The Nuclear Physics of Neutron Stars

The Nuclear Physics of Neutron Stars

Editor

Charalampos Moustakidis

MDPI

Basel • Beijing • Wuhan • Barcelona • Belgrade • Novi Sad • Cluj • Manchester

Editor
Charalampos Moustakidis
Department of Physics
Aristotle University of Thessaloniki
Thessaloniki
Greece

Editorial Office
MDPI AG
Grosspeteranlage 5
4052 Basel, Switzerland

This is a reprint of articles from the Special Issue published online in the open access journal *Symmetry* (ISSN 2073-8994) (available at: https://www.mdpi.com/journal/symmetry/special_issues/Nuclear_Physics_Neutron_Stars).

For citation purposes, cite each article independently as indicated on the article page online and as indicated below:

Lastname, A.A.; Lastname, B.B. Article Title. *Journal Name* **Year**, *Volume Number*, Page Range.

ISBN 978-3-7258-1599-9 (Hbk)
ISBN 978-3-7258-1600-2 (PDF)
doi.org/10.3390/books978-3-7258-1600-2

Contents

About the Editor

Charalampos Moustakidis

Dr. Charalampos Moustakidis is a theoretical nuclear physicist. He obtained his PhD in Theoretical Nuclear Physics from Aristotle University of Thessaloniki in 2001 and worked as a postdoctoral researcher at the University of Tübingen in Germany. Currently, Dr. Moustakidis is a Professor at the Physics Department of Aristotle University where he has worked since 2006. His main research interests concern the study of the structure and basic properties of compact stars (neutron stars, quark stars, hybrid stars, dark matter stars, etc.) where the knowledge of the equation of state that describes them plays a decisive role. His research group focuses their studies on how open problems concerning dense nuclear matter can be answered and how this can be achieved with the simultaneous combination of theoretical approaches and astrophysical observations (neutron star mergers, gravitational waves, etc.).

Preface

Neutron stars are considered extraordinary astronomical laboratories for the physics of nuclear matter as they have the most fascinating constitution of energy and matter in the universe. Recently, the detection of gravitational waves from the merging of two neutron stars in a binary neutron star system has created a new opportunity for exploring the physics of neutron stars. It is widely acknowledged that the majority of the static and dynamic processes of neutron stars are sensitively dependent on the equation of the state of dense nuclear matter employed. However, knowledge of the equation of state is very uncertain, especially at high densities. The main purpose of the present Special Issue, which comprises a collection of various contributions, is to shed light on some of the open problems concerning the nuclear physics of neutron stars and how they can possibly be addressed or even answered. In particular, as a part of this Special Issue, topics such as modern applications of the theory of nuclear matter in neutron stars, ideas for constraining the equation of state for both cold and hot nuclear matter with the help of recent observations, applications of recent modified theories of gravity to the properties of neutron stars, the astrophysical jets emerging from a wide variety of astrophysical compact objects, the possible existence of exotic particles, including, for example, the hypothetical X17 boson in the interior of neutron stars, applications of the so-called "pseudo-conformal model" that addresses dense compact star matter and is confronted with the astrophysical observables available at present, and ideas for a possible existence of compact objects called "Ghost stars" endowed with an arbitrarily small mass. In any case, the study of neutron stars still has many open problems to address. These problems arise from different aspects of physics including nuclear physics, particle physics, the theory of gravity, and statistical physics. Solving these problems requires not only the expansion of theoretical study via the introduction of new ideas and models but also the procurement of observational and experimental data via more systematic and extensive methods. The set of open problems–issues is quite large and only a few of them have been addressed in this Special Issue. Regardless, new observations or the results of new experiments may provide answers to these problems and may also lead to the creation of new open issues.

Charalampos Moustakidis
Editor

symmetry

Editorial

The Nuclear Physics of Neutron Stars

Charalampos Moustakidis

Department of Theoretical Physics, Aristotle University of Thessaloniki, 54124 Thessaloniki, Greece; moustaki@auth.gr

Neutron stars are considered extraordinary astronomical laboratories for the physics of nuclear matter as they have the most fascinating constitution of energy and matter in the Universe [1–3]. Recently, the detection of gravitational waves from the merging of two neutron stars in a binary neutron star system created a new opportunity for exploring the physics of neutron stars [4,5]. In particular, the majority of the static and dynamical processes of neutron stars are sensitively dependent on the equation of state of dense nuclear matter employed [6,7]. However, knowledge of the equation of state is very uncertain, especially at high densities; therefore, relevant predictions and estimations suffer. For example, there are strong speculations that a phase transformation from hadronic to quark matter takes place inside stars [8]. This theory needs to be thoroughly researched. Moreover, a long-standing goal in astrophysics is the determination of the maximum mass of neutron stars (both non-rotating and rotating) [9]. Neutron stars are directly related to the formation of black holes (Kerr black holes), connecting two of the most important astrophysical objects. As a consequence, the maximum mass of neutron stars is of great interest in studying the effects of both neutron stars and black holes on the dynamics of supernovae explosions. Furthermore, neutron stars, due to their compactness, may rotate very quickly compared to other astrophysical objects. In particular, the measurement of specific properties of rapidly rotating neutron stars (including their mass and radius, frequency, moment of inertia, and quadrupole moment) may lead to robust constraints on the equation of state as well as on the star's nuclear matter constitution at very high densities [10].

This Special Issue is a collection of various contributions dedicated to (a) modern applications of the theory of nuclear matter in neutron stars, (b) proposing ideas for constraining the equation of state for both and hot nuclear matter (low/high densities) with the help of recent observations as well as the detection of gravitational waves originating from neutron stars mergers, (c) relating the application of recent modified theories of gravity to the properties of neutron stars, (d) discussing some issues in relation to the astrophysical jets emerging from a wide variety of astrophysical compact objects, (e) discussing the possibility of the existence of exotic particles, including, for example, the hypothetical X17 boson in the interior of neutron stars, (f) presenting and analyzing the application of the so-called "pseudo-conformal model" that addresses dense compact star matter and is confronted with the astrophysical observables available at presents, and (g) exploring the existence of compact objects called "Ghost stars" endowed with arbitrarily small mass.

In contribution 1, the authors address the GW190814 Puzzle. In particular, the LIGO/Virgo collaboration observed a compact object with a mass of $2.59^{+0.08}_{-0.09}\ M_{\odot}$ as a component of a system in which the main companion was a black hole with a mass of $23\ M_{\odot}$ [11]. This observation immediately invited speculation as to whether this object falls into the neutron star–black hole mass gap. In any case, understanding nature of the GW190814 event will offer rich information concerning open issues, the speed of sound and possible phases transition into other degrees of freedom. The authors made an effort to examine possible constraints on the equation of state which were inferred from the consideration that the low-mass companion is a slow or rapidly rotating neutron star, also paying attention on the study of the tidal deformability and the radius of a possible

Citation: Moustakidis, C. The Nuclear Physics of Neutron Stars. *Symmetry* **2024**, *16*, 658. https://doi.org/10.3390/sym16060658

Received: 18 May 2024
Accepted: 22 May 2024
Published: 27 May 2024

high-mass candidate existing as an individual or component star in a binary neutron star system. They concluded that similar isolated neutron stars or systems may exist in the universe, and their possible future observation will shed light on the maximum neutron star mass problem.

The authors of contribution 2 proposed several constraints on the nuclear equation of state (EOS) currently available from neutron star (NS) observations and laboratory experiments and studied the existence of possible correlations among the properties of nuclear matter at saturation density using NS observables [12]. In particular, they used a set of different models that included several phenomenological EOSs based on Skyrme and relativistic mean field models as well as microscopic calculations based on different many-body approaches, i.e., the (Dirac–)Brueckner–Hartree–Fock theories, quantum Monte Carlo techniques, and the variational method. They concluded that while no correlation exists between tidal deformability and the value of nuclear symmetry energy at saturation for any NS mass value, very weak correlations seem to exist with the derivative of the nuclear symmetry energy and with the nuclear symmetry energy and with the nuclear incompressibility.

To overcome the problem that some Gogny-type [13] interactions lead to soft equations of state, the authors of contribution 3 built new Gogny parametrizations by modifying the density dependence of the symmetry energy predicted by the force in such a way that they could be applied to the neutron star domain and could also reproduce the properties of finite nuclei as well as their predecessors. These new parametrizations allowed the authors to obtain stiffer EOSs based on Gogny interactions which predict the maximum masses of neutron stars around two solar masses. Moreover, other global properties of stars determined, such as their moment of inertia and deformability, were in harmony with those obtained using other well-tested EOSs based on the SLy4 Skyrme force or the Barcelona–Catania–Paris–Madrid (BCPM) energy density functional.

Neutron stars are perfect candidates for investigating the effects of a modified gravity theory since curvature effects are significant and, more importantly, potentially testable. In most cases studied in the literature in the context of massive scalar–tensor theories, inflationary models were examined. The most important scalar–tensor model is the Higgs model, which depends on the values of the scalar field [14]. In view of the above, the author of contribution 4 investigated which potential form of the Higgs model is more appropriate for consistently describing a static neutron star. He proved numerically that the non-inflationary Higgs potential, which is valid for certain values of the scalar field in the Jordan frame, leads to extremely large maximum neutron star masses. Finally, he concluded that these results show the uniqueness of the inflationary Higgs potential since it is the only approximation for the Higgs model that provides self-consistent results.

The authors of contribution 5 examined collimated outflows of magnetized astrophysical plasma known as astrophysical jets, which have been observed to emerge from a wide variety of astrophysical compact objects. These systems can be considered either hydrodynamic (HD) or magnetohydrodynamic (MHD) in nature, which means that they are governed by non-linear partial differential equations [15]. The authors mainly focused on appropriate numerical solutions for the MHD (and/or RMHD) equations as well as a transfer equation for inside the jet and simulated multi-messenger emissions from specific astrophysical compact objects. They performed numerical simulations for neutrino, gamma ray, and secondary particle emissions. As concrete examples, they chose the galactic Cygnus X-1 and extragalactic LMC X-1 systems.

The author of contribution 6 described the mapping at high density of topological structure of baryonic matter to a nuclear effective field theory that implements hidden symmetries emergent from strong nuclear correlations [16]. The results are found to be consistent with the presently available observations in both normal nuclear matter and compact-star matter. The hidden symmetries involved are "local flavor symmetry" of the vector mesons identified to be (Seiberg-)dual to the gluons of QCD and hidden "quantum scale symmetry" with an IR fixed point with a "genuine dilaton (GD)" characterized by non-vanishing pion and dilaton decay constants. Both the skyrmion topology for $N_f \geq 2$ baryons and the fractional quantum Hall (FQH) droplet topology for $N_f = 1$ baryons are unified in the "homogeneous/hidden" Wess–Zumino term in the hidden local symmetry (HLS) Lagrangian. The possible indispensable role of the FQH droplets in going beyond the density regime of compact stars approaching scale-chiral restoration is explored by moving toward the limit where both the dilaton and the pion go massless [17].

The authors of contribution 7 complemented the nuclear equation of state (EOS) with a hypothetical 17 MeV boson [18] and observed that only instances with an admixture of 30–40% satisfied all the relevant constraints. The successful EOS resulted in a radius of around 13 km for a neutron star with a mass of $M_{NS} \simeq 1.4\ M_{\odot}$ and a maximum mass of around $M_{NS} \simeq 2.5\ M_{\odot}$. They found that the value of the radius is in agreement with the recent measurement by NICER, while the maximum mass is also in agreement with the mass of the remnant of the gravitational wave event GW190814. They concluded that it appears that these EOSs satisfy all the existing experimental constraints and can be considered universal nuclear equations of state.

The author of contribution 8 discussed and analyzed the so-called "pseudo-conformal model" that addresses dense, compact star matter and is confronted with the astrophysical observables available at present, with a focus on those obtained from gravity waves [17]. Their predictions were made nearly free of parameters as the model involving "topology change" remained more or less intact and "un-torpedoed" by the data.

The authors of contribution 9 focused on computing the saturation properties of symmetric and asymmetric nuclear matter using the finite-range simple effective interaction with the Yukawa form factor [19]. The results of higher-order derivatives of the energy per particle and the symmetry energy computed at saturation were compared with corresponding values extracted from studies involving theory, experiments, and astrophysical observations. In particular, the ability of the resulting equations of state to predict the threshold mass for prompt collapse in a binary neutron star merger and gravitational redshift was examined in terms of the compactness of the neutron star and the level of incompressibility at the central density of the maximum-mass star. Finally, they analyzed and compared the correlations existing between neutron star properties and nuclear matter saturation properties with the predictions of other models.

In contribution 10, the authors explored an idea proposed many years ago by Zeldovich and Novikov concerning the existence of compact objects endowed with arbitrarily small mass [20]. The energy density of such objects, which are called "Ghost stars", is negative in some regions of the fluid distribution, producing a vanishing total mass. Thus, on the boundary surface, the interior is matched to Minkowski space-time [21]. The authors provided some exact analytical solutions and analyzed their properties. With the help of observational data, they confirmed or dismissed the existence of this type of stellar object.

In general, the study of neutron stars still has many open problems to address. These problems arise from different aspects of physics including nuclear physics, particle physics, the theory of gravity, and statistical physics. Solving these problems requires not only the expansion of theoretical study via the introduction of new ideas and models but also procurement of observational and experimental data via more systematic and extensive methods. Since the set of open problems–issues is quite large, we only some below. Regardless, new observations or the results of new experiments may provide answers to these problems and may also lead to the creation of new open issues. In summary, some of the key problems associated with the study of neutron stars and the physics behind them are as follows (also see the relevant Ref. [22]):

1. The experimental determination of nuclear symmetry energy close to and above the nuclear saturation density.
2. The hyperon "puzzle": the problem of the strong softening of the equation of state of dense matter induced by the presence of hyperons, which leads to a maximum mass value incompatible with observations.

3. The Bose condensation in nuclear matter: the effects of pion and kaon condensation in the interior of neutron stars.
4. Hadron–quark phase transitions in dense nuclear matter, which have implications for the structure of neutron stars.
5. Determining what other phases exist in the phase diagram of dense matter at low temperatures and how we can use neutron star observations to learn about these phases.
6. Hybrid stars as confirmations of phase transitions in dense nuclear matter: the twin star and backbending phenomena.
7. The possible existence of a mass gap between neutron stars and black holes and its implications for the formation of neutron stars.
8. The maximum and minimum masses of neutron stars; the maximum mass has implications for the minimum mass of a black hole and, consequently, the total number of stellar-mass black holes in our Universe, the progenitor mass, and the EOS of dense matter. The minimum mass is related to its formation through stellar evolution.
9. The accurate measurement of the radius of a neutron star. If possible, simultaneous measurements of the masses and radii of several individual stars could pin down an EOS free from the applied nuclear model.
10. Determining what limits the spin frequencies of millisecond pulsars and why; additionally, determining how effective mechanisms are for reducing the rotation speed of pulsars (r-modes, f-modes, etc.).
11. Determining how rich information from a neutron star cooling curve can be used, which microscopic mechanisms are responsible for this process, and what their roles are.
12. Investigating the mystery of the appearance of glitches and starquakes. What are the roles of superfluidity and the crust–core interface? What are the relevant dissipative processes?
13. Studying the neutron star–dark matter admixture and its application to the existence and possible determination of dark matter in the Universe.
14. Determining the origin of the strong magnetic field in neutron stars and elucidating the physics of magnetars.
15. Investigating neutron star mergers as a major source of gravitational wave radiation and the roles of star structure and deformability.
16. Investigating neutron star binary mergers: can they explain the creation (nucleosynthesis) and existence of heavy elements in the universe?
17. The lifetime and final-stage possibilities of binary neutron star merger remnants.
18. Determining the origin of X-rays on the surfaces of rapidly rotating neutron stars and the role of the strong magnetic field; investigating accreting neutron stars in binary star systems as the strongest sources of X-rays in our galaxy.
19. Investigating collisions between neutron stars as sources of short gamma-ray bursts, some of the most powerful and violent explosions in the known universe. What we can learn from the interiors of neutron stars?
20. Investigating exotic stars (quark stars, strange stars, pion stars, preon stars, Thorne–Zytkow objects, and gravastars): their origin, structure, observation, and verification.

Although the above list is quite extensive, we have robust indications that in the coming years, both improvements in experimental methods and the accuracy of astrophysical observations, in close cooperation with theoretical research, will provide solutions for the majority of the aforementioned open problems.

Funding: This research received no external funding.

Conflicts of Interest: The author declares no conflicts of interest.

Abbreviations

The following abbreviations were used in this manuscript:

NS	Neutron star
EOS	Equation of state

List of Contributions

1. Kanakis-Pegios, A.; Koliogiannis, P.; Moustakidis, C.C. Probing the Nuclear Equation of State from the Existence of a $\sim 2.6\ M_{\odot}$ Neutron Star: The GW190814 Puzzle. *Symmetry* **2021**, *13*, 183.
2. Burgio, G.F.; Schulze, H.J.; Vidaña, I.; We, J.B. A Modern View of the Equation of State in Nuclear and Neutron Star Matter. *Symmetry* **2021**, *13*, 400.
3. Viñas, X.; Boquera, C.G.; Centelles, M.; Mondal, C.; Robledo, L.M. Unified Equation of State for Neutron Stars Based on the Gogny Interaction. *Symmetry* **2021**, *13*, 1613.
4. Oikonomou, V.K. Uniqueness of the Inflationary Higgs Scalar for Neutron Stars and Failure of Non-Inflationary Approximations. *Symmetry* **2022**, *14*, 32.
5. Papavasileiou, T.; Kosmas, O.; Sinatkas, I. Relativistic Magnetized Astrophysical Plasma Outflows in Black-Hole Microquasars. *Symmetry* **2022**, *14*, 485.
6. Rho, M. Mapping Topology of Skyrmions and Fractional Quantum Hall Droplets to Nuclear EFT for Ultra-Dense Baryonic Matter. *Symmetry* **2022**, *15*, 994.
7. Veselský, M.; Petousis, V.; Leja, J.; Navarro, L. Universal Nuclear Equation of State Introducing the Hypothetical X17 Boson. *Symmetry* **2023**, *15*, 49.
8. Rho, M. Dense Baryonic Matter Predicted in "Pseudo-Conformal Model". *Symmetry* **2023**, *15*, 1271.
9. Viñas, X.; Bano, P.; Naik, Z.; Routray, T.R. Nuclear Matter Properties and Neutron Star Phenomenology Using the Finite Range Simple Effective Interaction. *Symmetry* **2024**, *16*, 215.
10. Herrera, L.; Prisco, A.D.; Ospino, J. Ghost Stars in General Relativity. *Symmetry* **2024**, *16*, 562.

References

1. Shapiro, S.L.; Teukolsky, S.A. *Black Holes, White Dwarfs, and Neutron Stars*; JohnWiley and Sons: Hoboken, NJ, USA; New York, NY, USA, 1983.
2. Glendenning, N.K. *Compact Stars: Nuclear Physics, Particle Physics, and General Relativity*; Springer: Berlin/Heidelberg, Germany, 2000.
3. Haensel, P.; Potekhin, A.Y.; Yakovlev, D.G. *Neutron Stars 1: Equation of State and Structure*; Springer: New York, NY, USA, 2007.
4. Abbott, B.P.; Abbott, R.; Abbott, T.D.; Acernese, F.; Ackley, K.; Adams, C.; Adams, T.; Addesso, P.; Adhikari, R.X.; Adya, V.B.; et al. GW170817: Observation of Gravitational Waves from a Binary Neutron Star Inspiral. *Phys. Rev. Lett.* **2017**, *119*, 161101. [CrossRef] [PubMed]
5. Abbott, B.P.; Abbott, R.; Abbott, T.D.; Acernese, F.; Ackley, K.; Adams, C.; Adams, T.; Addesso, P.; Adhikari, R.X.; Adya, V.B.; et al. Properties of the Binary Neutron Star Merger GW170817. *Phys. Rev. X* **2019**, *9*, 011001. [CrossRef]
6. Oertel, M.; Hempel, M.; Klähn, T.; Typel, S. Equations of state for supernovae and compact stars. *Rev. Mod. Phys.* **2017**, *89*, 015007. [CrossRef]
7. Burgio, G.F.; Fantina, A.F. Nuclear Equation of State for Compact Stars and Supernovae. In *the Physics and Astrophysics of Neutron Stars*; Springer: Cham, Switzerland, 2018; Volume 457, p. 255.
8. Annala, E.; Gorda, T.; Kurkela, A.; Nättilä, J.; Vuorinen, A. Evidence for quark-matter cores in massive neutron stars. *Nat. Phys.* **2020**, *16*, 907–910. [CrossRef]
9. Lattimer, J.M. The Nuclear Equation of State and Neutron Star Masses. *Ann. Rev. Nucl. Sci.* **2012**, *62*, 485–515. [CrossRef]
10. Koliogiannis, P.S.; Moustakidis, C.C. Effects of the equation of state on the bulk properties of maximally rotating neutron stars. *Phys. Rev. C* **2020**, *101*, 015805. [CrossRef]
11. Abbott, R.; Abbott, T.D.; Abraham, S.; Acernese, F.; Ackley, K.; Adams, C.; Adhikari, R.X.; Adya, V.B.; Affeldt, C.; Agathos, M.; et al. GW190814: GravitationalWaves from the Coalescence of a 23 Solar Mass Black Hole with a 2.6 Solar Mass *Compact Object. Astrophys. J. Lett.* **2020**, *896*, L44. [CrossRef]
12. Baldo, B.; Burgio, G.F. The Nuclear Symmetry Energy. *Prog. Part. Nucl. Phys.* **2016**, *91*, 203. [CrossRef]
13. Dechargé, J.; Gogny, D. Hartree-Fock-Bogoliubov calculations with the D1 effective interaction on spherical nuclei. *Phys. Rev. C* **1980**, *21*, 1568. [CrossRef]

14. Odintsov, S.D.; Oikonomou, V.K. Neutron stars in scalar–tensor gravity with quartic order scalar potential. *Ann. Phys.* **2022**, *440*, 168839. [CrossRef]
15. Papavasileiou, T.V.; Kosmas, O.T.; Sinatkas, J. Simulations of Neutrino and Gamma-Ray Production from Relativistic Black-Hole Microquasar Jets. *Galaxies* **2021**, *9*, 67. [CrossRef]
16. Weinberg, S. What is quantum field theory, and what did we think it is? In *Conceptual Foundations of Quantum Field Theory*; Cambridge University Press: Boston, MA, USA, 1996; pp. 241–251.
17. Ma, Y.L.; Rho, M. Towards the hadron–quark continuity via a topology change in compact stars. *Prog. Part. Nucl. Phys.* **2020**, *113*, 103791. [CrossRef]
18. Krasznahorkay, A.J.; Csatlós, M.; Csige, L.; Gácsi, Z.; Gulyás, J.; Hunyadi, M.; Kuti, I.; Nyakó, B.M.; Stuhl, L.; Timár, J.; et al. Observation of Anomalous Internal Pair Creation in 8Be: A Possible Indication of a Light, Neutral Boson. *Phys. Rev. Lett.* **2016**, *116*, 042501. [CrossRef] [PubMed]
19. Behera, B.; Viñas, X.; Routray, T.R.; Centelles, M. Study of spin polarized nuclear matter and finite nuclei with finite range simple effective interaction. *J. Phys. G Nucl. Part. Phys.* **2015**, *42*, 045103. [CrossRef]
20. Zeldovich, Y.B.; Novikov, I.D. *Relativistic Astrophysics. Vol. I. Stars and Relativity*; University of Chicago Press: Chicago, IL, USA, 1971.
21. Herrera, L.; Prisco, A.D.; Ospino, J.; Fuenmayor, E. Conformally flat anisotropic spheres in general relativity. *J. Math. Phys.* **2001**, *42*, 2129. [CrossRef]
22. Prakash, M. Neutron stars as probes of extreme energy density matter. *Pramana* **2015**, *84*, 927–941. [CrossRef]

symmetry

MDPI

Article

Probing the Nuclear Equation of State from the Existence of a $\sim$2.6 $M_\odot$ Neutron Star: The GW190814 Puzzle

Alkiviadis Kanakis-Pegios †,‡, Polychronis S. Koliogiannis *,†,‡ and Charalampos C. Moustakidis †,‡

Department of Theoretical Physics, Aristotle University of Thessaloniki, 54124 Thessaloniki, Greece; alkanaki@auth.gr (A.K.P.); moustaki@auth.gr (C.C.M.)
* Correspondence: pkoliogi@physics.auth.gr
† Current address: Aristotle University of Thessaloniki, 54124 Thessaloniki, Greece.
‡ These authors contributed equally to this work.

Abstract: On 14 August 2019, the LIGO/Virgo collaboration observed a compact object with mass $\sim 2.59^{+0.08}_{-0.09}\ M_\odot$, as a component of a system where the main companion was a black hole with mass $\sim 23\ M_\odot$. A scientific debate initiated concerning the identification of the low mass component, as it falls into the neutron star–black hole mass gap. The understanding of the nature of GW190814 event will offer rich information concerning open issues, the speed of sound and the possible phase transition into other degrees of freedom. In the present work, we made an effort to probe the nuclear equation of state along with the GW190814 event. Firstly, we examine possible constraints on the nuclear equation of state inferred from the consideration that the low mass companion is a slow or rapidly rotating neutron star. In this case, the role of the upper bounds on the speed of sound is revealed, in connection with the dense nuclear matter properties. Secondly, we systematically study the tidal deformability of a possible high mass candidate existing as an individual star or as a component one in a binary neutron star system. As the tidal deformability and radius are quantities very sensitive on the neutron star equation of state, they are excellent counters on dense matter properties. We conjecture that similar isolated neutron stars or systems may exist in the universe and their possible future observation will shed light on the maximum neutron star mass problem.

Keywords: equation of state; neutron star; GW190814; maximum mass; tidal deformability

Citation: Kanakis-Pegios, A.; Koliogiannis, P.S.; Moustakidis. C.C. Probing the Nuclear Equation of State from the Existence of a $\sim$2.6 $M_\odot$ Neutron Star: the GW190814 Puzzle. *Symmetry* **2021**, *13*, 183. https://doi.org/10.3390/sym13020183

Academic Editor: Andrea Lavagno
Received: 17 December 2020
Accepted: 19 January 2021
Published: 24 January 2021

Publisher's Note: MDPI stays neutral with regard to jurisdictional clai-ms in published maps and institutio-nal affiliations.

1. Introduction

In Ref. [1] the authors reported the observation of a compact binary coalescence involving a 22.2–24.3 $M_\odot$ black hole and a compact object with a mass of a 2.50–2.67 $M_\odot$ (all measurements quoted at the 90% credible level). The announcement of the GW190814 event [1] triggered various theoretical studies concerning the equation of state (EoS) of dense nuclear matter, in order to explain the possibility of the second partner to be a very massive neutron star (NS). It is worth pointing out that the authors in Ref. [1] did not exclude the possibility that the second partner to be a NS or an exotic compact object, i.e., quark star, boson star or gravastar.

It is worth to point out the observation of the GW190814 event has some additional general benefits apart from the measurement of 2.6 $M_\odot$ of the second partner [1]. Firstly, this binary system has the most unequal mass ratio yet measured with gravitational waves close to the value of 0.112. Secondly, the dimensionless of the spin of the primary black hole is constrained $\leq$0.07, where various tests of general relativity confirm this value, as well as its predictions of higher-multiple emission at high confidence interval. Moreover, the GW190814 event poses a challenge for the understanding of the population of merging compact binaries. It was found after systematic analysis that the merger rate density of GW190814-like binary system to be 7^{+16}_{-6} Gpc^{-3} year^{-1} [1]. More relevant to the present study, the observation of the GW190814 event led to the following conclusion: due to the source's asymmetric masses, the lack of detection of an electromagnetic counterpart and of clear signature of tides or spin-induced quadrupole effect in the waveform of the gravitational waves we are not able to distinguish between a black hole-black hole and black hole-neutron star system [1]. In this case, one must count only to the comparison between the mass of the second partner with the estimation of maximum NSs mass $M_{\rm max}$ [2]. This is one of the subjects of the present work. It should be emphasized that the measurements of NSs mass can also inform us about a bound on the $M_{\rm max}$ independently of the assumptions of specific EoS. For example, Alsing et al. [3] fitting the known population of NSs in binaries to double-Gaussian mass distribution obtained the empirical constraint that $M_{\rm max} \leq 2.6\ M_{\odot}$ (with 90% confidence interval). Moreover, Farr and Chatziionannou [4] updated the previous study including recent measurements. Their study constraint the maximum mass $M_{\rm max} = 2.25^{+0.81}_{-0.26}\ M_{\odot}$ leading to the conclusion that the posterior probability (for the mass of the second partner $m_2 \leq M_{\rm max}$) is around only 29%. However, the prediction of $M_{\rm max}$ is sensitive on the selection rules mass of NSs (not only on binary systems but also isolated) as well as on the discovery of new events and consequently remains an open problem. Finally, the conclusion of the recent GW190814 event in comparison with previous ones (for example the GW170817 event [5]) may shed light on the problem of the $M_{\rm max}$. For example, the spectral EoSs which are conditioned by the GW170817 event, are once more elaborated to include the possibility that of the prediction of $M_{\rm max}$ at least equal to m_2. This approach lead to significant constraints on the radius and tidal deformability of a NS with mass of 1.4 $M_{\odot}$ ($R_{1.4} = 12.9^{+0.8}_{-0.7}$ km and $\Lambda_{1.4} = 616^{+273}_{-158}$ respectively [1]).

The consideration of a NS as the second partner has been studied in recent Refs. [6–17]. On the other hand, the case that the second partner is a quark or hybrid star has been explored in Refs. [18–24]. Finally, some modified theories of gravity have also been applied as possible solutions to the problem [25–27]. In particular, Tsokaros et al. [6] showed using viable equations that rapid uniform rotation is adequate to explain the existence of a stable 2.6 $M_{\odot}$ NS for moderately stiff EoSs but may not be adequate for soft ones. Huang et al. [7] concluded that using the density-dependent relativistic mean field model one cannot exclude the possibility of the secondary object to be a NS composed of hadronic matter. Zevin et al. [8] performing a systematic study led to the conclusion that the formation of GW190814-like systems at any measurable rate requires a supernova engine model that acts on longer timescales such that the proto-compact object can undergo substantial accretion immediately before the explosion. This conclusion hinds that if GW190814 is the result of a massive star binary evolution, the mass gap between NSs and black holes may be narrower or nonexistent. Fattoyev et al. [9] speculated that the maximum NS mass cannot be significantly higher than the existing observational limit and also the 2.6 $M_{\odot}$ compact object is likely to be the lightest black hole ever discovered. Essick and Landry [10] found that there is a $\leq$6% chance that GW190814 involved a slowly spinning NS, regardless of their assumed population model (considering no overlap between the NS and black hole mass distributions). Safarzadeh and Loeb [11] suggested that the secondary partner was born as a NS where a significant amount of the supernova ejecta mass from its formation remained bound to the binary due to the presence of the massive black hole companion. The bound mass forms a circumbinary accretion disk and its accretion onto the NS created a mass gap object. Godzieba et al. [12], showed how a lower limit on the maximum mass of NSs, in combination with upcoming measurements of NS radii by LIGO/Virgo and NICER, would constrain the EoS of dense matter and discussed the implications for the GW190814 event. Sedrakian et al. [13], allowing the hyperonization of dense matter, found that the maximal masses of hypernuclear stars, even for maximally rotating configurations, are inconsistent with a stellar nature interpretation of the light companion in GW190814. They concluded that the GW190814 event involved two black holes rather than a NS and a black hole. Biswas et al. [14] concluded that the odds of the secondary object in GW190814 being a NS improved by considering a stiff high-density EoS or a large rotation. Zhang and Li [15] showed that one possible explanation for GW190814's secondary component is a super-fast pulsar spinning faster than 971 Hz. Most et al. [16] stated that based on our current understanding of the nuclear matter EoS, it can be a rapidly rotating NS that collapsed to a rotating black hole at some point before the merger. Tan et al. [17] constructed heavy NSs by introducing non-trivial structure in the speed of sound sourced by deconfined Quantum Chromodynamics (QCD) matter. Within this approach they can explain the high mass of the second partner.

Zhang and Mann [18] indicated a new possibility that the currently observed compact stars, including the recently reported GW190814's secondary component can be quark stars composed of interacting up-down quark matter. Bombaci et al. [19] investigated the possibility that the low mass companion of the black hole in the source of GW190814 was a strange quark star. This possibility is viable within the so-called two-families scenario in which NSs and strange quark stars coexist. Demircik et al. [20] studied rapidly spinning compact stars with EoSs featuring a first-order phase transition between strongly coupled nuclear matter and deconfined quark matter and also compatible with the interpretation that the secondary component in GW190814 is an NS. Cao et al. [21] provided circumstantial evidence suggesting the recently reported GW190814's secondary component could be an up-down quark star. Dexheimer et al. [22] showed that state-of-the-art relativistic mean field models can generate massive stars reaching $\geq 2.05\ M_{\odot}$, while being in good agreement with gravitational-wave events and x-ray pulsar observations when quark vector interactions and higher-order self-vector interactions are introduced. Roupas et al. [23] showed not only that a color-flavor locked quark star with this mass is viable, but also they calculated the range of the model parameters, namely the color superconducting gap Δ and the bag constant B , that satisfies the strict LIGO constraints on the EoS.

In the study of Moffat [25], the modified gravity (MOG) theory is applied to the gravitational wave binary merger GW190814. He demonstrated that the modified Tolman–Oppenheimer–Volkoff (TOV) equation for a NS can produce a mass $M = 2.5$–$2.7\ M_{\odot}$ allowing for the binary secondary component to be identified as a heavy NS in the hypothesized mass gap $M = 2.5$–$5\ M_{\odot}$. Astashenok et al. [26] showed that a NS with this observed mass can be consistently explained with the mass-radius relation obtained by extended theories of gravity. Nunes et al. [27] found that from an appropriate and reasonable combination of modified gravity, rotation effects, and realistic soft EoSs, it is possible to achieve high masses and explain GW190814 secondary component.

Another issue worth mentioning is the transition of hadron matter to unconfined quark matter at a sufficiently high density (a few times the saturation density, see previous discussion). Recently, Annala et al. [28] claimed that the recent observation of gravitational waves from NSs merger could shed light on the possibility of hadrons to quark phase transition. Moreover, the emergence of strange hadrons (hyperons, etc.) around twice the nuclear saturation density, leads to an appreciable softness of the EoS, and consequently, in most cases the observed values of high mass NSs never reached. As NSs provide a rich testing ground for microscopic theories of dense nuclear matter, combining this study with the experimental data from ultra-relativistic heavy ion collisions may help significantly to improve our knowledge on phase transition theory in hadronic matter. In particular, the so-called hyperon puzzle may be addressed (or even more solved) in NS studies. To be more specific, following the discussion of Ref. [29], the hyperon puzzle is related to the difficulty to reconcile the measured masses of NSs with the presence of hyperons in their interiors. The presence of hyperons in the interior of NSs is due to the fermionic nature of nucleons. The chemical potential of neutron and proton increasing rapidly, as a function of the density. When the chemical potential of neutrons becomes sufficiently large, the most energetic neutrons can decay via the weak interaction into Λ hyperons and form and new Fermi sea for this hadronic species. Other hyperons can be formed with similar weak processes [29]. However, the inclusion of hyperons in NS matter was found that leads to an appreciable decrease of maximum NS mass, incompatible with the recent observations. It is stated that this is a common feature of various hyperon star structure calculations (see Ref. [29] and reference therein). Thus, although the presence of hyperons in NSs seems unavailable, their presence leads to low values of NS mass, far from observation. This problem is briefly summarized as hyperon puzzle. Of course, there are other studies where the authors stated that hyperon consideration on the EoS is not in contradiction with the predictions of very high NS masses (see Ref. [13]). In the the present study, we do not consider the case of additional degrees of freedom (hyperons, quarks, etc.) in the interior of NSs. In any case, additional theoretical calculation in combination with specific observation may lead to the solution of the hyperon puzzle and the reveal of the existence of free quark matter in the interior of NSs.

In the present work we concentrate our study on the case where the EoS of NS matter is pure hadronic and the hydrostatic equilibrium is described by the general relativistic equations (TOV equations) [30–34]. Moreover, we focus on two possible cases, that the second partner to be a slow rotating NS and, in the extreme case, to be a very rapidly rotating one (even close to the Kepler limit). Firstly, we consider that the dense nuclear matter properties are described by the MDI-APR [35] (MDI: momentum dependent interaction, APR: Akmal, Pandharipande and Ravenhall) nuclear model. This model has recently been applied successfully in similar studies and, according to our opinion, is a robust guide for NS studies. However, since the behavior of dense nuclear matter at high densities remains uncertain, the parametrization of the EoS via the speed of sound is almost inevitable, at least in the framework of hadronic EoSs. To be more specific, we construct large number of EoSs where for low densities (concerning the NS crust), we employ well established results, but for the EoS concerning the core, we apply a twofold consideration. For the outer part we employ the EoS predicted by the MDI-APR model and for the inner one, a parametrization based on the speed of sound upper limits is applied. In this study, the transition density and the upper limit of the speed of sound are the two free parameters.

In the first part of our study we concentrate on the effect of the speed of sound and transition density on the bulk NS (non-rotating and rapidly rotating) properties including the maximum mass, the Kepler frequency, the Kerr parameter, and the maximum central density. We explore under which circumstances the prediction of the mass range 2.5–2.67 $M_\odot$ of the second partner is possible. More importantly, we provide the constraints which are inferred by the above consideration.

In the second part of the paper we systematically study the tidal deformability of NSs by employing the large set of EoSs. We mainly focus in the case of high mass candidates existing as an individual star or as a partner in binary NSs system. Until now, there are no observations of an individual, or as a partner of binary system, very massive NS (close to 2.5 $M_\odot$). However, we consider that it is worth to examine this possibility, by focusing on the predictions of the tidal deformability and the radius, quantities that are very sensitive on the NS EoS. These quantities are excellent counters on dense matter properties. In the present work, we provide predictions about both the individual and averaged tidal deformability of a hypothetical binary NS system where the most massive partner has a mass in the region 2.5–2.67 $M_\odot$.

The article is organized as follows: in Section 2, we present the MDI-APR nuclear model along with quantities at the rotating configuration while in Section 3, we present the speed of of sound parametrization of the EoS. In Section 4 we provide the basic formalism for the tidal deformability. The results and the discussion are provided in Section 5 while Section 6 includes the concluding remarks of the present study. Finally, Section 7 contains information about the rotating configuration code.

2. The MDI-APR Model and the Rapidly Rotating Neutron Star

The structure of the EoS and the properties of NSs are studied under the MDI model. In this model, the energy per particle is available through the form [36,37]

$$
\begin{aligned}
E(n,I) &= \frac{3}{10}E_F^0 u^{2/3}\left[(1+I)^{5/3}+(1-I)^{5/3}\right]+\frac{1}{3}A\left[\frac{3}{2}-X_0I^2\right]u+\frac{\frac{2}{3}B\left[\frac{3}{2}-X_3I^2\right]u^{\sigma}}{1+\frac{2}{3}B'\left[\frac{3}{2}-X_3I^2\right]u^{\sigma-1}} \\
&+ \frac{3}{2}\sum_{i=1,2}\left[C_i+\frac{C_i-8Z_i}{5}I\right]\left(\frac{\Lambda_i}{k_F^0}\right)^3\left(\frac{((1+I)u)^{1/3}}{\frac{\Lambda_i}{k_F^0}}-\tan^{-1}\frac{((1+I)u)^{1/3}}{\frac{\Lambda_i}{k_F^0}}\right) \\
&+ \frac{3}{2}\sum_{i=1,2}\left[C_i-\frac{C_i-8Z_i}{5}I\right]\left(\frac{\Lambda_i}{k_F^0}\right)^3\left(\frac{((1-I)u)^{1/3}}{\frac{\Lambda_i}{k_F^0}}-\tan^{-1}\frac{((1-I)u)^{1/3}}{\frac{\Lambda_i}{k_F^0}}\right),
\end{aligned}
\tag{1}
$$

where $u = n/n_s$, with n_s denoting the saturation density ($n_s = 0.16$ fm^{-3}), $I = (n_n - n_p)/n$ is the asymmetry parameter, $X_0 = x_0 + 1/2$, and $X_3 = x_3 + 1/2$. The parameters A, B, σ, C_1, C_2, and B' appear in the description of symmetric nuclear matter (SNM) and are determined so that the relation $E(n_s, 0) = -16$ MeV holds. Λ_1 and Λ_2 are finite range parameters equal to $1.5k_F^0$ and $3k_F^0$, respectively, with k_F^0 being the Fermi momentum at the saturation density. The rest of the parameters, x_0, x_3, Z_1, and Z_2 appear in the description of asymmetric nuclear matter and, with a suitable parametrization, are used in order to obtain different forms for the density dependence of symmetry energy as well as the value of slope parameter L and the value of symmetry energy S_2 at the saturation density, defined as [35]

$$
L = 3n_s \frac{dS_2(n)}{dn}\bigg|_{n_s} \quad \text{and} \quad S_2(n) = \frac{1}{2}\frac{\partial^2 E(n,I)}{\partial I^2}\bigg|_{I=0}, \tag{2}
$$

and as a consequence different parametrizations of the EoS stiffness. In fact, for a specific value of L, the density dependence of symmetry energy is adjusted so that the energy of pure neutron matter is comparable with those of the existing state-of-the-art calculations.

The MDI model, that combines both density and momentum dependent interaction among the nucleons, is suitable for studying NS matter at zero (present study) as well as at finite temperature. In particular, although it was introduced by Gale et al. [38,39] to examine the influence of momentum dependent interactions on the momentum flow of heavy-ion collisions, the model has been modified, elaborated, and applied also in the study of the properties of nuclear matter at NSs. The advantages of the MDI model are: (a) reproduces with high accuracy the properties of SNM at the saturation density, including isovector quantities, (b) reproduces the microscopic properties of the Chiral model for pure neutron matter and the results of state-of-the-art calculations of Akmal et al. [40], (c) predicts maximum NS mass higher than the observed ones [41–43], and (d) maintains the causal behavior of the EoS even at densities higher than the ones at the maximum mass configuration.

In this work we apply the EoS produced in Ref. [35], where for the construction of the EoS, the MDI model and data from Akmal et al. [40] had been used (for more details see Ref. [35]). This EoS not only has the mentioned advantages, but also reproduces the mass of the second component of GW190814 event.

In addition, as a possible scenario is the rotation, we apply rotating configuration in the EoS. In fact, we are interested about the Kepler frequency and the maximum mass of the NS at this configuration. This frequency is considered as the one where the star would shed matter at its equator and consequently is the maximum one (mass-shedding limit). An interesting quantity, which connects the gravitational mass with the angular momentum of the star, is the Kerr parameter defined as

$$
\mathcal{K} = \frac{cJ}{GM^2}, \tag{3}
$$

where M and J are the gravitational mass and angular momentum, respectively. For the construction of the rotating equilibrium model we used the RNS code [44].

Furthermore, in Ref. [45] had been found an analytical relation which connects the Kerr parameter with the gravitational mass of the non-rotating case (TOV), expressed as

$$M_{\text{rot}} = M_{\text{TOV}}\left(1 + a_1\left(\frac{\mathcal{K}}{\mathcal{K}_{\max}}\right)^2 + a_2\left(\frac{\mathcal{K}}{\mathcal{K}_{\max}}\right)^4\right), \tag{4}$$

with $a_1 = 0.132$, $a_2 = 0.071$, and $\mathcal{K}_{\max}$ being the Kerr parameter at mass-shedding limit, which is used by the authors to imply constraints on the possible Kerr parameter of the second component, as well as the upper limit of a NS mass [16].

3. Speed of Sound Formalism and Stiffness of Equation of State

Another scenario that is followed to reproduce the mass of the second component is the possible stiffness of the EoS. This is achievable by studying the upper and lower limit on the speed of sound, as well as the possible transition density. In this consideration, we have parametrized the EoS, according to Refs. [3,46–52], as

$$P(\mathcal{E}) = \begin{cases} P_{\text{crust}}(\mathcal{E}), & \mathcal{E} \le \mathcal{E}_{\text{c-edge}} \\ P_{\text{NM}}(\mathcal{E}), & \mathcal{E}_{\text{c-edge}} \le \mathcal{E} \le \mathcal{E}_{\text{tr}} \\ \left(\frac{v_s}{c}\right)^2(\mathcal{E} - \mathcal{E}_{\text{tr}}) + P_{\text{NM}}(\mathcal{E}_{\text{tr}}), & \mathcal{E}_{\text{tr}} \le \mathcal{E}, \end{cases} \tag{5}$$

where P and $\mathcal{E}$ denote the pressure and energy density, respectively, and $\mathcal{E}_{\text{tr}}$ is the transition energy density. For the construction of the EoSs, we adopted the following: (a) in region $\mathcal{E} \le \mathcal{E}_{\text{c-edge}}$, we used the equation of Feynman et al. [53] and also of Baym et al. [54] for the crust and low densities of NS, (b) in the intermediate region, $\mathcal{E}_{\text{c-edge}} \le \mathcal{E} \le \mathcal{E}_{\text{tr}}$, we employed a specific EoS based on the MDI model and data from Akmal et al. [40], and (c) for $\mathcal{E}_{\text{tr}} \ge \mathcal{E}$ region, the EoS is maximally stiff with the speed of sound, defined as $v_s = c\sqrt{(\partial P/\partial \mathcal{E})_S}$ (where S is the entropy) fixed in the present work in the range $[c/\sqrt{3}, c]$. The lowest allowed value of the speed of sound, that is $(v_s/c)^2 = 1/3$, is introduced in order to be consistent with the possibility of a phase transition in quark matter. In this case, all the theoretical predictions lead to this value as an upper bound of the speed of sound. The implementation of speed of sound values between the limited ones will lead to results well constrained by the two mentioned limits. Although the energy densities below the $\mathcal{E}_{\text{c-edge}}$ have negligible effects on the maximum mass configuration, we used them in calculations for the accurate estimation of the tidal deformability. The cases which took effect in this study can be divided into two categories based on the fiducial baryon transition density, $n_{\text{tr}} = pn_s$, and the speed of sound as: (a) the ones where p takes the values $[1.5, 2, 3, 4, 5]$, while the speed of sound is parametrized in the two limiting cases, $(v_s/c)^2 = 1/3$ and $(v_s/c)^2 = 1$ and (b) the ones where p takes the values $[1.5, 2]$, while the speed of sound is parametrized in the range $(v_s/c)^2 = [1/3, 1]$. The predicted EoSs are functional of n_{tr} and (v_s/c) and implemented to study the possibly existence of a NS with $\sim$2.6 $M_\odot$, either non-rotating or a rotating one.

In approach followed in Equation (5), while the continuity on the EoS is well ensured, the continuity in the speed of sound at the transition density, due to its artificial character, is not. Therefore, in order to ensure the continuity and a smooth phase transition, we employ a method presented in Ref. [55]. We proceeded with the matching of the EoSs on the transition density by considering that, above this value, the speed of sound is parametrized as follows (for more details see Ref. [55])

$$\frac{v_s}{c} = \left(a - c_1 \exp\left[-\frac{(n - c_2)^2}{w^2}\right]\right)^{1/2}, \quad a \in [1/3, 1] \tag{6}$$

where the parameters c_1 and c_2 are fit to the speed of sound and its derivative at n_{tr}, and also to the demands $v_s(n_{\text{tr}}) = [c/\sqrt{3}, c]$ [46] according to the value of α. The remaining parameter w controls the width of the curve, where in our case is equal to 10^{-3} fm^{-3}

in order to preserve the NS properties. Using Equation (6), the EoS for $n \geq n_{\rm tr}$ can be constructed with the help of the following recipe [55]

$$\mathcal{E}_{i+1} = \mathcal{E}_i + \Delta\mathcal{E}, \quad P_{i+1} = P_i + \left(\frac{v_s}{c}(n_i)\right)^2 \Delta\mathcal{E}, \tag{7}$$

$$\Delta\mathcal{E} = \Delta n \left(\frac{\mathcal{E}_i + P_i}{n_i}\right), \tag{8}$$

$$\Delta n = n_{i+1} - n_i. \tag{9}$$

The treatment with both discontinuity and continuity in the speed of sound is presented in Table V of Ref. [46]. The outline was that the two approaches converge and consequently the effects of the discontinuity are negligible.

In Figure 1 we present the pressure as a function of the rest mass density ($\rho_{\rm rest} = n_b m_n$) and the square speed of sound in units of speed of light as a function of the transition density for the EoSs constructed in cases (a) and (b). In addition, we display the credibility intervals proposed by Ref. [5] from LIGO/Virgo collaboration for the GW170817 event. It is clear from these figures that the pure MDI-APR EoS is well-defined in the proposed limits of LIGO/Virgo collaboration and also fulfills the speed of light limit at high densities.

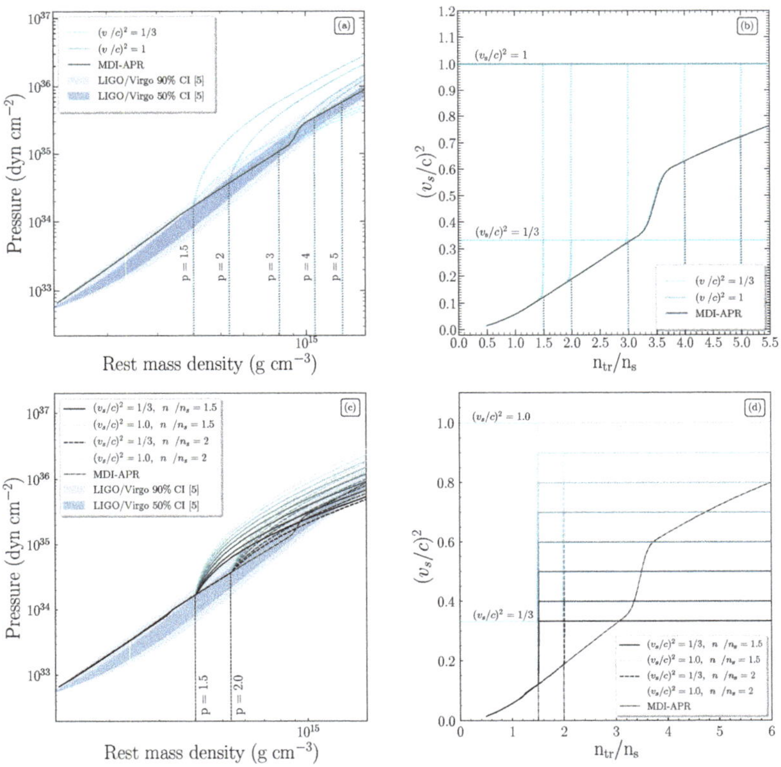

Figure 1. **(a)** Pressure as a function of rest mass density and **(b)** square speed of sound in units of speed of light as a function of the transition density, where p takes the values [1.5, 2, 3, 4, 5], while the speed of sound is parametrized in the two limiting cases, $(v_s/c)^2 = 1/3$ and $(v_s/c)^2 = 1$. **(c)** Pressure as a function of rest mass density and **(d)** square speed of sound in units of speed of light as a function of the transition density, where p takes the values [1.5, 2], while the speed of sound is parametrized in the range $(v_s/c)^2 = [1/3, 1]$ (As the speed of sound is getting higher values, the curves' color lightens). The vertical lines display the transition cases, while the shaded regions show the credibility interval extracted from Ref. [5].

4. Tidal Deformability

It has mentioned that the gravitational waves emitted from the final stages of an inspiraling binary NS system are one of the most important sources for the terrestrial gravitational waves detectors [56–64]. In such case, properties like the mass of the component stars can be measured. As Flanagan and Hinderer [58] articulated, the tidal effects can be measurable during this final stage of the inspiral.

The response of a NS to the presence of the tidal field, is described by a dimensionless tidal parameter, the tidal Love number k_2. This parameter depends on the NS structure; hence its mass and EoS. The tidal Love number k_2 is the coefficient of proportionality between the induced quadrupole moment Q_{ij} and the applied tidal field E_{ij} [58,65], given below

$$Q_{ij} = -\frac{2}{3}k_2\frac{R^5}{G}E_{ij} \equiv -\lambda E_{ij}, \tag{10}$$

where R is the NS radius and $\lambda = 2R^5k_2/3G$ is the other tidal parameter that we use in our study, the so-called tidal deformability. The tidal Love number k_2 is given by [58,59]

$$\begin{aligned} k_2 &= \frac{8\beta^5}{5}(1-2\beta)^2[2-y_R+(y_R-1)2\beta] \times \Big[2\beta(6-3y_R+3\beta(5y_R-8)) \\ &+ 4\beta^3\Big(13-11y_R+\beta(3y_R-2)+2\beta^2(1+y_R)\Big) \\ &+ 3(1-2\beta)^2[2-y_R+2\beta(y_R-1)]\ln(1-2\beta)\Big]^{-1}, \end{aligned} \tag{11}$$

where $\beta = GM/Rc^2$ is the compactness parameter of a NS. The quantity y_R is determined by solving the following differential equation

$$r\frac{dy(r)}{dr} + y^2(r) + y(r)F(r) + r^2Q(r) = 0, \tag{12}$$

with the initial condition $y(0) = 2$ [61]. $F(r)$ and $Q(r)$ are functionals of $\mathcal{E}(r)$, $P(r)$ and $M(r)$ defined as [56,61]

$$F(r) = \left[1 - \frac{4\pi r^2 G}{c^4}(\mathcal{E}(r) - P(r))\right]\left(1 - \frac{2M(r)G}{rc^2}\right)^{-1}, \tag{13}$$

and

$$\begin{aligned} r^2Q(r) &= \frac{4\pi r^2 G}{c^4}\left[5\mathcal{E}(r) + 9P(r) + \frac{\mathcal{E}(r)+P(r)}{\partial P(r)/\partial\mathcal{E}(r)}\right] \times \left(1 - \frac{2M(r)G}{rc^2}\right)^{-1} \\ &- 6\left(1-\frac{2M(r)G}{rc^2}\right)^{-1} - \frac{4M^2(r)G^2}{r^2c^4}\left(1 + \frac{4\pi r^3 P(r)}{M(r)c^2}\right)^2\left(1 - \frac{2M(r)G}{rc^2}\right)^{-2}. \end{aligned} \tag{14}$$

The numerical solution requires that the Equation (12) must be integrated self consistently with the TOV equations using the boundary conditions $y(0) = 2$, $P(0) = P_c$ and $M(0) = 0$ [56,59]. From the solution of TOV equations the mass M and radius R of the NS can be extracted, while the corresponding solution of the differential Equation (12) provides the value of $y_R = y(R)$. This parameter along with the quantity β are the basic ingredients of the tidal Love number k_2.

One parameter that is well constrained by the gravitational waves detectors is the chirp mass $\mathcal{M}_c$, which is a combination of the component masses of a binary NS system [66,67]

$$\mathcal{M}_c = \frac{(m_1m_2)^{3/5}}{(m_1+m_2)^{1/5}} = m_1\frac{q^{3/5}}{(1+q)^{1/5}}, \tag{15}$$

where m_1 is the mass of the heavier component star and m_2 is the lighter's one. Hence, the binary mass ratio $q = m_2/m_1$ is within the range $0 \leq q \leq 1$.

In addition, another binary parameter that can be measured from the analysis of the gravitational wave signal is the effective tidal deformability [66,67]

$$\tilde{\Lambda} = \frac{16}{13}\frac{(12q+1)\Lambda_1 + (12+q)q^4\Lambda_2}{(1+q)^5}, \quad (16)$$

where the key quantity q characterizes the mass asymmetry, and Λ_i is the dimensionless deformability defined as [66,67]

$$\Lambda_i = \frac{2}{3}k_2\left(\frac{R_i c^2}{M_i G}\right)^5 \equiv \frac{2}{3}k_2\beta_i^{-5}, \quad i = 1,2. \quad (17)$$

By combining Equations (17) and (11), one can find that Λ_i depends both on star's compactness and the value of $y(R)$.

We notice that Λ_i depends directly on the stiffness of the EoS through the compactness β and indirectly through the speed of sound which appears in Equation (14). The dependence of Λ_i on the behavior of $y(R)$ can lead to useful estimations or constraints on the tidal deformability itself. To be more specific, the applied EoS affects also the behavior of Λ regarding the NS's mass M and radius R . In our study we use the secondary very massive component of GW190814 system (see Ref. [1]) to examine the tidal deformability and the behavior of dense nuclear matter in the extreme scenario of such a massive NS.

5. Results and Discussion

The merger of a very massive black hole ($\sim$23 $M_\odot$) with a $\sim$2.6 $M_\odot$ compact object has recently been announced by the LIGO/Virgo collaboration, as the GW190814 event. The scenarios that follow the second merger component are that of (a) the lightest black hole, (b) the most compact NS, (c) a rapidly rotating NS, and (d) an exotic compact object. In the present work we studied only the second and third case scenarios, that is either a compact non-rotating NS or a rapidly rotating one.

5.1. Slow/Rapid Rotation: Implications to Neutron Star Properties

In Figure 2 we display the gravitational mass as a function of the Kerr parameter for the pure MDI-APR EoS. In addition, we note the universal relation Equation (4) for two limiting cases: (a) $M_{\rm TOV} = 2.08\ M_\odot$ and (b) $M_{\rm TOV} = 2.3\ M_\odot$ [16], and $\mathcal{K}_{\rm max} = 0.68$. The limiting cases were the minimum and maximum possible mass, respectively, for a NS based on the calculations provided in Ref. [16]. In accordance, the maximum value of the Kerr parameter was also calculated in Ref. [16] with respect to the minimum possible mass. With regard to the pure MDI-APR EoS, the relevant dependence was constructed through the RNS code having as input the angular momentum of the star until it reached its mass-shedding limit. This figure represents the limited area where the compact object should lie. The area was marked by the intersection of the gravitational mass, $M = 2.59^{+0.08}_{-0.09}\ M_\odot$, with the Kerr parameter, $\mathcal{K} = [0.49, 0.68]$ [16]. We note that the pure MDI-APR EoS was in the range of the described limits for the gravitational mass and Kerr parameter, as well as the ones introduced in Figure 1, being a suitable hadronic EoS to simulate the compact object of $\sim$2.6 $M_\odot$.

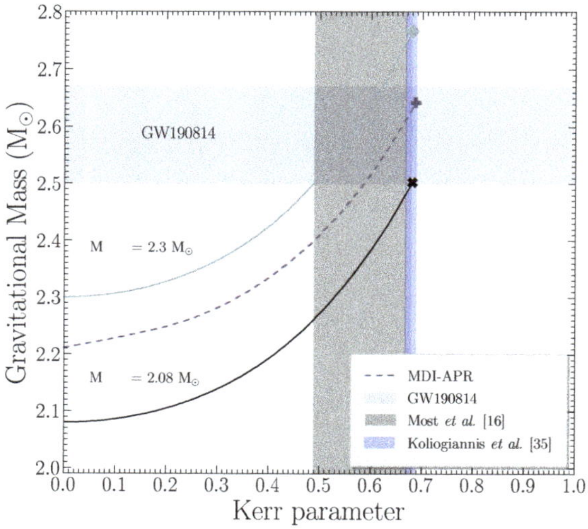

Figure 2. Gravitational mass as a function of Kerr parameter for the MDI-APR (MDI: momentum dependent interaction, APR: Akmal, Pandharipande and Ravenhall) equation of state (EoS). The solid lines from bottom to top represent the Equation (4) with $M_{\rm TOV} = 2.08\ M_\odot$ and $M_{\rm TOV} = 2.3\ M_\odot$. The mass range of the second component of GW190814 is noted with the horizontal shaded region, while with the vertical one (**left**), the possible region of Kerr parameter $\mathcal{K} = [0.49, 0.68]$ from Ref. [16] is shown. In addition, the region for the Kerr parameter $\mathcal{K}_{\rm max} = [0.67, 0.69]$ from Ref. [35], if the low mass component was rotating at its mass-shedding limit, is presented with the vertical shaded region (**right**). The markers point the maximum mass configuration at the mass-shedding limit.

In addition, taking into consideration the limiting case that the compact object was rotating at its mass-shedding limit, then constraints on the maximum value of the Kerr parameter, the corresponding equatorial radius, and the central energy density were possible. In particular, firstly we employed the relation found in Ref. [35]

$$\mathcal{K}_{\rm max} = 0.488 + 0.074\left(\frac{M_{\rm max}}{M_\odot}\right), \tag{18}$$

for the observable gravitational mass. For the mass of the second component, the Kerr parameter lay in the range $\mathcal{K}_{\rm max} = [0.67, 0.69]$, which is also noted in Figure 2. Secondly, using the derived relation from the recent Ref. [68], which connected the maximum value of Kerr parameter with the one of compactness parameter, as

$$\mathcal{K}_{\rm max} = 1.34\sqrt{\beta_{\rm max}} \quad \text{with} \quad \beta_{\rm max} = \frac{G}{c^2}\frac{M_{\rm max}}{R_{\rm max}}, \tag{19}$$

it was possible to extract a specific range for the corresponding equatorial radius. In this case, the corresponding equatorial radius lay in the range $R_{\rm max} = [14.77, 14.87]$ km.

We concentrated now on the microscopic properties of the NS, the speed of sound and the transition density. In Figure 3 we display the gravitational mass and Kerr parameter as a function of the transition density for the two limiting cases of the speed of sound based on Ref. [46]. From Figure 3a, the intersection of GW190814 mass area with the extracted curves provided us two regions of the possible transition density with respect to the applied speed of sound. By employing the formula from Ref. [46]

$$\frac{M_{\rm max}}{M_\odot} = \alpha_1 \coth\left[\alpha_2\left(\frac{n_{\rm tr}}{n_s}\right)^{1/2}\right], \tag{20}$$

where the coefficients a_1 and a_2 are given in Table 1, we were able to restrict the transition density with respect to the speed of sound. More precisely, we took under consideration two possible cases: (a) non-rotating and (b) maximally-rotating NS. In the first case, as Figure 3a shows, the possible transition density region was restricted between the two limiting cases of the speed of sound. However, as the lower limit in the speed of sound was not able to represent the gravitational mass of the low mass component, we found the lower possible speed of sound value that reproduced this mass at the specific value of the transition density. Consequently, the transition density was constrained in the range $n_{\rm tr} = [1.5, 3.2]\ n_s$ and the speed of sound in the range $(v_s/c)^2 = [0.45, 1]$.

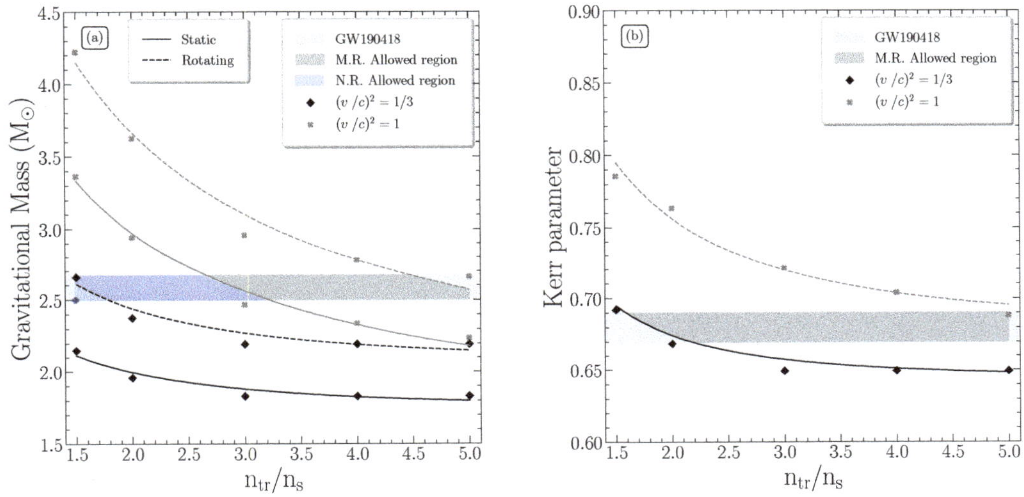

Figure 3. (**a**) Gravitational mass and (**b**) Kerr parameter as a function of transition density at the maximum mass configuration for the two limiting speed of sound bounds. The data at the maximum mass configuration is presented with diamonds for the $(v_s/c)^2 = 1/3$ bound and crosses for the $(v_s/c)^2 = 1$ bound. The plus marker denotes the lower bound in the speed of sound, $(v_s/c)^2 = 0.45$, assuming that the second component was a non-rotating neutron star (NS). The mass range of the second component of GW190814 is noted with the horizontal shaded region. (**a**) The lighter shaded region marks the allowed range for the transition density at the maximally-rotating (M.R.) configuration, while the darker one, marks the allowed region at the non-rotating (N.R.) configuration. (**b**) The darker shaded region marks the allowed range for the transition density at the maximally-rotating (M.R.).

In the second case, the possible transition density could be constrained both from the gravitational mass and the Kerr parameter. From Figure 3a it is clear that the transition density could take all the values in the area under consideration. However, from Figure 3b the transition density was constrained from the lower limit of the speed of sound. According to the derived formula from Ref. [46]

$$\mathcal{K}_{\rm max} = \alpha_3 \coth\left[\alpha_4 \left(\frac{n_{\rm tr}}{n_s}\right)^{1/2}\right], \quad (21)$$

where the coefficients a_3 and a_4 are given in Table 1, the rotating case allowed the existence of the low mass component at the whole range of speed of sound values and at transition densities in the range $n_{\rm tr} = [1.6, 5]\ n_s$.

Table 1. Coefficients of Equations (20) and (21) for the two speed of sound bounds. The abbreviation "N.R." corresponds to the non-rotating configuration and the "M.R." to the maximally-rotating one.

Speed of Sound Bounds	α_1		α_2		α_3		α_4	
	N.R.	M.R.	N.R.	M.R.	N.R.	M.R.	N.R.	M.R.
c	1.665	1.689	0.448	0.352	-	0.683	-	1.053
$c/\sqrt{3}$	1.751	2.069	0.964	0.883	-	0.645	-	1.348

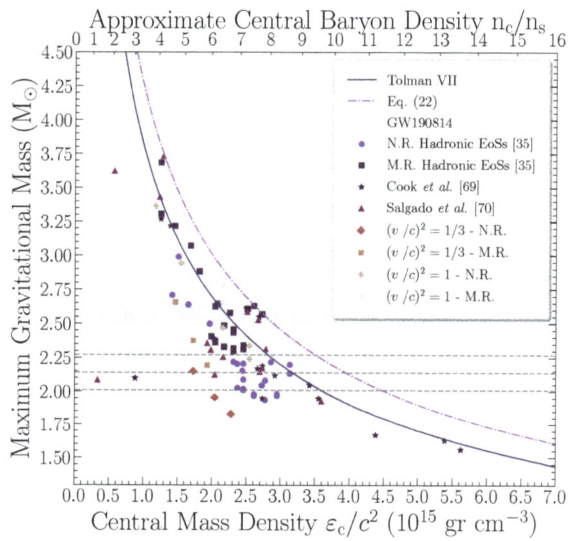

Figure 4. Gravitational mass as a function of the central energy/baryon density at the maximum mass configuration both at non-rotating and maximally-rotating case. Circles correspond to 23 hadronic EoSs [35] at the non-rotating case (N.R.), squares to the corresponding maximally-rotating (M.R.) one, stars to data of Cook et al. [69], and triangles to data of Salgado et al. [70]. In addition, rhombus and pluses mark the non-rotating configuration at the two limiting values of the sound speed, while crosses and polygons marks the maximally-rotating one. The horizontal dashed lines correspond to the observed NS mass limits (2.01 $M_\odot$ [41], 2.14 $M_\odot$ [42], and 2.27 $M_\odot$ [43]). Equation (22) is noted with the dashed-dotted line, while for comparison the Tolman VII analytical solution [35] is shown with the solid line. The mass range of the second component of GW190814 is noted with the horizontal shaded region.

One more interesting property of NSs is the central energy density, as it is connected with the study of the time evolution of pulsars and the appearance of a possible phase transition. In Ref. [35] a relation was found describing the upper bound for the density of cold baryonic matter, as

$$\frac{M}{M_\odot} = 4.25\sqrt{\frac{10^{15}\ \text{gr cm}^{-3}}{\varepsilon_c/c^2}}. \tag{22}$$

Figure 4 presents the maximum gravitational mass as a function both of the central energy density and the central baryon density. In particular, we present the results of 23 hadronic EoSs [35], for the non-rotating and maximally-rotating case, Tolman VII analytical solution, Equation (22), data from Cook et al. [69] and Salgado et al. [70], as well as the newly added data for the non-rotating and maximally-rotating case, both in $(v_s/c)^2 = 1/3$ and $(v_s/c)^2 = 1$, at the transition densities under consideration. Adopting the Equation (22) for the range of the gravitational mass of the low mass component in the GW190814 event, the central energy density could be constrained in the range $\varepsilon_c/c^2 = [2.53, 2.89]\ 10^{15}$ gr cm^{-3}, meaning that NSs with higher values of central energy density could not exist. Furthermore, from Figure 4, we could also extract the corresponding region for the central baryon density, that is $n_c = [7.27, 8.09]\ n_s$. Finally, all the extracted EoSs met the limit for the central energy/baryon density as they are included in the region described under Equation (22).

5.2. Tidal Effects and Speed of Sound: A Very Massive Neutron Star Hypothesis

5.2.1. Isolated Non-Rotating Neutron Star

Firstly, we concentrated our study of tidal deformability on the isolated non-rotating NS case, by using two transition densities $n_{\rm tr} = [1.5, 2]n_s$ and eight values of speed of sound bounds $(v_s/c)^2 = [1/3, 0.4, 0.5, 0.6, 0.7, 0.8, 0.9, 1]$. The values of transition density were taken to be close to the constraints of Ref. [71]. The numerical solution of TOV equations' system, by using the previous bounds for sound speed, provided the mass-radius diagram presented in Figure 5.

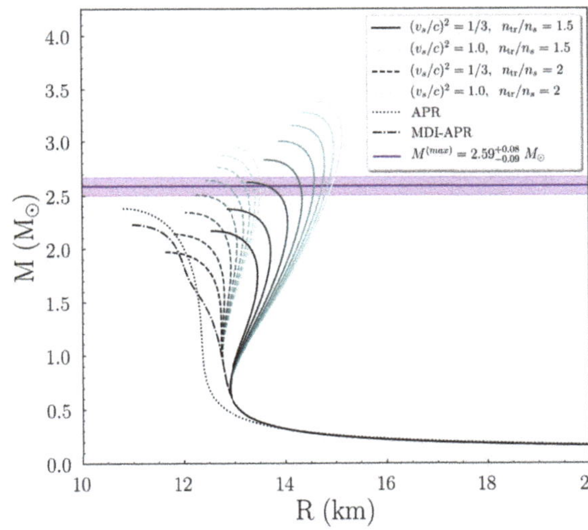

Figure 5. Mass vs. radius for an isolated non-rotating NS, for each transition density $n_{\rm tr}$ and all speed of sound cases. The higher values of speed of sound correspond to lighter curves' color. The purple horizontal line and region indicate the mass estimation of the massive compact object of Ref. [1]. The dashdot (dotted) curve corresponds to the MDI-APR (APR) EoS.

In Figure 5 one can observe two main branches, related to the transition density; the solid (dashed) curves correspond to the $n_{\rm tr} = 1.5n_s$ ($n_{\rm tr} = 2n_s$) case. In each branch, there are bifurcations in the families of EoSs, in analog to each speed of sound boundary condition. The higher the speed of sound, the lighter the representing color of the curves in the figure. The purple solid horizontal line, with the shaded region, indicates the estimation of the recently observed massive compact object of Ref. [1]. As Figure 5 shows, the branch of EoSs with $n_{\rm tr} = 1.5n_s$ provides stiffer EoSs compare to the $n_{\rm tr} = 2n_s$ branch. The EoSs of the $n_{\rm tr} = 1.5n_s$ case are more likely to provide such a massive non-rotating NS, than the $n_{\rm tr} = 2n_s$ case in which three EoSs of the total sum lay outside of the shaded region. Especially, between the same kind of transition density $n_{\rm tr}$ the EoSs with higher speed of sound bounds led to higher values of NS mass and radius, hence a high bound of the speed of sound (even more close to the causality as the transition density is getting higher) was needed for the description of such a massive compact object.

From the observation of Figure 5 a trend across the maximum masses contained in each branch of EoSs, characterized by the speed of sound bound, seemed to be inherent.

Therefore, we constructed the appropriate diagram of Figure 6. The cross (star) marks represent the maximum masses of $n_{\rm tr} = 1.5n_s$ ($n_{\rm tr} = 2n_s$) case. As the speed of sound bound was getting higher, the marks' color lightened. Similarly to the previous figure, the purple solid horizontal line, with the shaded region, indicates the estimation of the recently observed massive compact object of Ref. [1]. The red (green) curves represent the following fit formula for the $n_{\rm tr} = 1.5n_s$ ($n_{\rm tr} = 2n_s$) case, given below

$$M_{max} = c_1 d^{c_2} + c_3 d + c_4, \tag{23}$$

where $d = (v_s/c)^2$. The coefficients are given in Table 2.

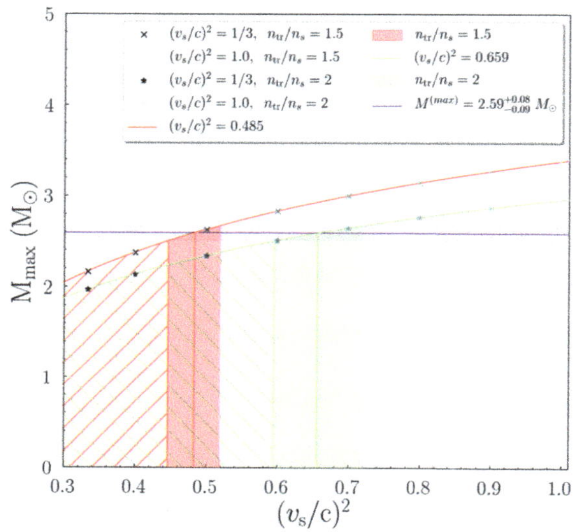

Figure 6. Dependence of a non-rotating NS's maximum mass M_{max} on the speed of sound values $(v_s/c)^2$ for each transition density $n_{\rm tr}$ (in units of saturation density n_s). The red vertical shaded region corresponds to the $n_{\rm tr} = 1.5n_s$ case, while the green one corresponds to the $n_{\rm tr} = 2n_s$ case. The red (green) vertical line indicates the corresponding value of the speed of sound for a massive object with $M = 2.59\ M_\odot$.

By using the mass estimation of the secondary component of GW190814 system, in combination with the fitting formulas mentioned above, we obtained estimations on the speed of sound values for each transition density $n_{\rm tr}$ scenario. In particular, for a non-rotating massive NS with $M = 2.59\ M_\odot$ the value of the speed of sound must be (a) $(v_s/c)^2 = 0.485$ ($n_{\rm tr} = 1.5n_s$), and (b) $(v_s/c)^2 = 0.659$ ($n_{\rm tr} = 2n_s$). The exact values' interval is given in Table 2. We observed that for higher values of transition density $n_{\rm tr}$ the fitted curve and marks were shifted downwards; hence the higher the point of the transition in density, the smaller the provided maximum mass. The higher values of speed of sound were more suitable to describe such massive NSs, until a specific boundary value of transition density $n_{\rm tr}$ in which even the causality would not be suitable. Therefore, a very massive non-rotating NS favored higher values of speed of sound than the $v_s = c/\sqrt{3}$ limit. We noticed that a lower bound on the transition density $n_{\rm tr}$ was needed to be able in the description of the observed NS mergers [71]. Therefore there was a contradiction since the transition density $n_{\rm tr}$ must be above a specific lower limit and not big enough to predict very massive masses. This kind of remark arose in the speed of sound value, respectively.

Table 2. Parameters of the Equation (23) and bounds of speed of sound value of Figure 6. The parameters c_1, c_3, and c_4 are in solar mass units $M_\odot$.

n_{tr}	c_1	c_2	c_3	c_4	$(v_s/c)^2_{min}$	$(v_s/c)^2$	$(v_s/c)^2_{max}$
$1.5n_s$	-1.6033×10^3	-7.56×10^{-4}	-1.64×10^{-1}	1.6068×10^3	0.448	0.485	0.52
$2n_s$	5.5754	0.2742	-0.6912	-1.9280	0.597	0.659	0.72

Moving on to the tidal parameters, we investigated the tidal Love number k_2 and the tidal deformability λ. In Figure 7 we display the two tidal parameters for the single NS case that we examined. In both diagrams, the vertical purple shaded region and line correspond to the GW190814 system's secondary component compact object. There were two main families of EoSs, distinguished by the transition density n_{tr}. In general, the EoSs with higher values of speed of sound bounds led to larger values on both tidal parameters. Therefore, a NS with a higher speed of sound more easily deformable, rather than a more compact star (smaller tidal deformation) with lower speed of sound. As Figure 7 shows, the EoSs with smaller transition density n_{tr} and higher $(v_s/c)^2$ values were more likely to predict a very massive NS of $M = 2.59\ M_\odot$. We postulate that a further study with higher transition density n_{tr} would lead to smaller values of tidal parameters, therefore to more compact stars and more difficult to be deformed. In this case, a very high value of speed of sound, even close to the causal limit, would be necessary to predict such a massive non-rotating NS.

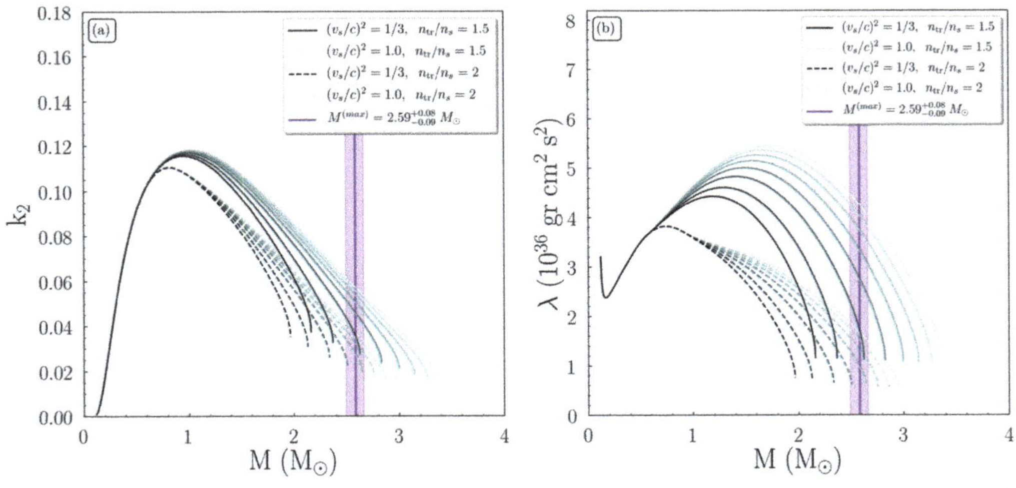

Figure 7. Tidal parameters (**a**) k_2 and (**b**) λ as a function of a NS's mass. The purple vertical line and shaded region indicate the estimation of the recently observed massive compact object of Ref. [1]. The solid (dashed) curves correspond to the $n_{tr} = 1.5n_s$ ($n_{tr} = 2n_s$) case. As the speed of sound is getting higher values, the curves' color lightens.

5.2.2. A Very Massive Neutron Star Component

Regarding the binary NS system case, we considered the scenario of a very heavy component NS, in agreement with the recent observation of GW190814 event [1]. Especially, we consider a heavy mass of $m_1 = 2.59\ M_\odot$ and we let the second star to fluctuate within the range $m_2 \in (1, 2.59)\ M_\odot$. By subtracting the component masses m_1, m_2 in Equation (15) we obtained the corresponding range for the values of $\mathcal{M}_c$. Since the masses were defined, from the Equations (16) and (17), the effective tidal deformability $\tilde{\Lambda}$ could be determined.

Figure 8a shows the effective tidal deformability $\tilde{\Lambda}$ as a function of the chirp mass $\mathcal{M}_c$, for all the possible binary NS systems with such a massive NS component. We noticed that from the total sum of EoSs that we studied in the single NS case above, in the binary case we used only those who can predict a NS with 2.59 $M_\odot$ mass. In general, there were two families of EoSs, distinguished by the transition density n_{tr}. Inside each family of EoSs, the EoSs with higher speed of sound value provided higher values of $\tilde{\Lambda}$. We noticed that for a binary system with $m_1 = 2.59\ M_\odot$ and $m_2 = 1.4\ M_\odot$ the chirp mass was $\mathcal{M}_c = 1.642\ M_\odot$. In addition, binary NS systems with both heavy components, therefore higher $\mathcal{M}_c$, led to smaller values of $\tilde{\Lambda}$. In such cases, possible limits on the lower bound of $\tilde{\Lambda}$ may be more suitable to extract constraints on the EoS.

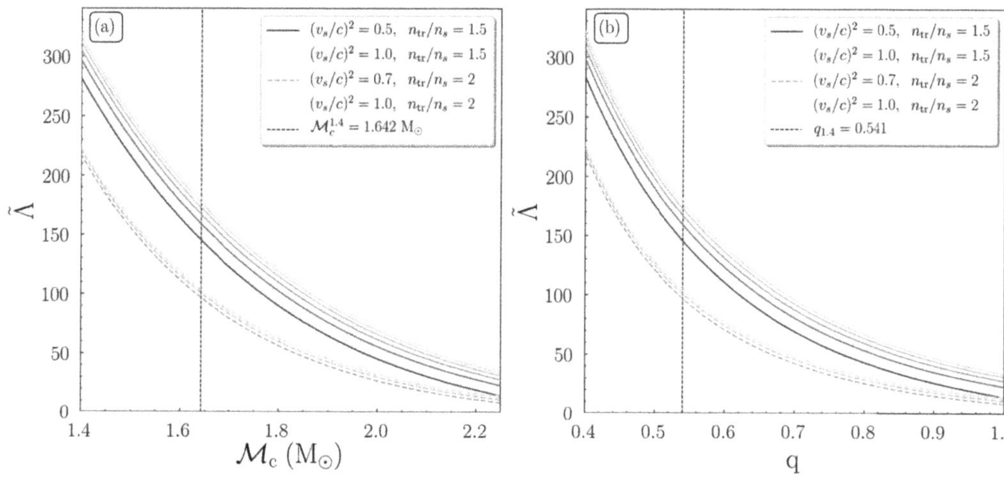

Figure 8. The effective tidal deformability $\tilde{\Lambda}$ as a function of (**a**) the chirp mass $\mathcal{M}_c$ and (**b**) binary mass ratio q, in the case of a very massive NS component, identical to Ref. [1]. As the speed of sound bound is getting higher, the color of EoSs lightens. The black dashed vertical line indicates (**a**) the corresponding chirp mass $\mathcal{M}_c$ and (**b**) mass ratio q, of a binary NS system with $m_1 = 2.59\ M_\odot$ and $m_2 = 1.4\ M_\odot$ respectively.

In the same way, Figure 8b shows the dependence of $\tilde{\Lambda}$ to the corresponding binary mass ratio q. We notice that this kind of $\tilde{\Lambda} - q$ diagram is different from the usual ones (see in comparison Figure 3 of Ref. [71]) because the chirp mass $\mathcal{M}_c$ has not a unique value. To be more specific, $\mathcal{M}_c$ is a variable and each point of Figure 8b corresponds to a different binary NS system with the heavier component in all cases to be a very massive NS of 2.59 $M_\odot$. One can observe a similar behavior of the curves, in analogue to Figure 8a; two main families and the EoSs with higher speed of sound provide higher values of $\tilde{\Lambda}$. As the binary NS systems were more symmetric ($q \to 1$), the binary tidal deformability got smaller. The highest values of $\tilde{\Lambda}$ corresponded to the most asymmetric binary NS systems. We notice that for a binary system with $m_1 = 2.59\ M_\odot$ and $m_2 = 1.4\ M_\odot$ the asymmetry ratio was $q = 0.541$.

Beyond the general behavior of $\tilde{\Lambda}$ that we studied above, it is in our interest to examine the radius and possible constraints that can be derived from it. Following the previous steps, we focused on the $R_{1.4}$ case of a $m_2 = 1.4\ M_\odot$ secondary component NS, as these values can be extracted from Figure 5. The heavier component NS is taken to be $m_1 = 2.59\ M_\odot$. By combining these values with Figure 7b and Equations (16) and (17), we obtained the $\tilde{\Lambda}$. In Figure 9 we display this dependence; the EoSs were in five main groups, characterized by the transition density n_{tr}. We noticed that we expanded our study to transition densities $n_{tr} = [1.25, 1.75, 2.25]$ to be more accurate in calculations and study in more detail the curves' behavior. The higher speed of sound values correspond to lighter marks' color. In analog to the remarks of the previous Figure 8, the high speed of sound bounds lead to higher $\tilde{\Lambda}$ and $R_{1.4}$. Moreover, we applied a fitting expression to the $(v_s/c)^2 = [0.8, 0.9, 1]$ cases. The expression was taken to be in the kind of the proposed relations of Refs. [72,73].

$$\tilde{\Lambda} = c_5 R_{1.4}^{c_6}, \tag{24}$$

where the coefficients for each case are given in Table 3. A recent study suggested a similar power-law relation that connects the tidal deformability of a single NS to the $R_{1.4}$ [74]. The significance of the tidal deformability $\Lambda_{1.4}$ and $R_{1.4}$ in order to extract information about microscopic quantities was studied in Ref. [75].

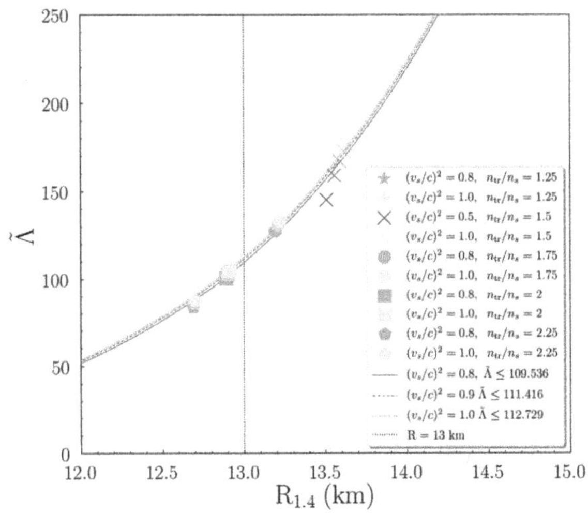

Figure 9. Effective tidal deformability $\tilde{\Lambda}$ vs. radius $R_{1.4}$ of a $m_2 = 1.4\ M_\odot$ NS. The heavier component of the system was taken to be $m_1 = 2.59\ M_\odot$. The lighter colors correspond to higher values of speed of sound bounds. The grey lines indicate the expression of Equation (24). The black dotted vertical line shows the proposed upper limit of Ref. [76].

By applying an upper limit on $R_{1.4}$ one can obtain an upper limit on $\tilde{\Lambda}$ for each case. We adopted the general limit of Ref. [76] that led us to the constraints of Table 3.

Table 3. Parameters of the Equation (24) and bounds of $\tilde{\Lambda}$ of Figure 9.

$(v_s/c)^2$	c_5 (km^{-1})	c_6	$\tilde{\Lambda}$
0.8	4.1897×10^{-9}	9.3518	109.536
0.9	5.3213×10^{-9}	9.2652	111.416
1	6.1109×10^{-9}	9.2159	112.729

6. Concluding Remarks

The GW190814 puzzle and its nature through the nuclear EoS has been addressed in this study. In particular, an effort to explain the existence of a $\sim$2.6 $M_\odot$ NS, which falls into the NS - black hole mass gap, had been made both for non-rotating and maximally-rotating NSs. For this reason, the MDI-APR EoS and its parametrization for various values of sound speed and transition density in the ranges $(v_s/c)^2 = [1/3, 1]$ and $n_{\rm tr} = [1.5, 5]\ n_s$, respectively, have been studied.

Firstly, we compare the MDI-APR EoS with the applicable range of the proposed Kerr parameter from Ref. [16], where the authors restraint it using an analytical relation connecting this property with the maximum mass of a non-rotating NS. The results shown that the MDI-APR EoS lies in the range of the gravitational mass of the low mass component, as well as the one for the Kerr parameter. In fact, the MDI-APR EoS proposes that the $\sim$2.6 $M_{\odot}$ compact star is a rapidly rotating NS, close to its mass-shedding limit.

By considering that the $\sim$2.6 $M_{\odot}$ compact star had been rotating at its mass-shedding limit, possible constraints can be extracted for the corresponding equatorial radius. The Kerr parameter at the mass-shedding limit can be expressed as a relation with the gravitational mass, and hereafter a region of $\mathcal{K}_{\rm max} = [0.67, 0.69]$ is extracted. This region also includes the upper limit of the relevant region from Ref. [16] in a narrow range. In addition, using a relation that connects the Kerr parameter and the compactness parameter at its mass-shedding limit, a possible tight region for the equatorial radius of the star is implied as $R_{\rm max} = [14.77, 14.87]$ km.

The upper limit on the central energy/baryon density is a very interesting property, as it is connected with the evolution of the NS and the possible appearance of a phase transition. From this analysis, the central energy density must be lower than the values in the range $\varepsilon_c/c^2 = [2.53, 2.89]\ 10^{15}$ gr cm^{-3}, while for the central baryon density, the corresponding range is $n_c = [7.27, 8.09]\ n_s$. The latter can inform us about the stability of the NS, as a NS with higher values of central energy/baryon density cannot exist, as well as the appearance of the back-bending process.

The transition density, along with the speed of sound, can infer various structures for the EoS. Assuming that the NSs is a non-rotating one, the transition density is constrained in the region $n_{\rm tr} = [1.5, 3.2]\ n_s$ while the corresponding value of sound speed must be in the range $(v_s/c)^2 = [0.45, 1]$. Otherwise, if the NS is considered as a maximally-rotating one, although the speed of sound implies no constraints, the transition density must be higher than 1.6 n_s.

In the case of non-rotating NS, the construction of the M-R diagram showed at first glance the cases that can describe the extreme scenario of our study. Moving to a more detailed diagram of the mass vs. the speed of sound bounds, it was feasible to extract stringent constraints on the speed of sound bounds for each case of transition density $n_{\rm tr}$. For $n_{\rm tr} = 1.5 n_s$ this bound is $(v_s/c)^2 \in [0.448, 0.52]$ while for $n_{\rm tr} = 2n_s$ is $(v_s/c)^2 \in [0.597, 0.72]$. We observe that the first lower bound is in agreement with the bound extracted above, which is a good validation of our result. We postulate that for higher transition densities $n_{\rm tr}$ it is more difficult to achieve such a massive non-rotating NS. As the transition density $n_{\rm tr}$ grows the speed of sound would need to be even close to the causal limit.

The study of the tidal parameters for a single non-rotating NS allowed us to examine the behavior of EoSs in each case. This lead to the general conclusion that the lower transition densities $n_{\rm tr}$ lead to higher tidal parameters. Therefore the transition density $n_{\rm tr} = 2n_s$ corresponds to a more compact and less deformable NS. Among the same kind of transition density $n_{\rm tr}$, the EoSs with higher speed of sound values provide higher tidal parameters. Hence, in a second level across the same kind of $n_{\rm tr}$ EoSs, the higher speed of sound bound signifies that the tidal deformation is higher and the star is less compact.

Concerning the binary NS system case, the adoption of a very massive component with $m_1 = 2.59\ M_{\odot}$ allowed us to investigate a variety of possible binary NS systems with such a heavy component. We notice that as the binary NS system consists of both heavy component stars, therefore high chirp mass $\mathcal{M}_c$, the effective tidal deformability $\tilde{\Lambda}$ is taking smaller values. Hence, the binary deformation is smaller in such systems. Similarly, the same behavior was noticed in the $\tilde{\Lambda} - q$ diagram in which the increasing binary mass symmetric ratio q leads to smaller values of $\tilde{\Lambda}$. In the case that the second component has a mass $m_2 = 1.4\ M_{\odot}$, the chirp mass of the system $\mathcal{M}_c$ and the ratio q are $\mathcal{M}_c = 1.642\ M_{\odot}$ and $q = 0.541$ respectively.

Lastly, we considered the case of a binary NS system with $m_1 = 2.59\ M_{\odot}$ with a secondary component $m_2 = 1.4\ M_{\odot}$. This selection permitted the study of the radius $R_{1.4}$ and the extraction of possible constraints. In general, the transition density $n_{\rm tr} = 1.5n_s$ provides higher values of $R_{1.4}$ and $\tilde{\Lambda}$ than the $n_{\rm tr} = 2n_s$ case. The examination of other transition densities $n_{\rm tr}$ permitted us to confirm this behavior. In addition, the high values of speed of sound $(v_s/c)^2$ exhibits a similar behavior; high speed of sound bounds provide higher values on both referred parameters. The adoption of an upper limit on the radius, allowed us to extract some upper limits on $\tilde{\Lambda}$ for each case of sound speed. Nevertheless the value of $n_{\rm tr}$, as the speed of sound bound is getting higher the upper limit on $\tilde{\Lambda}$ performs an analogous course. We conclude that the existence of such a massive non-rotating NS would require a significant differentiation from all the so far known cases, consisting in any case a unique and very interesting challenge for physics.

7. Materials and Methods

The numerical integration of the equilibrium equations for NSs is under the publicly available RNS code [44] by Stergioulas and Friedman [77]. This code was developed based on the method of Komatsu, Eriguchi, and Hachisu (KEH) [78], while modifications were introduced by Cook, Shapiro, and Teukolsky [79]. The input of the code was the EoS in a tabulated form which included the energy density, the pressure, the enthalpy, and the baryon density.

Author Contributions: Conceptualization, A.K.-P., P.S.K. and C.C.M.; Methodology, A.K.-P. and P.S.K.; Software, A.K.-P. and P.S.K.; Validation, A.K.-P., P.S.K. and C.C.M.; Formal Analysis, A.K.-P. and P.S.K.; Investigation, A.K.-P., P.S.K. and C.C.M.; Data Curation, A.K.-P. and P.S.K.; Writing—Original Draft Preparation, A.K.-P., P.S.K. and C.C.M.; Writing—Review & Editing, A.K.-P., P.S.K. and C.C.M.; Visualization, A.K.-P., P.S.K. and C.C.M.; Supervision, C.C.M. All authors contributed equally to this work. All authors have read and agreed to the published version of the manuscript.

Funding: This research received no external funding.

Institutional Review Board Statement: Not applicable.

Informed Consent Statement: Not applicable.

Data Availability Statement: Not applicable.

Acknowledgments: The authors thank K. Kokkotas for his constructive comments on the preparation of the manuscript and also L. Rezzolla for useful correspondence and clarifications.

Conflicts of Interest: The authors declare no conflict of interest.

Abbreviations

The following abbreviations are used in this manuscript:

EoS	Equation of state
NS	Neutron star
QCD	Quantum chromodynamics
MDI	Momentum dependent interaction
APR	Akmal, Pandharipande and Ravenhall
SNM	Symmetric Nuclear Matter
N.R.	Non-rotating configuration
M.R.	Maximally-rotating configuration

References

1. Abbott, R.; Abbott, T.D.; Abraham, S.; Acernese, F.; Ackley, K.; Adams, C.; Adhikari, R.X.; Adya, V.B.; Affeldt, C.; Agathos, M.; et al. GW190814: Gravitational Waves from the Coalescence of a 23 Solar Mass Black Hole with a 2.6 Solar Mass Compact Object. *Astrophys. J. Lett.* **2020**, *896*, L44. [CrossRef]
2. Datta, S; Phukon, K.S.; Bose, S. Recognizing black holes in gravitational-wave observations: Telling apart impostors in mass-gap binaries. *arXiv* **2020**, arXiv:gr-qc/2004.05974.
3. Alsing, J; Silva, H.O.; Berti, E. Evidence for a maximum mass cut-off in the neutron star mass distribution and constraints on the equation of state. *Mon. Not. R. Astron. Soc.* **2018**, *478*, 1377–1391. [CrossRef]
4. Farr, W.M.; Chatziioannou, K. A Population-Informed Mass Estimate for Pulsar J0740+6620. *Res. Notes AAS* **2020**, *4*, 65. [CrossRef]
5. Abbott, B.P.; Abbott, R.; Abbott, T.D.; Acernese, F.; Ackley, K.; Adams, C.; Adams, T.; Addesso, P.; Adhikari, R.X.; Adya, V.B.; et al. (The LIGO Scientific Collaboration and the Virgo Collaboration) GW170817: Measurements of Neutron Star Radii and Equation of State. *Phys. Rev. Lett.* **2018**, *121*, 161101. [CrossRef]
6. Tsokaros, A.; Ruiz, M.; Shapiro, S.L. GW190814: Spin and equation of state of a neutron star companion. *Astrophys. J.* **2020**, *905*, 48. [CrossRef]

7. Huang, K.; Hu, J.; Zhang, Y.; Shen, H. The Possibility of the Secondary Object in GW190814 as a Neutron Star. *Astrophys. J.* **2020**, *904*, 39. [CrossRef]
8. Zevin, M.; Spera, M.; Berry, C.; Kalogera, V. Exploring the Lower Mass Gap and Unequal Mass Regime in Compact Binary Evolution. *Astrophys. J. Lett.* **2020**, *899*, L1. [CrossRef]
9. Fattoyev, F.J.; Horowitz, C.J.; Piekarewicz, J.; Reed, B. GW190814: Impact of a 2.6 solar mass neutron star on nucleonic equations of state. *Phys. Rev. C* **2020**, *102*, 065805. [CrossRef]
10. Essick, R.; Landry, P.; Discriminating between Neutron Stars and Black Holes with Imperfect Knowledge of the Maximum Neutron Star Mass. *Astrophys. J.* **2020**, *904*, 80. [CrossRef]
11. Safarzadeh, M.; Loeb, A. Formation of Mass Gap Objects in Highly Asymmetric Mergers. *Astrophys. J. Lett.* **2020**, *899*, L15. [CrossRef]
12. Godzieba, D.A.; Radice, D.; Bernuzzi, S. On the maximum mass of neutron stars and GW190814. *arXiv* **2020**, arXiv:astro-ph.HE/2007.10999.
13. Sedrakian, A.; Weber, F.; Li, J.J. Confronting GW190814 with hyperonization in dense matter and hypernuclear compact stars. *Phys. Rev. D* **2020**, *102*, 041301. [CrossRef]
14. Biswas, B.; Nandi, R.; Bose, S.; Stergioulas, N. GW190814: On the properties of the secondary component of the binary. *arXiv* **2020**, arXiv:astro-ph.HE/2010.02090.
15. Zhang, N.B.; Li, B.A. GW190814's Secondary Component with Mass 2.50–2.67 $M_{\odot}$ as a Superfast Pulsar. *Astrophys. J.* **2020**, *902*, 38. [CrossRef]
16. Most, E.R.; Papenfort, L.J.; Weih, L.R.; Rezzolla, L. A lower bound on the maximum mass if the secondary in GW190814 was once a rapidly spinning neutron star. *Mon. Not. R. Astron. Soc.* **2020**, *499*, L82–L86. [CrossRef]
17. Tan, H.; Hostler-Noronha, J.; Yunes, N. Neutron Star Equation of State in light of GW190814. *Phys. Rev. Lett.* **2020**, *125*, 261104. [CrossRef]
18. Zhang, C.; Mann, R.B. Unified Interacting Quark Matter and its Astrophysical Implications. *arXiv* **2020**, arXiv:astro-ph.HE/2009.07182.
19. Bombaci, I.; Drago, A.; Logoteta, D.; Pagliara, G.; Vidana, I. Was GW190814 a black hole–strange quark star system? *arXiv* **2020**, arXiv:nucl-th/2010.01509.
20. Demircik, T.; Ecker, C.; Jarvinen, M. Rapidly Spinning Compact Stars with Deconfinement Phase Transition. *arXiv* **2020**, arXiv:astro-ph.HE/2009.10731.
21. Cao, Z.; Chen, L.W.; Chu, P.C.; Zhou, Y. GW190814: Circumstantial Evidence for Up-Down Quark Star. *arXiv* **2020**, arXiv:astro-ph.HE/2009.00942.
22. Dexheimer, V.; Gomes, R.O.; Klahn, T.; Han, S.; Salinas, M. GW190814 as a massive rapidly-rotating neutron star with exotic degrees of freedom. *arXiv* **2020**, arXiv:astro-ph.HE/2007.08493.
23. Roupas, Z.; Panotopoulos, G.; Lopes, I. QCD color superconductivity in compact stars: color-flavor locked quark star candidate for the gravitational-wave signal GW190814. *arXiv* **2020**, arXiv:astro-ph.HE/2010.11020.
24. Rather, I.A.; Usmani, A.A; Patra, S.K. Hadron-Quark phase transition in the context of GW190814. *arXiv* **2020**, arXiv:nucl-th/2011.14077.
25. Moffat, J.W. Modified Gravity (MOG) and Heavy Neutron Star in Mass Gap. *arXiv* **2020**, arXiv:gr-qc/2008.04404.
26. Astashenok, A.V.; Capozziello, S.; Odintsov, S.D.; Oikonomou, V.K. Extended gravity description for the GW190814 supermassive neutron star. *Phys. Lett. B* **2020**, *811*, 135910. [CrossRef]
27. Nunes, R.C.; Coelho, G.G.; Araujo, J.C.N. Weighing massive neutron star with screening gravity: a look on PSR J0740 + 6620 and GW190814 secondary component. *Eur. Phys. J. C* **2020**, *80*, 1115. [CrossRef]
28. Annala, E.; Gorda, T.; Kurkela, A.; Nättilä, J.; Vuorinen, A. Evidence for quark-matter cores in massive neutron stars. *Nat. Phys.* **2020**, *16*, 907–910. [CrossRef]
29. Bombaci, I. The Hyperon Puzzle in Neutron Stars. *JPS Conf. Proc.* **2017**, *17*, 101002.
30. Shapiro, S.L.; Teukolsky, S.A. *Black Holes, White Dwarfs, and Neutron Stars*; John Wiley and Sons: Hoboken, NJ, USA; New York, NY, USA, 1983.
31. Glendenning, N.K. *Compact Stars: Nuclear Physics, Particle Physics, and General Relativity*; Springer: Berlin/Heidelberg, Germany, 2000.
32. Haensel, P.; Potekhin, A.Y.; Yakovlev, D.G. *Neutron Stars 1: Equation of State and Structure*; Springer: New York, NY, USA, 2007.
33. Weinberg, S. *Gravitational and Cosmology: Principle and Applications of the General Theory of Relativity*; Wiley: New York, NY, USA, 1972.
34. Zel'dovich, Y.B.; Novikov, I.D. *Stars and Relativity*; Dover Publications, Inc.: New York, NY, USA, 1971.
35. Koliogiannis, P.S.; Moustakidis, C.C. Effects of the equation of state on the bulk properties of maximally rotating neutron stars. *Phys. Rev. C* **2020**, *101*, 015805. [CrossRef]
36. Prakash, M.; Bombaci, I.; Prakash, Manju; Ellis, P.J.; Lattimer, J.M.; Knorren, R. Composition and structure of protoneutron stars. *Phys. Rep.* **1997**, *280*, 1. [CrossRef]
37. Moustakidis, C.C. Effects of the nuclear equation of state on the r-mode instability and evolution of neutron stars. *Phys. Rev. C* **2015**, *91*, 035804. [CrossRef]
38. Gale, C.; Bertsch, G.; Das Gupta, S. Heavy-ion collision theory with momentum-dependent interactions. *Phys. Rev. C* **1987**, *35*, 1666. [CrossRef] [PubMed]

39. Gale, C.; Welke, G.M.; Prakash, M.; Lee, S.J.; Das Gupta, S. Transverse momenta, nuclear equation of state, and momentum-dependent interactions in heavy-ion collisions. *Phys. Rev. C* **1990**, *41*, 1545. [CrossRef] [PubMed]
40. Akmal, A.; Pandharipande, V.R.; Ravenhall, D.G. Equation of state of nucleon matter and neutron star structure. *Phys. Rev. C* **1998**, *58*, 1804. [CrossRef]
41. Antoniadis, J.; Freire, P.C.; Wex, N.; Tauris, T.M.; Lynch, R.S.; van Kerkwijk, M.H.; Kramer, M.; Bassa, C.; Dhillon, V.S.; Driebe, T.; et al. A Massive Pulsar in a Compact Relativistic Binary. *Science* **2013**, *340*, 1233232. [CrossRef]
42. Cromartie, H.T.; Fonseca, E.; Ransom, S.M.; Demorest, P.B.; Arzoumanian, Z.; Blumer, H.; Brook, P.R.; DeCesar, M.E.; Dolch, T.; Ellis, J.A.; et al. Relativistic Shapiro delay measurements of an extremely massive millisecond pulsar. *Nat. Astron.* **2020**, *4*, 72. [CrossRef]
43. Linares, M.; Shahbaz, T.; Casares, J. Peering into the Dark Side: Magnesium Lines Establish a Massive Neutron Star in PSR J2215+5135. *Astrophys. J.* **2018**, *859*, 54. [CrossRef]
44. Stergioulas, N. 1996. Available online: http://www.gravity.phys.uwm.edu/rns/ (accessed on 15 December 2020).
45. Breu, C.; Rezzolla, L. Maximum mass, moment of inertia and compactness of relativistic stars. *Mon. Not. R. Astron. Soc.* **2016**, *459*, 646. [CrossRef]
46. Margaritis, C.; Koliogiannis, P.S.; Moustakidis, C.C. Speed of sound constraints on maximally rotating neutron stars. *Phys. Rev. D* **2020**, *101*, 043023. [CrossRef]
47. Rhoades, C.E.; Ruffini, R. Maximum Mass of a Neutron Star. *Phys. Rev. Lett.* **1974**, *32*, 324. [CrossRef]
48. Kalogera, V.; Baym, G. The Maximum Mass of a Neutron Star. *Astrophys. J.* **1996**, *470*, L61. [CrossRef]
49. Koranda, S.; Stergioulas, N.; Friedman, J.L. Upper Limits Set by Causality on the Rotation and Mass of Uniformly Rotating Relativistic Stars. *Astrophys. J.* **1997**, *488*, 799. [CrossRef]
50. Chamel, N.; Haensel, P.; Zdunik, J.L.; Fantina, A.F. On the maximum mass of neutron stars. *Int. J. Mod. Phys. E* **2013**, *22*, 1330018. [CrossRef]
51. Podkowka, D.M.; Mendes, R.F.P.; Poisson, E. Trace of the energy-momentum tensor and macroscopic properties of neutron star. *Phys. Rev. D* **2018**, *98*, 064057. [CrossRef]
52. Xia, C.; Zhu, Z.; Zhou, X.; Li, A. Sound velocity in dense stellar matter with strangeness and compact stars. *arXiv* **2019**, arXiv:nucl-th/1906.00826.
53. Feynman, R.P.; Metropolis, N.; Teller, E. Equations of State of Elements Based on the Generalized Fermi-Thomas Theory. *Phys. Rev.* **1949**, *75*, 1561. [CrossRef]
54. Baym, G.; Pethik, C.; Sutherland, P. The Ground State of Matter at High Densities: Equation of State and Stellar Models. *Astrophys. J.* **1971**, *170*, 299. [CrossRef]
55. Tews, I.; Carlson, J.; Gandolfi, S.; Reddy, S. Constraining the Speed of Sound inside Neutron Stars with Chiral Effective Field Theory Interactions and Observations. *Astrophys. J.* **2018**, *860*, 149. [CrossRef]
56. Postnikov, S.; Prakash, M.; Lattimer, J.M. Tidal Love numbers of neutron and self-bound quark stars. *Phys. Rev. D* **2010**, *82*, 024016. [CrossRef]
57. Baiotti, L. Gravitational waves from neutron star mergers and their relation to the nuclear equation of state. *Prog. Part. Nucl. Phys.* **2019**, *109*, 103714. [CrossRef]
58. Flanagan, E.E.; Hinderer, T. Constraining neutron-star tidal Love numbers with gravitational-wave detectors. *Phys. Rev. D* **2008**, *77*, 021502(R). [CrossRef]
59. Hinderer, T. Tidal Love Numbers of Neutron Stars. *Astrophys. J.* **2008**, *677*, 1216. [CrossRef]
60. Damour, T.; Nagar, A. Relativistic tidal properties of neutron stars. *Phys. Rev. D* **2009**, *80*, 084035. [CrossRef]
61. Hinderer, T.; Lackey, B.D.; Lang, R.N.; Read, J.S. Tidal deformability of neutron stars with realistic equations of state and their gravitational wave signatures in binary inspiral. *Phys.Rev. D* **2010**, *81*, 123016. [CrossRef]
62. Fattoyev, F.J.; Carvajal, J.; Newton, W.G.; Li, B.A. Constraining the high-density behavior of the nuclear symmetry energy with the tidal polarizability of neutron stars. *Phys. Rev. C* **2013**, *87*, 015806. [CrossRef]
63. Lackey, B.D.; Wade, L. Reconstructing the neutron-star equation of state with gravitational-wave detectors from a realistic population of inspiralling binary neutron stars. *Phys. Rev. D* **2015**, *91*, 043002. [CrossRef]
64. Takátsy, J.; Kovács, P. Comment on "Tidal Love numbers of neutron and self-bound quark stars". *Phys. Rev. D* **2020**, *102*, 028501. [CrossRef]
65. Thorne, K.S.; Tidal stabilization of rigidly rotating, fully relativistic neutron stars. *Phys. Rev. D* **1998**, *58*, 124031. [CrossRef]
66. Abbott, B.P.; Abbott, R.; Abbott, T.D.; Acernese, F.; Ackley, K.; Adams, C.; Adams, T.; Addesso, P.; Adhikari, R.X.; Adya, V.B.; et al. GW170817: Observation of Gravitational Waves from a Binary Neutron Star Inspiral. *Phys. Rev. Lett.* **2017**, *119*, 161101. [CrossRef]
67. Abbott, B.P.; Abbott, R.; Abbott, T.D.; Acernese, F.; Ackley, K.; Adams, C.; Adams, T.; Addesso, P.; Adhikari, R.X.; Adya, V.B.; et al. Properties of the Binary Neutron Star Merger GW170817. *Phys. Rev. X* **2019**, *9*, 011001. [CrossRef]
68. Koliogiannis, P.S.; Moustakidis, Ch.C. Thermodynamical description of hot rapidly rotating neutron stars and neutron stars merger remnant. *arXiv* **2020**, arXiv:astro-ph.HE/2007.10424 .
69. Cook, G.B.; Shapiro, S.L.; Teukolsky, S.A. Rapidly Rotating Neutron Stars in General Relativity: Realistic Equations of State. *Astrophys. J.* **1994**, *424*, 823. [CrossRef]
70. Salgado, M.; Bonazzola, S.; Gourgoulhon, E.; Haensel, P. High precision rotating neutron star models 1: Analysis of neutron star properties. *Astron. Astrophys.* **1994**, *291*, 155.

71. Kanakis-Pegios, A.; Koliogiannis, P.S.; Moustakidis, C.C. Speed of sound constraints from tidal deformability of neutron stars. *Phys. Rev. C* **2020**, *102*, 055801. [CrossRef]
72. Zhao, T.; Lattimer, J.M. Tidal deformabilities and neutron star mergers. *Phys. Rev. D* **2018**, *98*, 063020. [CrossRef]
73. De, S.; Finstad, D.; Lattimer, J.M.; Brown, D.A.; Berger, E.; Biwer, C.M. Tidal Deformabilities and Radii of Neutron Stars from the Observation of GW170817. *Phys. Rev. Lett.* **2018**, *121*, 091102. [CrossRef]
74. Tsang, C.Y.; Tsang, M.B.; Danielewicz, P.; Fattoyev, F.J.; Lynch, W.G. Insights on Skyrme parameters from GW170817. *Phys. Lett. B* **2019**, *796*, 1. [CrossRef]
75. Wei, J.-B.; Lu, J.-J.; Burgio, G.F.; Li, Z.-H.; Schulze, H.-J. Are nuclear matter properties correlated to neutron star observables? *Eur. Phys. J. A* **2020**, *56*, 63. [CrossRef]
76. Raithel, C.A.; Özel, F.; Psaltis, D. Tidal Deformability from GW170817 as a Direct Probe of the Neutron Star Radius. *Astrophys. J. Lett.* **2018**, *857*, L23. [CrossRef]
77. Stergioulas, N.; Friedman, J.L. Comparing Models of Rapidly Rotating Relativistic Stars Constructed by Two Numerical Methods. *Astrophys. J.* **1995**, *444*, 306. [CrossRef]
78. Komatsu, H.; Eriguchi, Y.; Hachisu, I. Rapidly rotating general relativistic stars. I - Numerical method and its application to uniformly rotating polytropes. *Mon. Not. R. Astron. Soc.* **1989**, *237*, 355. [CrossRef]
79. Cook, G.B.; Shapiro, S.L.; Teukolsky, S.A. Rapidly Rotating Polytropes in General Relativity. *Astrophys. J.* **1994**, *422*, 227. [CrossRef]

symmetry

MDPI

Article

A Modern View of the Equation of State in Nuclear and Neutron Star Matter

G. Fiorella Burgio *, Hans-Josef Schulze, Isaac Vidaña and Jin-Biao Wei

INFN Sezione di Catania, Dipartimento di Fisica e Astronomia, Università di Catania, Via Santa Sofia 64, 95123 Catania, Italy; schulze@ct.infn.it (H.-J.S.); isaac.vidana@ct.infn.it (I.V.); jinbiao.wei@ct.infn.it (J.-B.W.)
* Correspondence: fiorella.burgio@ct.infn.it

Abstract: Background: We analyze several constraints on the nuclear equation of state (EOS) currently available from neutron star (NS) observations and laboratory experiments and study the existence of possible correlations among properties of nuclear matter at saturation density with NS observables. **Methods:** We use a set of different models that include several phenomenological EOSs based on Skyrme and relativistic mean field models as well as microscopic calculations based on different many-body approaches, i.e., the (Dirac–)Brueckner–Hartree–Fock theories, Quantum Monte Carlo techniques, and the variational method. **Results:** We find that almost all the models considered are compatible with the laboratory constraints of the nuclear matter properties as well as with the largest NS mass observed up to now, $2.14^{+0.10}_{-0.09}\,M_\odot$ for the object PSR J0740+6620, and with the upper limit of the maximum mass of about 2.3–2.5 $M_\odot$ deduced from the analysis of the GW170817 NS merger event. **Conclusion:** Our study shows that whereas no correlation exists between the tidal deformability and the value of the nuclear symmetry energy at saturation for any value of the NS mass, very weak correlations seem to exist with the derivative of the nuclear symmetry energy and with the nuclear incompressibility.

Keywords: nuclear matter; neutron star; equation of state; gravitational waves; binary mergers

Citation: Burgio, G.F.; Schulze, H.-J.; Vidaña, I.; Wei, J.-B. A Modern View of the Equation of State in Nuclear and Neutron Star Matter. *Symmetry* **2021**, *13*, 400. https://doi.org/10.3390/sym13030400

Academic Editors: Charalampos Moustakidis and Sergei D. Odintsov

Received: 15 December 2020
Accepted: 24 February 2021
Published: 28 February 2021

1. Introduction

A neutron star (NS) is the collapsed core of a massive star ($8-25M_\odot$, with $M_\odot \approx 2\times10^{33}$ g the mass of the sun), which at the end point of its evolution cannot be supported by hydrostatic pressure and collapses, producing a supernova explosion. A huge amount of gravitational energy is released, mainly in the form of neutrino radiation, and this leads to the complete destruction of the progenitor star. NSs may have masses in the range $M \sim 1-2M_\odot$ and radii of about 10–15 km. A huge amount of data has been collected from more than 50 years of NS observations, performed with ground-based and on-board telescopes covering all bands of the electromagnetic spectrum. In 2017, the observations witnessed an important breakthrough thanks to the direct detection (GW170817) of gravitational waves from such an event by the Advanced LIGO and Virgo collaborations Ref. [1–3]. In fact, it has been found that the observations of NS mergers can potentially provide strong constraints on the nuclear equation of state (EOS), as discussed in Refs. [4,5].

The EOS of isospin-asymmetric nuclear matter plays a major role not only in the study of NS structure and composition but also in the evolution of core-collapse supernovae and binary compact-star mergers [6,7]. Additionally, matter flows generated in heavy ion collisions (HIC) and the properties of nuclei in their ground state are strongly affected by the relevant features of the EOS, in particular the symmetry energy and its first derivative and the compressibility. In principle, it can be expected that in high-energy HICs as well as in supernova explosions and binary NS mergers, thermal effects are quite important, and thus they should be correctly included in the EOS. Besides that, the large density reached in the inner core of a NS can pose several theoretical problems because a complete theory of nuclear interactions based on QCD cannot be solved yet on the lattice for arbitrarily large values of density, temperature and isospin asymmetry, because of the well-known sign problem. Therefore, one has to rely on theoretical models and methods of the nuclear many-body theory in order to build the nuclear EOS.

Among possible observables regarding NSs, the mass and radius are the most promising ones, and they could ideally be used to infer the NS EOS within a certain observational uncertainty. While the masses of several NSs are known with good precision [8–12], information on their radii is less accurate [13,14]. However, simultaneous measurement of both quantities for several objects is required in order to constrain the EOS of NS matter and allow robust conclusions. The recent observation of gravitational waves (GWs) emitted during the merger of two corotating NSs [1–3] has opened the door to new possibilities of obtaining information on their masses and radii, by means of the measurement of the tidal deformability [15,16], and us allowed to deduce upper and lower limits on it [2,17]. The tidal deformability measures the linear response of the quadrupole deformation to a (weak) external gravitational field, and thus could be well constrained by the new data. It is therefore of interest to examine these quantities and their relations with other observables in theoretical calculations of the EOS. Moreover, the improved accuracy for the radius reached in recent observations by NICER (Neutron Star Interior Composition Explorer) Ref. [18,19], and the future planned missions like eXTP [20], will allow us to statistically infer NS mass and radius to better accuracy.

In this work we analyze the available constraints on the nuclear EOS, and compare with those derived from both phenomenological and ab-initio theoretical models. Possible correlations among the properties of nuclear matter close to saturation density with related quantities deduced from NS observations and nuclear physics experiments will be analyzed. We limit ourselves to the description of NS matter considering only nucleonic degrees of freedom, thus ignoring the possible appearance of hyperons [21] and a phase transition to the quark phase [22].

The article is organized as follows. In Section 2 we briefly review the different theoretical approaches for the nuclear EOS and illustrate the ones we adopt in the present study. We then discuss the experimental constraints on the nuclear EOS in Section 3. A brief overview of different EOSs of betastable matter is given in Section 4, and in Section 5 we discuss the comparison and correlations between NS EOS, nuclear physics and astrophysical constraints. Conclusions are drawn in Section 6.

2. Equations of State

The most commonly used theoretical approaches to determine the nuclear EOS can be classified into phenomenological and microscopic ones. (Non-)relativistic phenomenological approaches are based on effective interactions that are built to describe finite nuclei in their ground state, and therefore predictions at high isospin asymmetries should be considered with care [23]. In fact, at larger densities no experimental data are available, and therefore their behaviour can be very different. Skyrme interactions [24,25] and relativistic mean-field (RMF) models [26] are among the most used ones.

In this work we use a limited sample of Skyrme forces, namely GS and Rs [27], SLy4 [28] of the Lyon group, the old SV [29], SkI4 [30] of the SkI family, SkMP [31] and SkO [32]. We also include three EOSs derived within the modern Brussels-Montreal family of unified models, i.e., BSk22,24,26, which are commonly used in NS calculations [33]. We also consider two types of RMF models, which are based on effective Lagrangian densities where the interaction between baryons is described in terms of meson exchanges. In particular, we adopt models with constant meson-baryon couplings described by the Lagrangian density of the nonlinear Walecka model (NLWM), and models with density-dependent couplings [hereafter referred to as density-dependent models (DDM)]. Within the first type, we consider the models GM1 and GM3 [34]. For the DDM, we consider the models DDME1, DDME2 [35] and TW99 [36]. A further phenomenological RMF EOS, the SFHO EOS [37], has been used for comparison. A larger sample of RMF models, consistent with the analysis of Ref. [38], has been studied in Refs. [39,40], where the Love number and corresponding tidal deformabilities show very good agreement with the recent data from the GW170817 merger event.

Realistic two- and three-nucleon forces, that describe nucleon scattering data in free space and the properties of the deuteron, are the essential input for the microscopic approaches. These interactions are based on meson-exchange theory [41,42], or the recent chiral perturbation theory [43–46]. The main theoretical challenge is the treatment of the short-range repulsive core, which characterizes the nucleon-nucleon interaction, and this makes the difference among the available many-body approaches. The most well-known are the Brueckner–Hartree–Fock (BHF) [47] and its relativistic version, the Dirac–Brueckner–Hartree–Fock (DBHF) [48–50] theories, the variational method [51], the self-consistent Green's function technique [52,53], the Quantum Monte Carlo techniques [54,55], the chiral effective field theory [56] and the $V_{\mathrm{low\,k}}$ approach [57].

In this paper we adopt several BHF EOSs based on different nucleon-nucleon potentials, namely the Bonn B (BOB) [41,58], the Nijmegen 93 (N93) [42,59], and the Argonne V_{18} (V18) [60]. In all those cases, the two-body forces are supplemented by nucleonic three-body forces (TBF), which are needed in all non-relativistic many-body methods in order to reproduce correctly the saturation properties of nuclear matter. Since a complete theory of TBF starting from first principles is not available yet, we adopt either phenomenological or microscopic models [61–64]. The phenomenological approach is based on the Urbana model (labelled as UIX) [62,65,66], whereas the microscopic TBF employes the same meson exchange as in the two-body force, as described in detail in Refs. [64,67]. Within the BHF framework, we also examine an EOS based on a potential model which includes explicitly the quark-gluon degrees of freedom, named FSS2 [68,69]. This reproduces correctly the saturation point of symmetric nuclear matter (SNM) and the binding energy of few-nucleon systems, and does not need TBF. We use two different EOS versions labelled respectively as FSS2CC and FSS2GC. Moreover, we compare these BHF EOSs with the often-used results of the Dirac-BHF method (DBHF) [49], which employs the Bonn A potential, in the following labelled DBHF(A), and a more recent calculation performed with the Bonn B potential [50], and labelled DBHF(B). We also compare with the APR EOS [51] based on the variational method and the V_{18} potential, and a parametrization of a recent Auxiliary Field Diffusion Monte Carlo (AFDMC) calculation [70].

3. Bulk Properties of Nuclear Matter

Around saturation density ρ_0 and isospin asymmetry $\delta \equiv (\rho_n - \rho_p)/\rho = 0$, being $\rho_n(\rho_p)$ the neutron (proton) density and ρ the total nucleonic density, the nuclear EOS can be characterized by a set of few isoscalar (E_0, K_0) and isovector $(S_0, L, K_{\mathrm{sym}})$ parameters, which can be constrained by nuclear experiments. The parameters are related to the coefficients of a Taylor expansion of the energy per particle of asymmetric nuclear matter as a function of density and isospin asymmetry,

$$E(\rho,\delta) = E_{\mathrm{SNM}}(\rho) + E_{\mathrm{sym}}(\rho)\delta^2\,, \quad (1)$$

$$E_{\mathrm{SNM}}(\rho) = E_0 + \frac{K_0}{2}x^2\,, \quad (2)$$

$$E_{\mathrm{sym}}(\rho) = S_0 + Lx + \frac{K_{\mathrm{sym}}}{2}x^2\,, \quad (3)$$

where $x \equiv (\rho - \rho_0)/3\rho_0$, E_0 is the energy per particle of symmetric nuclear matter at ρ_0, K_0 the incompressibility and $S_0 \equiv E_{\mathrm{sym}}(\rho_0)$ is the symmetry energy coefficient at saturation. These parameters are defined as

$$K_0 \equiv 9\rho_0^2 \frac{d^2 E_{\rm SNM}}{d\rho^2}(\rho_0)\,, \tag{4}$$

$$S_0 \equiv \frac{1}{2}\frac{\partial^2 E}{\partial \delta^2}(\rho_0, 0)\,, \tag{5}$$

$$L \equiv 3\rho_0 \frac{dE_{\rm sym}}{d\rho}(\rho_0)\,, \tag{6}$$

$$K_{\rm sym} \equiv 9\rho_0^2 \frac{d^2 E_{\rm sym}}{d\rho^2}(\rho_0)\,. \tag{7}$$

The values of these parameters at ρ_0 for the various considered EOSs are listed in Table 1.

Table 1. Saturation properties and NS observables predicted by the considered EOSs.

Model	**EOS**	ρ_0 **[fm^{-3}]**	$-E_0$ **[MeV]**	S_0 **[MeV]**	L **[MeV]**	K_0 **[MeV]**	$M_{\rm max}$ **[$M_\odot$]**	$\Lambda_{1.2}$	$\Lambda_{1.4}$	$\Lambda_{1.6}$
Micro.	BOB	0.170	15.4	33.6	70	238	2.50	1366	570	252
	V18	0.178	13.9	32.3	67	207	2.36	1082	442	188
	N93	0.185	16.1	36.5	77	229	2.25	1234	473	190
	UIX	0.171	14.9	33.5	61	171	1.96	848	309	112
	APR	0.159	15.9	33.4	51	233	2.20	720	274	110
	DBHF(A)	0.181	16.2	34.4	69	218	2.31	1635	681	295
	DBHF(B)	0.186	16.2	32.8	67	272	-	830	327	133
	FSS2CC	0.157	16.3	31.8	52	219	1.94	814	295	106
	FSS2GC	0.170	15.6	31.0	51	185	2.08	697	262	101
	AFDMC	0.160	16.0	31.3	60	239	2.21	822	293	109
Skyrme	Gs	0.158	15.6	31.2	94	239	2.13	1769	659	253
	Rs	0.158	15.1	30.8	86	248	2.12	1652	618	238
	SLy4	0.160	16.0	31.8	45	232	2.05	756	287	111
	SV	0.155	16.0	33.0	97	305	2.43	2224	914	393
	SkI4	0.158	16.2	33.7	106	245	2.17	1203	474	194
	SkMP	0.158	15.6	34.3	82	244	2.11	1295	487	188
	SkO	0.157	15.8	29.7	70	230	2.01	1252	451	164
	BSk22	0.158	16.1	32.0	69	246	2.26	1553	632	268
	BSk24	0.158	16.1	30.0	46	246	2.28	1260	523	227
	BSk26	0.159	16.1	30.0	38	241	2.17	830	327	133
NLWM	SFHO	0.157	16.2	32.8	53	244	2.06	862	334	132
	GM1	0.153	16.3	32.5	94	300	2.36	2223	913	393
	GM3	0.153	16.4	32.5	90	241	2.02	1688	617	228
DDM	DDME1	0.152	16.2	33.1	55	245	2.47	1765	773	355
	DDME2	0.152	16.1	32.3	51	251	2.51	1834	806	374
	TW99	0.153	16.2	32.8	55	240	2.08	1041	404	162
Exp.		$\sim$ 0.14–0.17	$\sim$ 15–17	28.5–34.9	30–87	220–260	$> 2.14^{+0.10}_{-0.09}$		70–580	
Ref.		[71]	[71]	[6,72]	[73,74]	[6,72]	[12]		[2]	

From the measurements of nuclear masses [75] and density distributions [76] the values $E_0 = -16 \pm 1$ MeV and $\rho_0 = 0.14 - 0.17$ fm^{-3} are obtained, whereas the value of K_0 can be extracted from the analysis of isoscalar giant monopole resonances in heavy nuclei. For the latter, results suggest $K_0 = 240 \pm 10$ MeV [77], or $K_0 = 248 \pm 8$ MeV [78], thus pointing to a rather soft EOS, as confirmed by HIC experiments [79].

Experimental information on the symmetry energy at saturation S_0 and its first derivative L can be obtained from the analysis of giant [80] and pygmy [81,82] dipole resonances, isospin diffusion measurements [83], isobaric analog states [84], measurements of the neutron skin thickness in heavy nuclei [85–88] and the meson production in HICs [89]. However, whereas S_0 is more or less well established ($\approx 3\,\text{MeV}$), the values of L ($30\,\text{MeV} < L < 87\,\text{MeV}$), and especially those of K_{sym} ($-400\,\text{MeV} < K_{sym} < 100\,\text{MeV}$) are still quite uncertain and poorly constrained [90,91], and therefore we disregard them in our analysis.

From Table 1, we notice that all the adopted EOSs in this work agree fairly well with the empirical values, except the slightly too low E_0 and K_0 for V18, too large S_0 for N93, and too low K_0 for UIX and FSS2GC. We notice that several phenomenological models predict too large L values, whereas all the microscopic EOSs are largely compatible. This is clearly displayed in Figure 1, where L is plotted as a function of the symmetry energy at saturation S_0. The shaded areas represent the experimental data currently available. In particular, we report the constraints inferred from the study of isospin diffusion in HICs [92] (blue band), electric dipole polarizability [93] (violet band), the neutron-skin thickness in Sn isotopes [94] (grey region), the finite-range droplet mass (FRDM) model [95] and the isobaric-analog-state (IAS) phenomenology combined with the skin-width data (green diagonal region) [96]. Moreover the horizontal (red) band is obtained from a Bayesian analysis of mass and radius measurements of NSs [97], and the dashed curve is the unitary gas bound on symmetry energy parameters [90]. Only values of (S_0, L) to the right of the curve are permitted, and therefore all the microscopic and some of the phenomenological models fulfill these constraints. We observe that there is no area of the parameter space where all constraints are simultaneously fulfilled, and this is likely due to the current uncertainties that plague the interpretation of the raw data. It should be stressed that the constraints on the EOS result from combining the raw data with theoretical models, and therefore they show some model dependence. On this basis, no theoretical model can be ruled out.

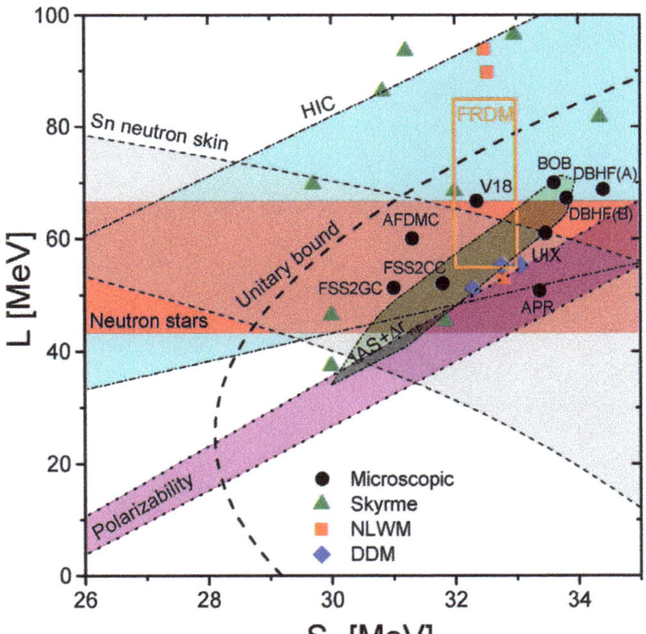

Figure 1. The relation between the symmetry energy at saturation density S_0 and its slope L. The full symbols represent the predictions of microscopic approaches (black circles), Skyrme EOSs (green triangles), NLWM models (red squares) and DDM approaches (blue diamonds), see Table 1 for the numerical values. The shaded areas represent experimental bands, see text for details.

We report in Figure 2 the symmetry energy E_{sym} as a function of the baryon density. The results in the left (right) panel are plotted for the microscopic (phenomenological) EOSs, and are compared with the experimental data displayed by the shaded areas. In particular, the grey area represents the diffusion data of HICs, the green area includes the flow data obtained by the FOPI-LAND collaboration [98] on the collective flow, and the blue area is the experimental region checked by the ASY-EOS collaboration [99]. The full orange contour shows the results on the isobaric analog states (IAS), obtained in Ref. [96]. We see that most of the EOSs, both microscopic and phenomenological ones, are compatible with experimental data up to around the saturation density, whereas for larger densities some EOSs tend to predict smaller values for the symmetry energy that are below the experimental areas. This is a clear sign of discrepancy, which results in a much larger difference at larger values of the baryon density, such as the ones characterizing the inner core of a NS. We stress once again that the inferred constraints are model dependent, since the data interpretation requires theoretical simulations.

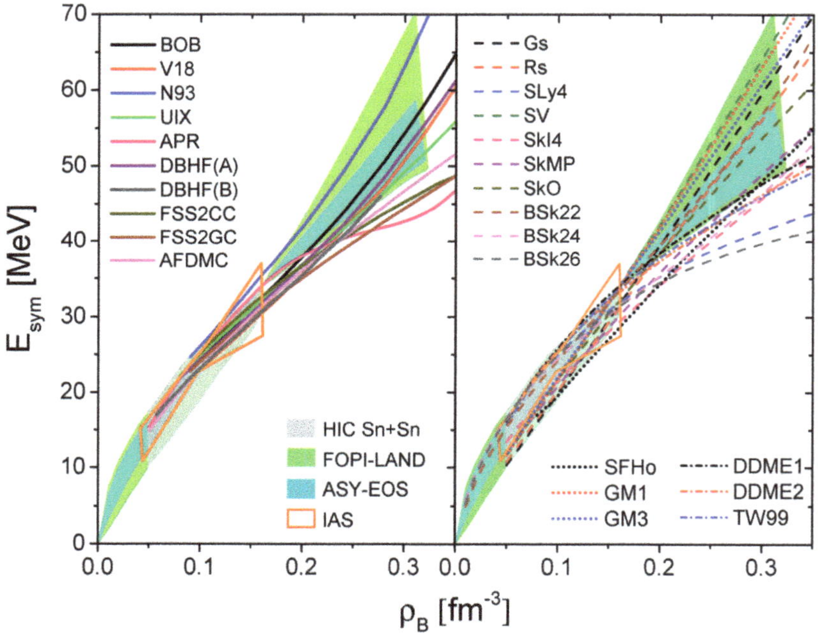

Figure 2. The symmetry energy vs. the baryon density for all the discussed EOSs. The green, blue and grey bands represent experimental data from HICs, whereas the orange contour represents the IAS calculations. See text for details.

4. EOS for Betastable Matter

Once the EOSs for symmetric and pure neutron matter are defined, one can calculate the composition and the EOS of cold, neutrino-free, catalyzed matter. For charge-neutral matter in beta-equilibrium with neutrons, protons, and leptons (e^-, μ^-), the EOS is computed in the following standard way [100]. One starts from the energy density of lepton/baryon matter as a function of the different partial densities ρ_i of the species $i = n, p, e, \mu$,

$$\varepsilon(\rho_n, \rho_p, \rho_e, \rho_\mu) = (\rho_n m_n + \rho_p m_p) + (\rho_n + \rho_p) E(\rho_n, \rho_p) + \varepsilon(\rho_e) + \varepsilon(\rho_\mu) , \tag{8}$$

where m_i are the corresponding masses, $E(\rho_n, \rho_p)$ is the energy per particle of asymmetric nuclear matter, and the leptonic contribution is calculated assuming ultrarelativistic and relativistic expressions for the energy densities of electrons $\varepsilon(\rho_e)$ and muons $\varepsilon(\rho_\mu)$, respectively [100]. We have used the parabolic approximation [101,102] of the energy per particle of asymmetric nuclear matter in Equation (1), with the symmetry energy calculated simply as the difference between the energy per particle of pure neutron matter and symmetric nuclear matter,

$$E_{\text{sym}}(\rho) \approx E(\rho_n = \rho, \rho_p = 0) - E(\rho_n = \rho/2, \rho_p = \rho/2) \,. \tag{9}$$

From the energy density, Equation (8), the various chemical potentials can be computed,

$$\mu_i = \frac{\partial \varepsilon}{\partial \rho_i} \,, \tag{10}$$

and imposing the beta-equilibrium conditions, $\mu_i = b_i \mu_n - q_i \mu_e$ (b_i and q_i denoting baryon number and charge of species i), along with the charge neutrality, $\sum_i \rho_i q_i = 0$, one can find the equilibrium composition ρ_i at fixed baryon density ρ, and finally the EOS,

$$p(\varepsilon) = \rho^2 \frac{d}{d\rho} \frac{\varepsilon(\rho_i(\rho))}{d\rho} = \rho \frac{d\varepsilon}{d\rho} - \varepsilon = \rho \mu_n - \varepsilon \,. \tag{11}$$

Once the EOS of betastable matter is known, one can solve the Tolman-Oppenheimer-Volkoff (TOV) [100] equations which describe the structure of a non-rotating spherically symmetric star in general relativity

$$\begin{aligned} \frac{dp}{dr} &= -G \frac{\varepsilon m}{r^2} \left(1 + \frac{p}{\varepsilon}\right) \left(1 + \frac{4\pi p r^3}{m}\right) \left(1 - \frac{2Gm}{r}\right)^{-1} , \\ \frac{dm}{dr} &= 4\pi r^2 \varepsilon \,, \end{aligned} \tag{12}$$

where G is the gravitational constant, p the pressure, ε the energy density and m the mass enclosed within a sphere of radius r. For each given central density, the integration of the TOV equations gives the mass and radius of the star corresponding to that density; this way one can construct an entire family of static configurations. It turns out that the NS mass has a maximum value as a function of the radius (or central density), above which the star is unstable against collapse to a black hole. The value of the maximum mass depends strongly on the nuclear EOS, hence the observation of a mass higher than the maximum mass allowed by a given EOS simply rules out that EOS.

We notice that the above mentioned theoretical methods cannot describe inhomogeneous and clusterized matter, and therefore for the low-density part $\rho < \rho_t \approx 0.08\ \text{fm}^{-3}$, one has to adopt the well-known Negele-Vautherin EOS [103] in the medium-density regime ($0.001\ \text{fm}^{-3} < \rho < \rho_t$), and the ones by Baym-Pethick-Sutherland [104] and Feynman-Metropolis-Teller [105] for lower densities $\rho < 0.001\ \text{fm}^{-3}$.

5. Constraints on the EOS from Terrestrial Laboratories and Astrophysical Observations

As already mentioned in Section 3, HICs at energies ranging from few tens to several hundreds MeV per nucleon have been exploited for extracting the gross properties of the nuclear EOS from the data. In fact, at a large enough energy, the two colliding nuclei produce flows of matter due to the large compression, resulting in a strong emission of nucleons and fragments of different sizes. The transverse flow, which is measured, depends sensitively on the pressure developed in the fireball at the moment of maximum compression during the collision. Additionally, the subthreshold K^+ production in heavy ion reactions has been demonstrated to probe the fireball density reached during the collision, with this being the ideal situation for exploring the EOS and its incompressibility.

In Ref. [106] the flow and kaon production analysis was summarized by plotting the region in the pressure versus density plane. A reasonable EOS should pass through it, and this is displayed in Figure 3 (left panels) as an orange area for the subthreshold kaon production [107], and as a grey area for the flow data [98]. Those results point in the direction of a soft EOS, i.e., values of the compressibility in the range $180 \leq K \leq 250$ MeV at density close to saturation. Those values are compatible with the ones extracted from the data on monopole oscillations [77]. In the upper (lower) panels of Figure 3 we plot the results for the microscopic (phenomenological) EOSs. We notice that most of the adopted EOSs are compatible with laboratory data, but some of them are too repulsive and therefore incompatible with experiments. In particular, among the considered microscopic EOSs, UIX, APR, and N93 are well compatible with the data extracted from HICs over the whole density range, whereas BOB, DBHF, and V18 are only marginally compatible large density (actually never reached in HICs), and those are characterized by a larger stiffness.

The EOS also rules the dynamics of NS mergers, in particular the final fate of the merger, a prompt or delayed collapse to a black hole or a single NS, as well as the amount of ejected matter which undergoes the nucleosynthesis of heavy elements, which strongly depends on the EOS. During the inspiral phase, the influence of the EOS is evident on the tidal polarizability, $\Lambda = \frac{2}{3}k_2\beta^{-5}$, where k_2 is the Love number and $\beta = GM/R$ is the compactness. An upper limit of $\Lambda < 800$ was initially given in the first GW170817 analysis for a $1.4\,M_\odot$ NS [1], but later on the analysis was improved by assuming that both NSs have the same EOS, thus giving different limits of $\Lambda = 190^{+390}_{-120}$, which translates into a measure of radius $R = 11.9^{+1.4}_{-1.4}$ km [2]. In this latter analysis, the values of the pressure as a function of density were extracted, and those are displayed as the blue hatched areas in Figure 3 (right panels). We notice that in this case the comparison has to be performed for the betastable case. We observe that almost all microscopic EOSs, except UIX, turn out to be compatible with the GW170817 data at density $\rho > 2\rho_0$, whereas the nuclear collision data look more restrictive. Additionally, for the phenomenological case some EOSs turn out to be marginally compatible with the observational data, as for the flow data in the symmetric case.

A very important constraint to be fulfilled is the value of the maximum mass for the different EOSs, which has to be compatible with the observational data. In Figure 4 we display the mass-radius relations obtained with microscopic and phenomenological EOSs, shown respectively as solid and broken curves. We observe that most models give values for the maximum mass larger than $2\,M_\odot$, except the soft microscopic UIX and FSS2GC, which therefore are compatible with current observational data [9–11], in particular with the largest mass observed up to now, $2.14^{+0.10}_{-0.09}\,M_\odot$ at 68% confidence interval for the object PSR J0740 + 6620 [12] (dark orange band). For completeness, we also display the observational limits at 95% confidence interval (light orange band). Analogous limits are plotted also for the object PSR J0348 + 0432 [11] (grey bands), which are more restrictive at high confidence level. Apart from these lower limits, some recent theoretical analyses of the GW170817 event indicate an upper limit on the maximum mass of about $2.33\,M_\odot$ (68%) or $2.5\,M_\odot$ (95%) (displayed by red horizontal lines) [108–111], with which several of both the microscopic and phenomenological EOSs would be compatible as well. We also display the Bayesian parameter estimation of the mass and equatorial radius of the millisecond pulsar PSR J0030 + 0451 [18,19], as recently reported by the NICER mission. The M, R values inferred from the analysis of the collected data (green and light grey zones) are $1.36^{+0.15}_{-0.16}\,M_\odot$ and $12.71^{+1.14}_{-1.19}$ km [18], or $1.44^{+0.15}_{-0.14}\,M_\odot$ and $13.02^{+1.24}_{-1.06}$ km [19].

Let us now turn to the discussion of the tidal deformability and its possible correlations with nuclear matter properties in its ground state. As already anticipated, the analysis of the GW170817 event [1–3] produced a value of $\tilde{\Lambda} < 730$, assuming equal mass merging. If both NSs have the same EOS, this leads to the constraints $70 < \Lambda_{1.4} < 580$ and $10.5 < R_{1.4} < 13.3$ km [2] for a $1.4\,M_\odot$ NS. A more stringent lower limit $\tilde{\Lambda} > 400$ [17] on the average tidal deformability was imposed by the high luminosity of the kilonova AT2017gfo following the NS merger event. This constraint could indicate that $R_{1.4} \gtrsim 12$ km [112–115], but it has to be taken with great care [116,117].

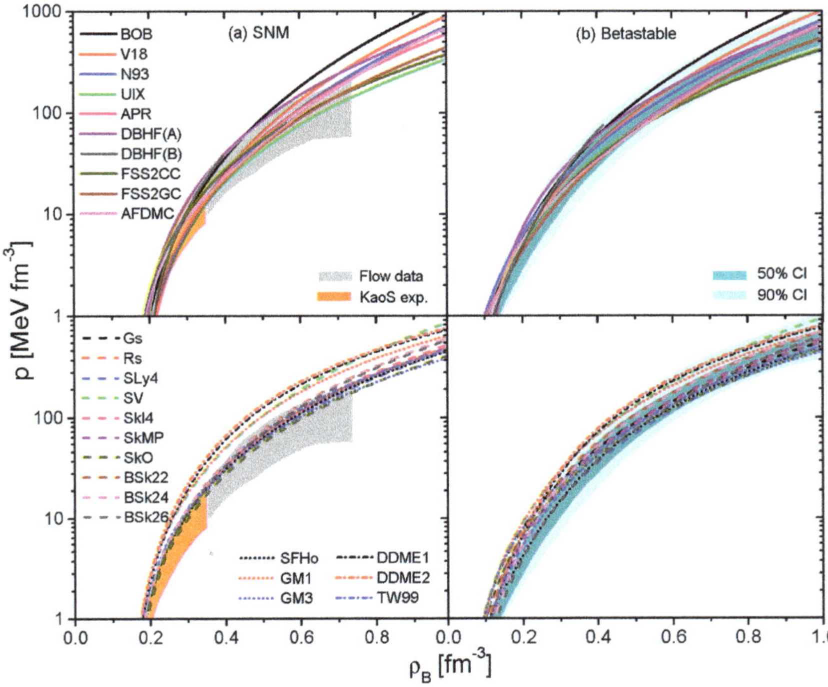

Figure 3. Pressure vs. baryon density for the symmetric case (**left** panels), and the beta-stable case (**right** panels). The upper (lower) panels display results for microscopic (phenomenological) EOSs. Constraints derived from HIC data are displayed in the left panels as orange (KaoS experiment) and grey (flow data) bands. Limits deduced by the GW170817 event are labelled by blue bands in the right panels. See text for details.

The possibility of finding correlations between properties of nuclear matter and NS observables has been recently explored [118,119]. In the following we further explore this issue, using the set of microscopic and phenomenological EOSs listed in Table 1. In Figure 5 we show the tidal deformability of a 1.2 (upper panels), 1.4 (central panels) and 1.6 (lower panels) solar-mass NS as a function of the symmetry energy at saturation S_0 (left panels), its first derivative L (central panels) and K_0 (right panels). The light- and dark-shaded bands in the central panels represent the limits inferred from the observational data of the GW170817 event [2] together with the experimental limits reported in Table 1. The degree of correlation is quantified by the correlation factor

$$r(x,y) = \frac{1}{n-1} \frac{\sum_x \sum_y (x-\bar{x})(y-\bar{y})}{\sigma_x \sigma_y}, \tag{13}$$

with n being the number of data pairs, $\bar{x}$ and $\bar{y}$ being the mean values of x and y, and σ_x and σ_y being their standard deviations. We obtain the following values

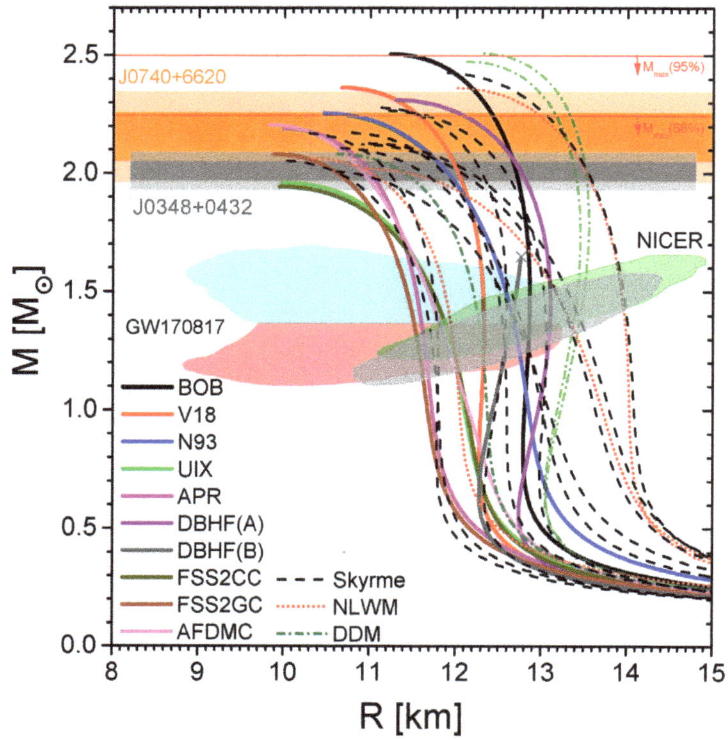

Figure 4. Mass-radius relations predicted by the different EOSs listed in Table 1. The observed masses of the millisecond pulsar PSR J0740 + 6620 [12] and of J0384-0432 [11] are also shown, as well as constraints inferred from the analysis of the GW170817 event and observations reported by the NICER mission [18,19]. See text for details.

$$r([S_0,\ L,\ K_0],\ \Lambda_{1.2}) = [0.006,\ 0.635,\ 0.709]\,, \tag{14}$$

$$r([S_0,\ L,\ K_0],\ \Lambda_{1.4}) = [0.206,\ 0.551,\ 0.702]\,, \tag{15}$$

$$r([S_0,\ L,\ K_0],\ \Lambda_{1.6}) = [0.145,\ 0.459,\ 0.682]\,. \tag{16}$$

As can be seen the correlation factor is rather small in the case of S_0 for the three values of the tidal deformability, indicating that no correlation at all exists between Λ and S_0 for any value of the NS mass. Instead, a very weak correlation of Λ seems to exist with L and a slightly stronger, although still weak, exists with K_0. We observe that in the case of $\Lambda_{1.4}$ all but two (SFHO and TW99) of the NLWM and DDME models are incompatible with the observational constraint [2]. On the contrary, most of the Skyrme models lie within the shaded bands, except for a few cases. Regarding the microscopic models, they are almost all in agreement with the GW observations and experimental constraints on S_0 and L, although only four are compatible with the constraints on K_0.

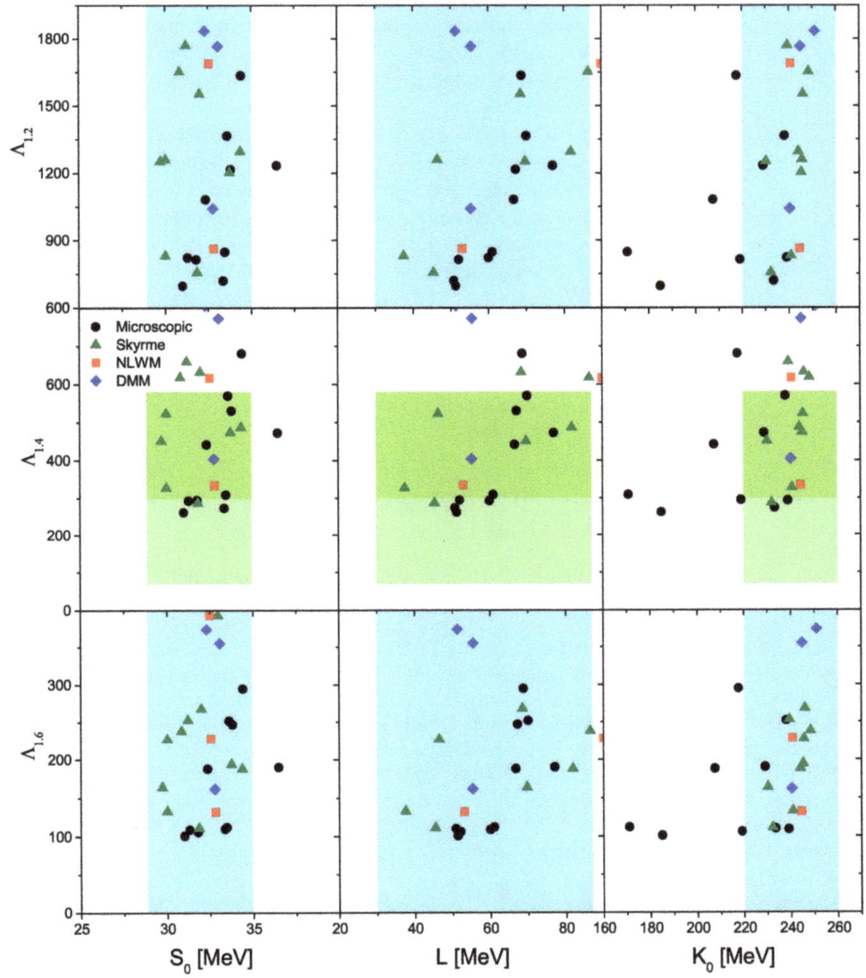

Figure 5. See Table 1 for the numerical values. displayed as a function of S_0, L and K_0 for the various EOSs. See Table 1 for the numerical values.

6. Conclusions

In this work, we have analyzed several constraints on the nuclear EOS currently available from NS observations and laboratory experiments. For this purpose, we have used a set of different models that include several phenomenological EOSs based on Skyrme and relativistic mean field models as well as microscopic calculations based on the (Dirac–) Brueckner–Hartree–Fock theories, the variational method and Quantum Monte Carlo techniques. To select the most compatible EOSs among the ones considered in this work, we have employed in particular the experimental constraints on several properties of nuclear matter at saturation density derived from different experiments as well as observational constraints on the mass, radius and tidal deformability imposed by recent measurements of the masses of millisecond pulsars [12], the data of the NICER mission [18,19] and the GW170817 NS merger event [1–3]. We have found that almost all considered models are compatible with the laboratory constraints of the nuclear-matter properties as well as with the largest masses observed up to now, $2.14^{+0.10}_{-0.09}\,M_\odot$ for the object PSR J0740 + 6620 [12], and with the upper limit of the maximum mass of about 2.3–2.5 $M_\odot$ [108–111] deduced from the analysis of the GW170817 event. Our study of possible correlations among properties of nuclear matter at saturation density with NS observables, particularly with the tidal deformability, has shown that no correlation exists between Λ and S_0 for any value of the NS mass, but weak correlations of Λ do exist with L and with K_0.

We would like to finish by noticing that while the isoscalar part of the nuclear EOS is rather well constrained by the major experimental, observational and theoretical advances, the isovector one is less well known mainly due to our still limited knowledge of the nuclear force and, particularly, of its in-medium modifications and its spin and isospin dependence. Future NS observations, such as the precise simultaneous measurement of the mass and radius of a single NS, together with laboratory experiments planned in next-generation radioactive ion beam facilities, are fundamental to provide more stringent constraints on the nuclear EOS, and are very much awaited for.

Author Contributions: Writing—review and editing, G.F.B., H.-J.S., I.V. and J.-B.W. The authors contributed equally to this work. All authors have read and agreed to the published version of the manuscript.

Funding: This work was partially funded by "PHAROS" COST Action CA16214 and the China Scholarship Council, No. 201706410092.

Data Availability Statement: Not applicable.

Conflicts of Interest: The authors declare no conflicts of interest.

References

1. Abbott, B.P.; Abbott, R.; Abbott, T.D.; Acernese, F.; Ackley, K.; Adams, C.; Adams, T.; Addesso, P.; Adhikari, R.X.; Adya, V.B.; et al. GW170817: Observation of Gravitational Waves from a Binary Neutron Star Inspiral. *Phys. Rev. Lett.* **2017**, *119*, 161101. [CrossRef]
2. Abbott, B.P.; Abbott, R.; Abbott, T.D.; Acernese, F.; Ackley, K.; Adams, C.; Adams, T.; Addesso, P.; Adhikari, R.X.; Adya, V.B.; et al. GW170817: Measurements of neutron star radii and equation of state. *Phys. Rev. Lett.* **2018**, *121*, 161101. [CrossRef] [PubMed]
3. Abbott, B.P.; Abbott, R.; Abbott, T.D.; Acernese, F.; Ackley, K.; Adams, C.; Adams, T.; Addesso, P.; Adhikari, R.X.; Adya, V.B.; et al. Properties of the Binary Neutron Star Merger GW170817. *Phys. Rev. X* **2019**, *9*, 011001. [CrossRef]
4. Blaschke, D.; Colpi, M.Y.; Horowitz, C.J.; Radice, D.E. First joint gravitational wave and electromagnetic observations: Implications for nuclear and particle physics. *Eur. Phys. J. A* **2019**, *55*.
5. Baiotti, L. Gravitational waves from neutron star mergers and their relation to the nuclear equation of state. *Prog. Part. Nucl. Phys.* **2019**, *109*, 103714. [CrossRef]
6. Oertel, M.; Hempel, M.; Klähn, T.; Typel, S. Equations of state for supernovae and compact stars. *Rev. Mod. Phys.* **2017**, *89*, 015007. [CrossRef]
7. Burgio, G.F.; Fantina, A.F. Nuclear Equation of state for Compact Stars and Supernovae. In *The Physics and Astrophysics of Neutron Stars*; Springer: Cham, Switzerland, 2018; Volume 457, p. 255.
8. Lattimer, J.M. The Nuclear Equation of State and Neutron Star Masses. *Ann. Rev. Nucl. Sci.* **2012**, *62*, 485–515. [CrossRef]
9. Demorest, P.B.; Pennucci, T.; Ransom, S.M.; Roberts, M.S.; Hessels, J.W. A two-solar-mass neutron star measured using Shapiro delay. *Nature* **2010**, *467*, 1081–1083. [CrossRef]
10. Fonseca, E.; Pennucci, T.T.; Ellis, J.A.; Stairs, I.H.; Nice, D.J.; Ransom, S.M.; Demorest, P.B.; Arzoumanian, Z.; Crowter, K.; Dolch, T. The NANOGrav Nine-year Data Set: Mass and Geometric Measurements of Binary Millisecond Pulsars. *Astrophys. J.* **2016**, *832*, 167. [CrossRef]
11. Antoniadis, J.; Freire, P.C.; Wex, N.; Tauris, T. M.; Lynch, R.S.; van Kerkwijk, M.H.; Kramer, M.; Bassa, C.; Dhillon, V.S.; Driebe, T.; et al. A Massive Pulsar in a Compact Relativistic Binary. *Science* **2013**, *340*, 6131. [CrossRef]
12. Cromartie, H.T.; Fonseca, E.; Ransom, S.M.; Demorest, P.B.; Arzoumanian, Z.; Blumer, H.; Brook, P.R.; DeCesar, M.E.; Dolch, T.; Ellis, J.A.; et al. Relativistic Shapiro delay measurements of an extremely massive millisecond pulsar. *Nat. Astron.* **2019**, *4*, 72–76. [CrossRef]
13. Özel, F.; Freire, P. Masses, Radii, and the Equation of State of Neutron Stars. *Ann. Rev. Astron. Astrophys.* **2016**, *54*, 401–440. [CrossRef]
14. Guillot, S.; Servillat, M.; Webb, N.A.; Rutledge, R.E. Measurement of the Radius of Neutron Stars with High Signal-to-noise Quiescent Low-mass X-Ray Binaries in Globular Clusters. *Astrophys. J.* **2013**, *772*, 7. [CrossRef]
15. Hartle, J.B. Slowly rotating relativistic stars. 1. Equations of structure. *Astrophys. J.* **1967**, *150*, 1005–1029. [CrossRef]
16. Flanagan, E.E.; Hinderer, T. Constraining neutron star tidal Love numbers with gravitational wave detectors. *Phys. Rev. D* **2008**, *77*, 021502. [CrossRef]
17. Radice, D.; Perego, A.; Zappa, F.; Bernuzzi, S. GW170817: Joint Constraint on the Neutron Star Equation of State from Multimessenger Observations. *Astrophys. J.* **2018**, *852*, L29. [CrossRef]

18. Riley, T.E.; Watts, A.L.; Bogdanov, S.; Ray, P.S.; Ludlam, R.M.; Guillot, S.; Arzoumanian, Z.; Baker, C.L.; Bilous, A.V.; Chakrabarty, D. A NICER View of PSR J0030+0451: Millisecond Pulsar Parameter Estimation. *Astrophys. J. Lett.* **2019**, *887*, L21. [CrossRef]
19. Miller, M.C.; Lamb, F.K.; Dittmann, A.J.; Bogdanov, S.; Arzoumanian, Z.; Gendreau, K.C.; Guillot, S.; Harding, A.K.; Ho, W.C.G.; Lattimer, J.M.; et al. PSR J0030+0451 Mass and Radius from NICER Data and Implications for the Properties of Neutron Star Matter. *Astrophys. J.* **2019**, *887*, L24. [CrossRef]
20. Watts, A.L.; Yu, W.; Poutanen, J.; Zhang, S.; Bhattacharyya, S.; Bogdanov, S.; Ji, L.; Patruno, A.; Riley, T.E.; Bakala, P.; et al. Dense matter with eXTP. *Sci. China Phys. Mech. Astron.* **2019**, *62*, 29503. [CrossRef]
21. Schulze, H.J.; Polls, A.; Ramos, A.; Vidana, I. Maximum mass of neutron stars. *Phys. Rev. C* **2006**, *73*, 058801. [CrossRef]
22. Annala, E.; Gorda, T.; Kurkela, A.; Nättilä, J.; Vuorinen, A. Evidence for quark-matter cores in massive neutron stars. *Nat. Phys.* **2020**, *16*, 907–910. [CrossRef]
23. Stone, J.R.; Reinhard, P.G. The Skyrme interaction in finite nuclei and nuclear matter. *Prog. Part. Nucl. Phys.* **2007**, *58*, 587–657. [CrossRef]
24. Vautherin, D.; Brink, D.M. Hartree-Fock Calculations with Skyrme's Interaction. I. Spherical Nuclei. *Phys. Rev. C* **1972**, *5*, 626–647. [CrossRef]
25. Quentin, P.; Flocard, H. Self-Consistent Calculations of Nuclear Properties with Phenomenological Effective Forces. *Annu. Rev. Nucl. Part. Sci.* **1978**, *28*, 523–594. [CrossRef]
26. Boguta, J.; Bodmer, A.R. Relativistic calculation of nuclear matter and the nuclear surface. *Nucl. Phys. A* **1977**, *292*, 413–428. [CrossRef]
27. Friedrich, J.; Reinhard, P.G. Skyrme-force parametrization: Least-squares fit to nuclear ground-state properties. *Phys. Rev. C* **1986**, *33*, 335–351. [CrossRef] [PubMed]
28. Chabanat, E. Report No. LYCENT 9501. Unpublished. Ph.D. Thesis, Universitè Claude Bernard Lyon-1, Villeurbanne, France, 1995.
29. Beiner, M.; Flocard, H.; Van Giai, N.; Quentin, P. Nuclear ground-state properties and self-consistent calculations with the skyrme interaction (I). Spherical description. *Nucl. Phys. A* **1975**, *238*, 29–69. [CrossRef]
30. Reinhard, P.G.; Flocard, H. Nuclear effective forces and isotope shifts. *Nucl. Phys. A* **1995**, *584*, 467–488. [CrossRef]
31. Bennour, L.; Heenen, P.H.; Bonche, P.; Dobaczewski, J.; Flocard, H. Charge distributions of ^{208}Pb, ^{206}Pb, and ^{205}Tl and the mean-field approximation. *Phys. Rev. C* **1989**, *40*, 2834–2839. [CrossRef]
32. Reinhard, P.G.; Dean, D.J.; Nazarewicz, W.; Dobaczewski, J.; Maruhn, J.A.; Strayer, M.R. Shape coexistence and the effective nucleon-nucleon interaction. *Phys. Rev. C* **1999**, *60*, 014316. [CrossRef]
33. Pearson, J.M.; Chamel, N.; Potekhin, A.Y.; Fantina, A.F.; Ducoin, C.; Dutta, A.K.; Goriely, S. Unified equations of state for cold non-accreting neutron stars with Brussels–Montreal functionals—I. Role of symmetry energy. *Mon. Not. R. Astron. Soc.* **2018**, *481*, 2994–3026. [CrossRef]
34. Glendenning, N.K.; Moszkowski, S.A. Reconciliation of neutron-star masses and binding of the Lambda in hypernuclei. *Phys. Rev. Lett.* **1991**, *67*, 2414–1417. [CrossRef]
35. Nikšić, T.; Vretenar, D.; Finelli, P.; Ring, P. Relativistic Hartree-Bogoliubov model with density-dependent meson-nucleon couplings. *Phys. Rev. C* **2002**, *66*, 024306. [CrossRef]
36. Typel, S.; Wolter, H.H. Relativistic mean field calculations with density-dependent meson-nucleon coupling. *Nucl. Phys. A* **1999**, *656*, 331–364. [CrossRef]
37. Steiner, A.W.; Hempel, M.; Fischer, T. Core-collapse supernova equations of state based on neutron star observations. *Astrophys. J.* **2013**, *774*, 17. [CrossRef]
38. Dutra, M.; Lourenço, O.; Avancini, S.S.; Carlson, B.V.; Delfino, A.; Menezes, D.P.; Providência, C.; Typel, S.; Stone, J.R. Relativistic mean-field hadronic models under nuclear matter constraints. *Phys. Rev. C* **2014**, *90*, 055203. [CrossRef]
39. Lourenço, O.; Dutra, M.; Lenzi, C.H.; Flores, C.V.; Menezes, D.P. Consistent relativistic mean-field models constrained by GW170817. *Phys. Rev. C* **2019**, *99*, 045202. [CrossRef]
40. Lourenço, O.; Dutra, M.; Lenzi, C. H.; Biswal, S. K.; Bhuyan, M.; Menezes, D. P. Consistent Skyrme parametrizations constrained by GW170817. *Eur. Phys. J. A* **2020**, *56*, 32.
41. Machleidt, R.; Holinde, K.; Elster, C. The Bonn Meson Exchange Model for the Nucleon Nucleon Interaction. *Phys. Rep.* **1987**, *149*, 1–89. [CrossRef]
42. Nagels, M.M.; Rijken, T.A.; de Swart, J.J. A Low-Energy Nucleon-Nucleon Potential from Regge Pole Theory. *Phys. Rev. D* **1978**, *17*, 768. [CrossRef]
43. Weinberg, S. Effective chiral lagrangians for nucleon-pion interactions and nuclear forces. *Nucl. Phys. B* **1991**, *363*, 3–18. [CrossRef]
44. Weinberg, S. Nuclear forces from chiral lagrangians. *Phys. Lett. B* **1990**, *251*, 288–292. [CrossRef]
45. Entem, D.R.; Machleidt, R. Accurate charge-dependent nucleon-nucleon potential at fourth order of chiral perturbation theory. *Phys. Rev. C* **2003**, *68*, 041001. [CrossRef]
46. Epelbaum, E.; Hammer, H.W.; Meißner, U.G. Modern theory of nuclear forces. *Rev. Mod. Phys.* **2009**, *81*, 1773–1825. [CrossRef]
47. Day, B.D. Elements of the Brueckner-Goldstone Theory of Nuclear Matter. *Rev. Mod. Phys.* **1967**, *39*, 719–744. [CrossRef]
48. Brockmann, R.; Machleidt, R. Relativistic nuclear structure. I. Nuclear matter. *Phys. Rev. C* **1990**, *42*, 1965–1980. [CrossRef] [PubMed]

49. Gross-Boelting, T.; Fuchs, C.; Faessler, A. Covariant representations of the relativistic Brueckner T-matrix and the nuclear matter problem. *Nucl. Phys. A* **1999**, *648*, 105–137. [CrossRef]
50. Müther, H.; Sammarruca, F.; Ma, Z. Relativistic effects and three-nucleon forces in nuclear matter and nuclei. *Int. J. Mod. Phys. E* **2017**, *26*, 1730001. [CrossRef]
51. Akmal, A.; Pandharipande, V.R.; Ravenhall, D.G. Equation of state of nucleon matter and neutron star structure. *Phys. Rev. C* **1998**, *58*, 1804–1828. [CrossRef]
52. Kadanoff, L.; Baym, G. *Quantum Statistical Mechanics*; W.A. Benjamin Inc.: New York, NY, USA, 1962.
53. Dickhoff, W.; Van Neck, D. *Many-Body Theory Exposed*! World Scientific: Singapore, 2005; p. 752.
54. Wiringa, R.B.; Pieper, S.C.; Carlson, J.; Pand haripande, V.R. Quantum Monte Carlo calculations of A=8 nuclei. *Phys. Rev. C* **2000**, *62*, 014001. [CrossRef]
55. Gandolfi, S.; Illarionov, A.Y.; Schmidt, K.E.; Pederiva, F.; Fantoni, S. Quantum Monte Carlo calculation of the equation of state of neutron matter. *Phys. Rev. C* **2009**, *79*, 054005. [CrossRef]
56. Drischler, C.; Furnstahl, R.J.; Melendez, J.A.; Phillips, D.R. How Well Do We Know the Neutron-Matter Equation of State at the Densities Inside Neutron Stars? A Bayesian Approach with Correlated Uncertainties. *Phys. Rev. Lett.* **2020**, *125*, 202702. [CrossRef] [PubMed]
57. Bogner, S.K.; Furnstahl, R.J.; Schwenk, A. From low-momentum interactions to nuclear structure. *Prog. Part. Nucl. Phys.* **2010**, *65*, 94–147. [CrossRef]
58. Machleidt, R. The Meson theory of nuclear forces and nuclear structure. *Adv. Nucl. Phys.* **1989**, *19*, 189–376.
59. Stoks, V.G.J.; Klomp, R.A.M.; Terheggen, C.P.F.; de Swart, J.J. Construction of high quality N N potential models. *Phys. Rev. C* **1994**, *49*, 2950–2962. [CrossRef] [PubMed]
60. Wiringa, R.B.; Stoks, V.G.J.; Schiavilla, R. An Accurate nucleon-nucleon potential with charge independence breaking. *Phys. Rev. C* **1995**, *51*, 38–51. [CrossRef] [PubMed]
61. Grangé, P.; Lejeune, A.; Martzolff, M.; Mathiot, J.F. Consistent three-nucleon forces in the nuclear many-body problem. *Phys. Rev. C* **1989**, *40*, 1040–1060. [CrossRef] [PubMed]
62. Baldo, M.; Bombaci, I.; Burgio, G.F. Microscopic nuclear equation of state with three-body forces and neutron star structure. *Astron. Astrophys.* **1997**, *328*, 274–282.
63. Zuo, W.; Lejeune, A.; Lombardo, U.; Mathiot, J.F. Microscopic three-body force for asymmetric nuclear matter. *Eur. Phys. J. A* **2002**, *14*, 469–475. [CrossRef]
64. Li, Z.H.; Lombardo, U.; Schulze, H.J.; Zuo, W. Consistent nucleon-nucleon potentials and three-body forces. *Phys. Rev. C* **2008**, *77*, 034316. [CrossRef]
65. Pudliner, B.S.; Pandharipande, V.R.; Carlson, J.; Wiringa, R.B. Quantum Monte Carlo calculations of A <= 6 nuclei. *Phys. Rev. Lett.* **1995**, *74*, 4396–4399.
66. Pudliner, B.S.; Pandharipande, V.R.; Carlson, J.; Pieper, S.C.; Wiringa, R.B. Quantum Monte Carlo calculations of nuclei with A <= 7. *Phys. Rev. C* **1997**, *56*, 1720–1750.
67. Li, Z.H.; Schulze, H.J. Neutron star structure with modern nucleonic three-body forces. *Phys. Rev. C* **2008**, *78*, 028801. [CrossRef]
68. Baldo, M.; Fukukawa, K. Nuclear Matter from Effective Quark-Quark Interaction. *Phys. Rev. Lett.* **2014**, *113*, 242501. [CrossRef]
69. Fukukawa, K.; Baldo, M.; Burgio, G.F.; Lo Monaco, L.; Schulze, H.J. Nuclear matter equation of state from a quark-model nucleon-nucleon interaction. *Phys. Rev. C* **2015**, *92*, 065802. [CrossRef]
70. Gandolfi, S.; Illarionov, A.Y.; Fantoni, S.; Miller, J.C.; Pederiva, F.; Schmidt, K.E. Microscopic calculation of the equation of state of nuclear matter and neutron star structure. *Mon. Not. Roy. Astron. Soc.* **2010**, *404*, L35–L39. [CrossRef]
71. Margueron, J.; Hoffmann Casali, R.; Gulminelli, F. Equation of state for dense nucleonic matter from metamodeling. I. Foundational aspects. *Phys. Rev. C* **2018**, *97*, 025805. [CrossRef]
72. Li, B.A.; Han, X. Constraining the neutron-proton effective mass splitting using empirical constraints on the density dependence of nuclear symmetry energy around normal density. *Phys. Lett. B* **2013**, *727*, 276–281. [CrossRef]
73. Shlomo, S.; Kolomietz, V.M.; Colò, G. Deducing the nuclear-matter incompressibility coefficient from data on isoscalar compression modes. *Eur. Phys. J. Hadron. Nucl.* **2006**, *30*, 23–30.
74. Piekarewicz, J. Do we understand the incompressibility of neutron-rich matter? *J. Phys. Nucl. Phys.* **2010**, *37*, 064038. [CrossRef]
75. Audi, G.; Wapstra, A.H.; Thibault, C. The AME2003 atomic mass evaluation . (II). Tables, graphs and references. *Nucl. Phys. A* **2003**, *729*, 337–676. [CrossRef]
76. de Vries, H.; de Jager, C.W.; de Vries, C. Nuclear Charge-Density-Distribution Parameters from Electron Scattering. *At. Data Nucl. Data Tables* **1987**, *36*, 495. [CrossRef]
77. Colò, G.; van Giai, N.; Meyer, J.; Bennaceur, K.; Bonche, P. Microscopic determination of the nuclear incompressibility within the nonrelativistic framework. *Phys. Rev. C* **2004**, *70*, 024307. [CrossRef]
78. Piekarewicz, J. Unmasking the nuclear matter equation of state. *Phys. Rev. C* **2004**, *69*, 041301. [CrossRef]
79. Fuchs, C.; Faessler, A.; Zabrodin, E.; Zheng, Y.M. Probing the Nuclear Equation of State by K^+ Production in Heavy-Ion Collisions. *Phys. Rev. Lett.* **2001**, *86*, 1974–1977. [CrossRef]
80. Garg, U. ; Li, T. ; Okumura, S. ; Akimune, H. ; Fujiwara, M. ; Harakeh, M.N. ; Itoh, M. ; Iwao, Y. ; Kawabata, T. ; Kawase, K.; et al. The Giant Monopole Resonance in the Sn Isotopes: Why is Tin so "Fluffy"? *Nucl. Phys. A* **2007**, *788*, 36–43. [CrossRef]

81. Klimkiewicz, A.; Paar, N.; Adrich, P.; Fallot, M.; Boretzky, K.; Aumann, T.; Cortina-Gil, D.; Datta Pramanik, U.; Elze, W.; Emling, H.; et al. Nuclear symmetry energy and neutron skins derived from pygmy dipole resonances. *Phys. Rev. C* **2007**, *76*, 051603. [CrossRef]
82. Carbone, A.; Colò, G.; Bracco, A.; Cao, L.G.; Bortignon, P.F.; Camera, F.; Wieland, O. Constraints on the symmetry energy and neutron skins from pygmy resonances in Ni68 and Sn132. *Phys. Rev. C* **2010**, *81*, 041301. [CrossRef]
83. Chen, L.W.; Ko, C.M.; Li, B.A. Determination of the Stiffness of the Nuclear Symmetry Energy from Isospin Diffusion. *Phys. Rev. Lett.* **2005**, *94*, 032701. [CrossRef] [PubMed]
84. Danielewicz, P.; Lee, J. Symmetry energy I: Semi-infinite matter. *Nucl. Phys. A* **2009**, *818*, 36–96. [CrossRef]
85. Brown, B.A. Neutron Radii in Nuclei and the Neutron Equation of State. *Phys. Rev. Lett.* **2000**, *85*, 5296–5299. [CrossRef]
86. Typel, S.; Brown, B.A. Neutron radii and the neutron equation of state in relativistic models. *Phys. Rev. C* **2001**, *64*, 027302. [CrossRef]
87. Horowitz, C.J.; Pollock, S.J.; Souder, P.A.; Michaels, R. Parity violating measurements of neutron densities. *Phys. Rev. C* **2001**, *63*, 025501. [CrossRef]
88. Roca-Maza, X.; Centelles, M.; Viñas, X.; Warda, M. Neutron Skin of Pb208, Nuclear Symmetry Energy, and the Parity Radius Experiment. *Phys. Rev. Lett.* **2011**, *106*, 252501. [CrossRef]
89. Fuchs, C. pp i Kaon production in heavy ion reactions at intermediate energies. *Prog. Part. Nucl. Phys.* **2006**, *56*, 1–103. [CrossRef]
90. Tews, I.; Lattimer, J.M.; Ohnishi, A.; Kolomeitsev, E.E. Symmetry Parameter Constraints from a Lower Bound on Neutron-matter Energy. *Astrophys. J.* **2017**, *848*, 105. [CrossRef]
91. Zhang, N.B.; Cai, B.J.; Li, B.A.; Newton, W.G.; Xu, J. How tightly is nuclear symmetry energy constrained by unitary Fermi gas? *Nucl. Sci. Tech.* **2017**, *28*, 181. [CrossRef]
92. Tsang, M.B.; Zhang, Y.; Danielewicz, P.; Famiano, M.; Li, Z.; Lynch, W.G.; Steiner, A.W. Constraints on the Density Dependence of the Symmetry Energy. *Phys. Rev. Lett.* **2009**, *102*, 122701. [CrossRef]
93. Roca-Maza, X.; Viñas, X.; Centelles, M.; Agrawal, B.K.; Colò, G.; Paar, N.; Piekarewicz, J.; Vretenar, D. Neutron skin thickness from the measured electric dipole polarizability in ^{68}Ni^{120}Sn and ^{208}Pb. *Phys. Rev. C* **2015**, *92*, 064304. [CrossRef]
94. Chen, L.W.; Ko, C.M.; Li, B.A.; Xu, J. Density slope of the nuclear symmetry energy from the neutron skin thickness of heavy nuclei. *Phys. Rev. C* **2010**, *82*, 024321. [CrossRef]
95. Möller, P.; Myers, W.D.; Sagawa, H.; Yoshida, S. New Finite-Range Droplet Mass Model and Equation-of-State Parameters. *Phys. Rev. Lett.* **2012**, *108*, 052501. [CrossRef]
96. Danielewicz, P.; Lee, J. Symmetry energy II: Isobaric analog states. *Nucl. Phys. A* **2014**, *922*, 1–70. [CrossRef]
97. Steiner, A.W.; Lattimer, J.M.; Brown, E.F. The Neutron Star Mass-Radius Relation and the Equation of State of Dense Matter. *Astrophys. J. Lett.* **2013**, *765*, L5. [CrossRef]
98. Ritman, J.L.; Herrmann, N.; Best, D.; Alard, J.P.; Amouroux, V.; Bastid, N.; Belyaev, I.; Berger, L.; Biegansky, J.; Buta, A.; et al. On the transverse momentum distribution of strange hadrons produced in relativistic heavy ion collisions. *Z. Phys. A* **1995**, *352*, 355–357. [CrossRef]
99. Russotto, P.; Gannon, S.; Kupny, S.; Lasko, P.; Acosta, L.; Adamczyk, M.; Al-Ajlan, A.; Al-Garawi, M.; Al-Homaidhi, S.; Amorini, F.; et al. Results of the ASY-EOS experiment at GSI: The symmetry energy at suprasaturation density. *Phys. Rev. C* **2016**, *94*, 034608. [CrossRef]
100. Shapiro, S.L.; Teukolsky, S.A. *Black Holes, White Dwarfs and Neutron Stars: The Physics of Compact Objects*; Wiley: New York, NY, USA, 2008.
101. Baldo, M.; Burgio, G.F.; Schulze, H.J. Hyperon stars in the Brueckner-Bethe-Goldstone theory. *Phys. Rev. C* **2000**, *61*, 055801. [CrossRef]
102. Bombaci, I.; Lombardo, U. Asymmetric nuclear matter equation of state. *Phys. Rev. C* **1991**, *44*, 1892–1900. [CrossRef]
103. Negele, J.W.; Vautherin, D. Neutron star matter at subnuclear densities. *Nucl. Phys. A* **1973**, *207*, 298–320. [CrossRef]
104. Baym, G.; Pethick, C.; Sutherland, P. The Ground state of matter at high densities: Equation of state and stellar models. *Astrophys. J.* **1971**, *170*, 299–317. [CrossRef]
105. Feynman, R.P.; Metropolis, N.; Teller, E. Equations of State of Elements Based on the Generalized Fermi-Thomas Theory. *Phys. Rev.* **1949**, *75*, 1561–1573. [CrossRef]
106. Danielewicz, P.; Lacey, R.; Lynch, W.G. Determination of the Equation of State of Dense Matter. *Science* **2002**, *298*, 1592–1596. [CrossRef]
107. Miśkowiec, D. Observation of enhanced subthreshold K^+ production in central collisions between heavy nuclei. *Phys. Rev. Lett.* **1994**, *72*, 3650–3653. [CrossRef]
108. Shibata, M.; Fujibayashi, S.; Hotokezaka, K.; Kiuchi, K.; Kyutoku, K.; Sekiguchi, Y.; Tanaka, M. Modeling GW170817 based on numerical relativity and its implications. *Phys. Rev. D* **2017**, *96*, 123012. [CrossRef]
109. Margalit, B.; Metzger, B.D. Constraining the Maximum Mass of Neutron Stars From Multi-Messenger Observations of GW170817. *Astrophys. J.* **2017**, *850*, L19. [CrossRef]
110. Rezzolla, L.; Most, E.R.; Weih, L.R. Using gravitational-wave observations and quasi-universal relations to constrain the maximum mass of neutron stars. *Astrophys. J.* **2018**, *852*, L25. [CrossRef]
111. Shibata, M.; Zhou, E.; Kiuchi, K.; Fujibayashi, S. Constraint on the maximum mass of neutron stars using GW170817 event. *Phys. Rev. D* **2019**, *100*, 023015. [CrossRef]

112. Most, E.R.; Weih, L.R.; Rezzolla, L.; Schaffner-Bielich, J. New constraints on radii and tidal deformabilities of neutron stars from GW170817. *Phys. Rev. Lett.* **2018**, *120*, 261103. [CrossRef] [PubMed]
113. Lim, Y.; Holt, J.W. Neutron star tidal deformabilities constrained by nuclear theory and experiment. *Phys. Rev. Lett.* **2018**, *121*, 062701. [CrossRef] [PubMed]
114. Malik, T.; Alam, N.; Fortin, M.; Providência, C.; Agrawal, B.K.; Jha, T.K.; Kumar, B.; Patra, S.K. GW170817: Constraining the nuclear matter equation of state from the neutron star tidal deformability. *Phys. Rev. C* **2018**, *98*, 035804. [CrossRef]
115. Burgio, G.F.; Drago, A.; Pagliara, G.; Schulze, H.J.; Wei, J.B. Are Small Radii of Compact Stars Ruled out by GW170817/AT2017gfo? *Astrophys. J.* **2018**, *860*, 139. [CrossRef]
116. Radice, D.; Dai, L. Multimessenger parameter estimation of GW170817. *Eur. Phys. J. Hadron. Nucl.* **2019**, *55*, 50. [CrossRef]
117. Kiuchi, K.; Kyutoku, K.; Shibata, M.; Taniguchi, K. Revisiting the Lower Bound on Tidal Deformability Derived by AT 2017gfo. *Astrophys. J. Lett.* **2019**, *876*, L31. [CrossRef]
118. Wei, J.B.; Lu, J.J.; Burgio, G.F.; Li, Z.H.; Schulze, H.J. Are nuclear matter properties correlated to neutron star observables? *Eur. Phys. J. A* **2020**, *56*, 63. [CrossRef]
119. Burgio, G.F.; Vidaña, I. The Equation of State of Nuclear Matter: From Finite Nuclei to Neutron Stars. *Universe* **2020**, *6*, 119. [CrossRef]

symmetry

MDPI

Article

Unified Equation of State for Neutron Stars Based on the Gogny Interaction

Xavier Viñas [1,*], Claudia Gonzalez-Boquera [1], Mario Centelles [1], Chiranjib Mondal [2] and Luis M. Robledo [3,4]

1 Departament de Física Quàntica i Astrofísica and Institut de Ciències del Cosmos (ICCUB), Facultat de Física, Universitat de Barcelona, Martí i Franquès 1, 08028 Barcelona, Spain; claugb@fqa.ub.edu (C.G.-B.); mario@fqa.ub.edu (M.C.)
2 Laboratoire de Physique Corpusculaire Caen, 6 Boulevard Marchal Juin, CEDEX, 14000 Caen, France; mondal@lpccaen.in2p3.fr
3 Departamento de Física Teórica and CIAFF, Universidad Autónoma de Madrid, 28049 Madrid, Spain; luis.robledo@uam.es
4 Center for Computational Simulation, Universidad Politécnica de Madrid, Campus de Montegancedo, Boadilla del Monte, 28660 Madrid, Spain
* Correspondence: xavier@fqa.ub.edu

Abstract: The effective Gogny interactions of the D1 family were established by D. Gogny more than forty years ago with the aim to describe simultaneously the mean field and the pairing field corresponding to the nuclear interaction. The most popular Gogny parametrizations, namely D1S, D1N and D1M, describe accurately the ground-state properties of spherical and deformed finite nuclei all across the mass table obtained with Hartree–Fock–Bogoliubov (HFB) calculations. However, these forces produce a rather soft equation of state (EoS) in neutron matter, which leads to predict maximum masses of neutron stars well below the observed value of two solar masses. To remove this limitation, we built new Gogny parametrizations by modifying the density dependence of the symmetry energy predicted by the force in such a way that they can be applied to the neutron star domain and can also reproduce the properties of finite nuclei as good as their predecessors. These new parametrizations allow us to obtain stiffer EoS's based on the Gogny interactions, which predict maximum masses of neutron stars around two solar masses. Moreover, other global properties of the star, such as the moment of inertia and the tidal deformability, are in harmony with those obtained with other well tested EoSs based on the SLy4 Skyrme force or the Barcelona–Catania–Paris–Madrid (BCPM) energy density functional. Properties of the core-crust transition predicted by these Gogny EoSs are also analyzed. Using these new Gogny forces, the EoS in the inner crust is obtained with the Wigner–Seitz approximation in the Variational Wigner–Kirkwood approach along with the Strutinsky integral method, which allows one to estimate in a perturbative way the proton shell and pairing corrections. For the outer crust, the EoS is determined basically by the nuclear masses, which are taken from the experiments, wherever they are available, or by HFB calculations performed with these new forces if the experimental masses are not known.

Keywords: unified equation of state; Gogny interaction; neutron star; symmetry energy; tidal deformability; moment of inertia

Citation: Viñas, X.; Gonzalez-Boquera, C.; Centelles, M.; Mondal, C.; Robledo, L.M. Unified Equation of State for Neutron Stars Based on the Gogny Interaction. *Symmetry* **2021**, *13*, 1613. https://doi.org/10.3390/sym13091613

Academic Editor: Charalampos Moustakidis

Received: 26 July 2021
Accepted: 18 August 2021
Published: 2 September 2021

1. Introduction

The standard Gogny interactions of the D1 family [1] consist of a finite-range part, which is modeled by two Gaussian form-factors including all the possible spin and isospin exchange terms, a zero-range density dependent term, which simulates the effect of the three-body forces, and a spin-orbit force, which is also of zero-range as in the case of Skyrme forces. Large-scale Hartree–Fock–Bogoliubov (HFB) calculations performed in a harmonic oscillator basis with the D1S parametrization [2,3] reveal that there is a systematic drift in the binding energy of neutron-rich nuclei (see [4] for more details). To overcome this deficiency, new parametrizations of the Gogny interaction, namely D1N [5] and D1M [6] were proposed. Unlike the D1S and D1N forces, whose parameters were obtained following the fitting protocol established in Ref. [1], the parameters of the D1M interaction were obtained by minimizing the energy *rms* deviation of 2149 measured nuclear masses of the AME2003 evaluation [7]. It is worthwhile to mention that, in the calibration of the D1N and D1M forces, in order to improve the description of neutron-rich nuclei, it was imposed that these interactions would follow the trend of the microscopic neutron matter EoS of Friedman and Pandharipande [8]. The D1M force reproduces the experimental nuclear masses of 2149 nuclei with an energy *rms* deviation of 798 keV. As an example, we display in the right panel of Figure 1 the binding energy differences between theoretical, computed with the D1M force at HFB level [9,10], and experimental binding energies, taken from the 2012 mass evaluation [11], of 620 even–even spherical and deformed nuclei. The theoretical binding energies include the HFB contribution and the rotational energy correction. However, the quadrupole zero point energy correction, which was included in the original fit, is approximated by a constant shift in the energy. We see that these differences are scattered around zero and do not show any energy drift for large neutron numbers. In the left panel of the same Figure, we display the same differences but computed with the D1S force. In this case the previously mentioned drift of binding energies can be clearly appreciated.

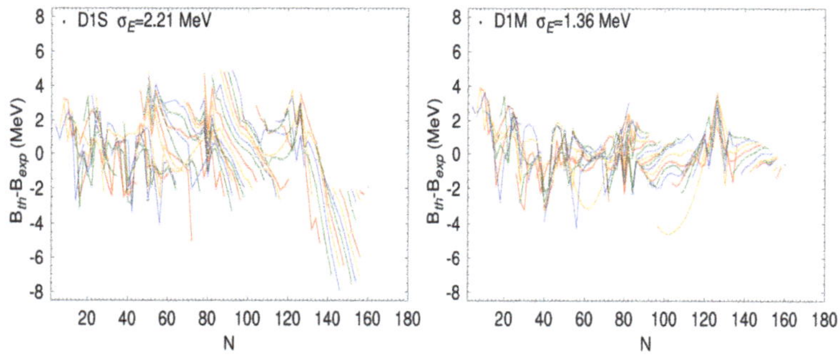

Figure 1. Differences between the computed and the experimental binding energies of 620 even–even nuclei. Theoretical calculations are performed with the Gogny D1S (**left panel**) and D1M (**right panel**) interactions. The experimental values are taken from [11].

However, the use of Gogny interactions in the neutron star (NS) domain does not work so well. In recent years it has been shown [12,13] that the most successful Gogny parametrizations, namely D1S, D1N and D1M, fall short in predicting a maximum NS mass of two solar masses ($M_\odot$), as required by some well contrasted astronomical observations [14–17]. A new extension of the Gogny force with a finite–range density–dependent term has been recently postulated [18]. This interaction, denoted as D2, has not been used much in finite nuclei calculations due to the complexity introduced by the finite range of the density–dependent term, but its EoS is able to reproduce the correct limit for the NS masses [19,20]. The structure of a standard NS composed by neutrons, protons and leptons (electrons and muons) in charge and in β–equilibrium is driven by its EoS, which allows the expression of the total pressure P of the system to be written as a function of the baryonic density ρ. The EoS is the essential input needed to solve the Tolman–Oppenheimer–Volkov (TOV) equations, whose solution provides the mass–radius relationship of the NS. Throughout this work we consider that the NS is non-rotating, cold and locally charge neutral and in absolute thermodynamic equilibrium. This is a reasonable picture for an NS that was created a long time ago and had enough time to cool down.

In the uniform core of the star, the total pressure is given by the sum of the baryonic (P_b) and leptonic (P_l) contributions:

$$P = P_b + \sum_l P_l = \rho^2 \frac{\partial E_b}{\partial \rho} + \sum_l \rho_l^2 \frac{\partial E_l}{\partial \rho_l}, \quad (1)$$

where $l = e, \mu$. In (1) E_b and E_l are the baryon and lepton energies per particle and $\rho = \rho_n + \rho_p$ is the total baryon density with ρ_n and ρ_p being the neutron and proton densities, respectively. The lepton densities ρ_l, owing to the charge equilibrium, are related to the proton density by $\rho_p = \rho_e + \rho_\mu$, where ρ_e and ρ_μ are the electron and muon densities. Changing from the neutron and proton densities to the total density ρ and to the isospin asymmetry $\delta = (\rho_n - \rho_p)/\rho$, each contribution to the total pressure (1) can also be written as

$$\begin{aligned} P_b &= \mu_n \rho_n + \mu_p \rho_p - \mathcal{H}_b(\rho, \delta) \\ P_l &= \mu_l \rho_l - \mathcal{H}_l(\rho_l), \end{aligned} \quad (2)$$

where $\mathcal{H}_b$ and $\mathcal{H}_l$ are the baryonic (b) and leptonic ($l = e, \mu$) energy densities and μ_n, μ_p, μ_e and μ_μ are the neutron, proton, electron and muon chemical potentials, respectively, which are defined as

$$\mu_n = \frac{\partial \mathcal{H}_b}{\partial \rho_n}; \quad \mu_p = \frac{\partial \mathcal{H}_b}{\partial \rho_p}; \quad \mu_e = \frac{\partial \mathcal{H}_e}{\partial \rho_e}; \quad \mu_\mu = \frac{\partial \mathcal{H}_\mu}{\partial \rho_\mu}. \quad (3)$$

In stable neutron star matter (NSM) the direct Urca processes

$$n \to p + l + \bar{\nu}_l \qquad \text{and} \qquad p + l \to n + \nu_l \quad (4)$$

take place simultaneously. Assuming that the neutrinos eventually leave the star, the β-equilibrium condition leads to

$$\mu_n - \mu_p = \mu_e = \mu_\mu. \quad (5)$$

The EoSs for NSM in logarithmic scale as a function of the baryonic density computed for some of the Gogny interactions used in this work and obtained previously in Refs. [12,13] are displayed in Figure 2 together with the EoS provided by the BCPM energy density functional [21], which we will use here as a benchmark, as well as the EoS obtained using the SLy4 [22] and BSk22 [23] Skyrme forces. The BCPM EoS, derived in the framework of the microscopic Brueckner–Bethe–Goldstone theory (see [21] and references therein), is in very good agreement with the EoS provided by the SLy4 force [22], which was specifically built for astrophysical calculations. We can also see that the EoS corresponding to the BSk22 Skyrme force obtained by the Brussels–Montreal group and reported in Ref. [23] (also see Ref. [24]) is stiffer than the EoSs computed with the SLy4 Skyrme force and the BCPM energy density functional. From this Figure we can see that the EoSs obtained with the D1N and D1M forces show an increasing trend with growing baryon density but softer than the behavior exhibited by the BCPM EoS. We can also see that the EoS for NSM calculated with the D1S force reaches a maximum value at around twice the normal saturation density and decreases for larger densities. As a consequence of this anomalous behavior, the TOV equations cannot be solved in the D1S case, which implies that the D1S interaction is not well suited for astrophysical calculations. The shaded area in Figure 2 depicts the region in the P-ρ plane consistent with the experimental collective flow data in Heavy-Ion Collisions (HIC) [25]. From this Figure we can see that none of the EoSs computed with the standard Gogny interactions are able to clearly pass through the region constrained by the collective flow in HIC.

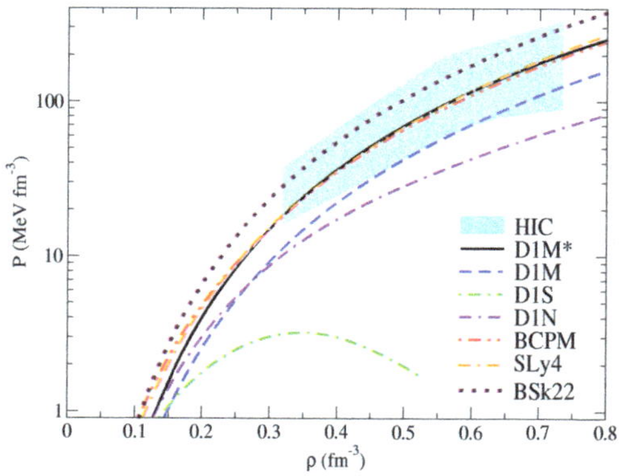

Figure 2. Equation of state (total pressure in logarithmic scale against baryon density) for neutron star matter computed with the D1M*, D1M, D1N and D1S Gogny interactions, with the BCPM energy density functional and with the SLy4 and BSk22 Skyrme forces. Constraints coming from collective flow in heavy-ion collisions are also included [25].

The baryonic part of the EoS is basically driven by the energy density of highly asymmetric nuclear matter (ANM) $\mathcal{H}_b(\rho,\delta)$, where the isospin asymmetry δ takes values around 0.9. To characterize this energy density, which is close to the pure neutron matter, it is extremely useful to introduce the symmetry energy, which can be understood as the energy cost to convert all protons into neutrons in symmetric nuclear matter. The energy per particle $E_b(\rho,\delta) = \mathcal{H}_b(\rho,\delta)/\rho$ in ANM can be written as a Taylor expansion with respect to the isospin asymmetry around $\delta = 0$:

$$E_b(\rho,\delta) = E_b(\rho,\delta=0) + \sum_{k=1}^{\infty} E_{sym,2k}(\rho)\delta^{2k}, \tag{6}$$

where we have assumed the charge symmetry of the strong interaction, which implies that only even powers of δ appear in (6). The first term of the expansion, $E_b(\rho,\delta=0)$ is the energy per baryon in symmetric nuclear matter and the coefficients of the Taylor expansion are given by:

$$E_{sym,2k} = \frac{1}{(2k)!}\frac{\partial^{2k}E_b(\rho,\delta)}{\partial\delta^{2k}}\bigg|_{\delta=0}. \tag{7}$$

The symmetry energy coefficient E_{sym} is usually defined as the second-order coefficient in the expansion (6), i.e., $E_{sym} \equiv E_{sym,2}$. In many cases the energy per particle in ANM is well approximated taking only the quadratic term in the expansion (6), that is,

$$E_b(\rho,\delta) = E_b(\rho,\delta=0) + E_{sym}(\rho)\delta^2. \tag{8}$$

Therefore, it is also possible to define the symmetry energy as the difference between the energy per particle in pure neutron matter and in symmetric nuclear matter,

$$E'_{sym} = E_b(\rho,\delta=1) - E_b(\rho,\delta=0). \tag{9}$$

Taking into account (6), it is clear that the definition (9) corresponds to the whole sum of the coefficients $E_{sym,2k}$. The difference between both definitions of the symmetry energy depends on the importance of the contribution of the terms higher than the quadratic one in the expansion (6). A detailed discussion about the higher-order symmetry energy contributions in the case of Gogny interactions can be found in Refs. [13,20]. In Figure 3 we display the symmetry energy, defined as Equation (7) with $k = 1$, as a function of the baryonic density computed with different Gogny forces available in the literature and taken from Refs. [12,13]. In the same Figure we also show the symmetry energy constraints extracted from the isobaric analog states (IAS) and from IAS combined with neutron skins [26], the constraints from the electric dipole polarizability α_D in ^{208}Pb [27] and from transport simulations in heavy-ion collisions in Sn isotopes [28].

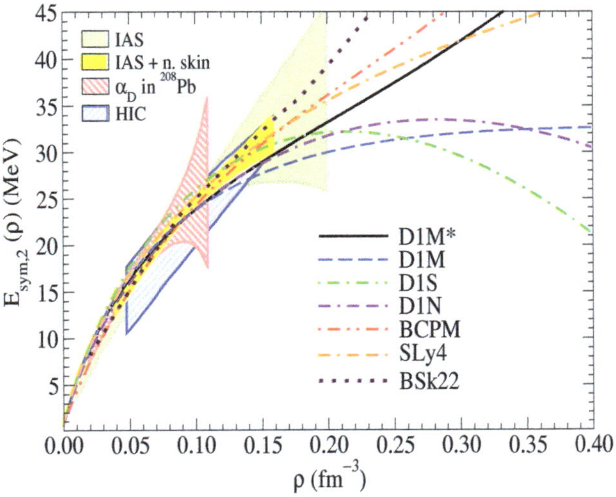

Figure 3. Symmetry energy, defined as Equation (7) with $k = 1$ against the baryon density predicted by the D1M*, D1M, D1S and D1N Gogny interactions, the BCPM energy density functional and the SLy4 and BSk22 Skyrme forces. Some constraints coming from isobaric analog states (IAS) (green), from IAS plus neutron skins (IAS + n.skin) (yellow), electric dipole polarizability α_D in ^{208}Pb (α_D in ^{208}Pb) (dashed red) and heavy-ion collisions (dashed blue) are also included [26–28].

From Figure 3 we can see that, below the saturation density, the symmetry energy behaves in a very similar way for all the considered forces taking values around 30 MeV at saturation. This is due to the fact that in this region the symmetry energy is well constrained by the nuclear masses to which the parameters of the different effective interactions have been fitted (see Refs. [29–31] for a review about the range of the symmetry energy obtained from different constraints). Above the saturation density, the symmetry energy predicted by the different interactions differ more among them. For example, we can see that the symmetry energy computed with the D1S and D1N parametrizations reaches maximum values of 30–40 MeV, and then decrease with increasing density until vanishing around 3–4 times the saturation density, where the isospin instability starts. In the case of the D1M force the symmetry energy also reaches a maximum value, which remains practically constant in the whole density range needed to solve the TOV equations. From the same Figure 3 we also observe that the symmetry energy computed with the BCPM energy density functional shows a different trend, growing with increasing density. The symmetry energy computed with the SLy4 and BSk22 Skyrme forces, which provide realistic EoSs, also shows an increasing trend with growing density, BSk22 being stiffer and SLy4 softer in the high-density domain above 0.20 fm^{-3}. These results show that the behavior of the symmetry energy as a function of the density above the saturation is crucial for describing properly the EoS of neutron-rich matter in the high-density regime, which, in turn, is the most relevant input for the study of many NS properties.

An important feature of the symmetry energy is its density content calculated at saturation density. This quantity is usually characterized by the slope of the symmetry energy L, which is defined as

$$L = 3\rho_0 \frac{\partial E_{sym}(\rho)}{\partial \rho}\bigg|_{\rho_0}. \tag{10}$$

The slope parameter is connected with different properties of finite nuclei, as for example the neutron skin thickness in heavy nuclei such as ^{208}Pb (see [32–35] and references therein). The numerical values of the slope parameter L predicted by different models span a very large range between 10 and 120 MeV, pointing out that this quantity is poorly constrained by the available experimental data. A compilation of possible L values extracted from different laboratory experiments and astronomical observations can be found in Refs. [19,36,37]. From the theoretical side, some recent microscopic calculations have estimated the slope parameter in the ranges L = 43.8–48.6 MeV [38], L = 20–65 MeV [39] and L = 45–70 MeV [40]. The values of the slope parameter predicted by the standard Gogny forces of the D1 family are relatively small, L = 22.43 MeV (D1S), L = 24.83 MeV (D1M) and L = 33.58 MeV (D1N) [12]. These values, which are clearly smaller than the value L = 52.96 MeV predicted by the BCPM energy density functional and those of the SLy4 and BSk22 Skyrme forces, clearly explain the soft behavior of the symmetry energy displayed in Figure 3 and consequently the softness of the EoS in NS matter predicted by such forces (see Figure 2). In Figure 4 we display some bounds of the symmetry energy at saturation $E_{sym}(\rho_0)$ and its slope L provided by recent laboratory data, astronomical observations and *ab initio* calculations using chiral interactions [30,36,41,42]. We see that the symmetry energy and its slope predicted by the Gogny forces D1M and D1N lie outside the constrained region in the $E_{sym}(\rho_0)$-L plane, while the point corresponding to the D1S interaction is at the lower edge of the region estimated from the measured electric dipole polarizability in ^{68}Ni, ^{120}Sn and ^{208}Pb [41].

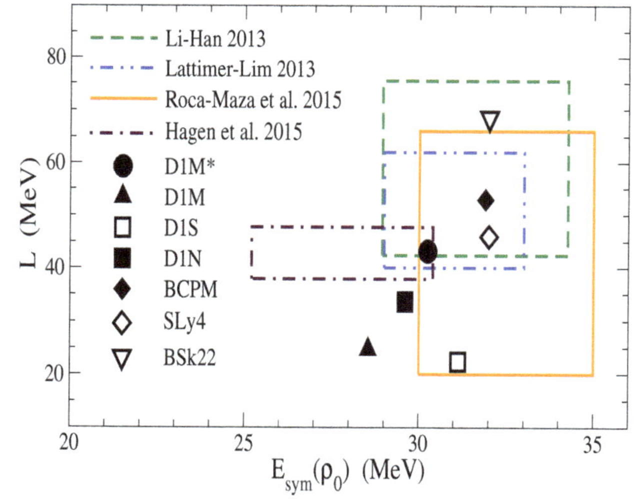

Figure 4. Slope of the symmetry energy L against the symmetry energy at saturation density for some Gogny interactions, the BCPM energy density functional and the SLy4 and BSk22 Skyrme forces. We have included some constraints extracted from the literature [30,36,41,42].

From this discussion it is clear that the standard Gogny interactions of the D1 family are not well suited for applications in the NS domain. To overcome this situation we designed some parametrizations of the Gogny type of forces starting from the D1M interaction [19,20,43,44] aimed to predict a maximum mass in NS of $2M_\odot$ without losing its ability to describe finite nuclei with a quality similar to those found using the D1M force. The purpose of this paper is to review those new parametrizations and compare them with previous results. The paper is organized as follows. In the second section we describe the method used to fit these new Gogny parametrizations, namely D1M* and D1M**. In the third section we describe how the EoS in the inner and outer crust using the D1M* interaction is obtained. In the same section the study of the core–crust transition using the thermodynamical and dynamical methods is briefly summarized. The fourth section is devoted to discussing some global NS properties such as the mass–radius relation, the moment of inertia, its crustal properties and the tidal deformability estimated with the new Gogny interaction D1M*. We also compare in this section the D1M* results with the predictions provided by other different models. Finally, our conclusions are presented in the last section.

2. Gogny Interactions Adapted for Astrophysical Calculations

The standard Gogny interaction of the D1 family consists of a finite range term, which is modeled by two form factors of Gaussian type and includes all possible spin and isospin exchange terms, plus a zero-range density-dependent contribution. To describe finite nuclei, a spin–orbit interaction—which is zero-range like in the case of Skyrme forces—is also added. With all these ingredients the Gogny interaction reads:

$$V(\mathbf{r}_1,\mathbf{r}_2) = \sum_{i=1,2}(W_i + B_iP_\sigma - H_iP_\tau - M_iP_\sigma P_\tau)e^{-\frac{r^2}{\mu_i^2}} + t_3(1 + x_3P^\sigma)\rho^\alpha(\mathbf{R})\delta(\mathbf{r}) + iW_{LS}(\sigma_1 + \sigma_2)\{\mathbf{k}' \times \delta(\mathbf{r})\mathbf{k}\}, \quad (11)$$

where $\mathbf{r}$ and $\mathbf{R}$ are the relative and the center of mass coordinates of the two nucleons, and $\mu_1 \simeq 0.5$–0.7 fm and $\mu_2 \simeq 1.2$ fm are the ranges of the two Gaussian form factors, which simulate the short- and long-range components of the force, respectively. The Skyrme-type t_3 and x_3 parameters control the density dependent part of the force.

To determine the parameters of the new Gogny interactions, denoted D1M* and D1M**, we start from the D1M force and modify the parameters of the finite-range part of the interaction, which are the ones that control the stiffness of the symmetry energy, keeping the binding energy and charge radius of finite nuclei predicted by these interactions as close as possible to the values obtained with the original D1M force. This way of proceeding is similar to the one used with some Skyrme forces and RMF parametrizations, such as SAMi-J [45], KDE0-J [46] or FSU-TAMU [47,48].

Therefore, we readjust the eight parameters W_i, B_i, H_i and M_i ($i = 1,2$) of the finite-range part of the Gogny interaction. The ranges of the two Gaussian form factors and the zero-range part of the force are kept fixed to the original values of D1M. The open parameters are constrained by imposing in symmetric nuclear matter the same values of the saturation density, energy per particle, incompressibility modulus and effective mass as the ones predicted by the original D1M force. It has been claimed in earlier literature that finite nuclei energies constrain the symmetry energy at a subsaturation density of about 0.1 fm^{-3} better than at saturation density [32,49]. Hence, we impose that the symmetry energy of the modified interaction at this particular density also equals the corresponding value provided by the D1M force. In order to preserve the pairing properties of D1M we also require that, in the new force the combinations $W_i - B_i - H_i + M_i$ (i = 1,2), which govern the strength of the pairing interaction, take the same value as in the original D1M force. There is still an open parameter, which we chose to be B_1. This parameter is used to modify the slope of the symmetry energy at saturation L, which in turn determines the maximum mass of the neutron star. We adjust this parameter B_1 in such a way that the maximum mass computed with the new parametrizations of the Gogny force are $2M_\odot$ (D1M*) and $1.91M_\odot$ (D1M**). Finally, we perform a fine tuning of the strength t_3 of the density-dependent term of the interaction in order to optimize the description of the masses of finite nuclei. To this end we compute the energies of 620 spherical and deformed even–even nuclei of the AME2012 database [11] at HFB level using the HFBaxial code [9]. As it is customary with Gogny forces, we carry out the HFB calculations in a harmonic oscillator basis. The parameters and size of the basis are chosen as to optimize the binding energies for each value of mass number A. An approximate second-order gradient is used to solve with confidence the HFB equations [10]. It has been known for a long time that some Skyrme parametrizations present numerical instabilities when the finite-nuclei calculations are performed on a mesh in a coordinate space, see e.g., [50] and references therein. It has been recently shown that the Gogny parameter sets may also display finite-size instabilities [51] that lead to diverging results in the coordinate-space calculations of finite nuclei [51,52]. This is the case of the D1N and D1M* forces [51,52] and, to a lesser extent, of D1M [52]. Therefore, the HFB calculations of finite nuclei with the new parameter set D1M* are to be performed in a harmonic oscillator basis [19,52]. The numerical values of the parameters of the new forces D1M* and D1M** were reported in Refs. [19,20,43]. For the sake of completeness, we collect them also here in Table 1, along with the parameters of D1M. In Table 2 we report the nuclear matter properties predicted by the D1M* and D1M** forces, as well as by the BCPM energy density functional, which is used in this work as a benchmark for comparison with the results provided by the new Gogny parametrizations D1M* and D1M**.

From Table 1, we observe that the finite-range parameters W_i, B_i, H_i and M_i of the modified D1M* and D1M** forces are larger in absolute value than the ones in the original D1M interaction. However, as can be seen in Table 2, the saturation properties of symmetric nuclear matter (namely, the saturation density ρ_0, the energy per particle E_0 at saturation, the incompressibility K_0, and the effective mass m^*/m) and the symmetry energy at a density 0.1 fm^{-3}, predicted by the D1M** interaction coincide with the values computed with the D1M force as a consequence of the fitting protocol used to obtain the parameters of the modified forces. In the case of the D1M* force we also slightly changed the t_3 parameter by an amount of 1 MeV to improve the finite nuclei description with this interaction. As a consequence of this small change in t_3, the symmetric nuclear matter properties involved in the reparametrization changes slightly compared to the corresponding values predicted by the D1M force, as can be seen in Table 2. The properties that differ significantly between the new parametrizations and D1M are the symmetry energy at saturation density ($E_{sym}(\rho_0)$) and, visibly, the density dependence of the symmetry energy, which governs the isovector part of the interaction. The latter is quantified by the slope parameter L, which varies from a value L = 24.84 MeV in the original D1M force to L = 43.18 MeV for D1M* and to L = 33.91 MeV for D1M**, as required to obtain a stiffer EoS in NS matter, which in turn allows predictions of the maximum mass of $2M_\odot$ and $1.91M_\odot$, respectively.

Table 1. Parameters of the D1M, D1M* and D1M** Gogny forces. The coefficients W_i, B_i, H_i and M_i are given in MeV, μ_i in fm and t_3 in MeV fm^4. The values of the other parameters of the modified interactions are the same as in the D1M force (namely, x_3 = 1, α = 1/3 and W_{LS} = 115.36 MeV fm^5).

D1M	W_i	B_i	H_i	M_i	μ_i
i = 1	−12,797.57	14,048.85	−15,144.43	11,963.81	0.50
i = 2	490.95	−752.27	675.12	−693.57	1.00
	t_3	x_3	α	W_{LS}	
	1562.22	1	1/3	115.36	
D1M*	W_i	B_i	H_i	M_i	μ_i
i = 1	−17,242.0144	19,604.4056	−20,699.9856	16,408.3344	0.50
i = 2	675.3860	−982.8150	905.6650	−878.0060	1.00
	t_3	x_3	α	W_{LS}	
	1561.22	1	1/3	115.36	
D1M**	W_i	B_i	H_i	M_i	μ_i
i = 1	−15,019.7922	16,826.6278	−17,922.2078	14,186.1122	0.50
i = 2	583.1680	−867.5425	790.3925	−785.7880	1.00
	t_3	x_3	α	W_{LS}	
	1562.22	1	1/3	115.36	

Let us now briefly discuss the main properties and predictions of these modified Gogny forces. As can be seen from Figure 3, the symmetry energy as a function of the baryon density obtained using D1M* shows a different behavior compared to the one exhibited by the standard Gogny interactions D1S, D1N and D1M. Above saturation the symmetry energy computed with D1M* increases with growing density and takes values close to the ones predicted by the BCPM energy density functional. As a consequence of this behavior, in the high–density domain the EoS predicted by the D1M* interaction follows closely the trend of the BCPM EoS, passing nicely through the region of the $P-\rho$ plane constrained by the experimental data of the heavy-ion collisions as can be seen in Figure 2. Finally let us point out that the representative points of the D1M* force lie within the region of the $E_{sym}(\rho_0)$-L plane constrained by the majority of the experimental data, as is seen in Figure 4. In order to check the ability of the D1M* force to describe finite nuclei, we plot in Figure 5 the differences between the binding energies of a set of 620 even–even nuclei computed with this new force and with the original D1M interaction along different isotopic chains covering the whole nuclear chart. We see that these differences are actually very small, lying within a window of ± 3 MeV for all the computed nuclei. As a general trend, the binding energy predicted by D1M* is larger than the one provided by D1M for neutron deficient nuclei of the isotopic chains and the opposite happens for neutron rich nuclei of the chain.

Table 2. Nuclear matter properties predicted by the D1M*, D1M** and D1M Gogny interactions and by the BCPM energy density functional.

	ρ_0 (fm^{-3})	E_0 (MeV)	K_0 (MeV)	m^*/m	$E_{sym}(\rho_0)$ (MeV)	$E_{sym}(0.1)$ (MeV)	L (MeV)
D1M	0.1647	−16.02	224.98	0.746	28.55	23.80	24.83
D1M*	0.1650	−16.06	225.38	0.746	30.25	23.82	43.18
D1M**	0.1647	−16.02	224.98	0.746	29.37	23.80	33.91
BCPM	0.1600	−16.00	213.75	1.000	31.92	24.20	52.96

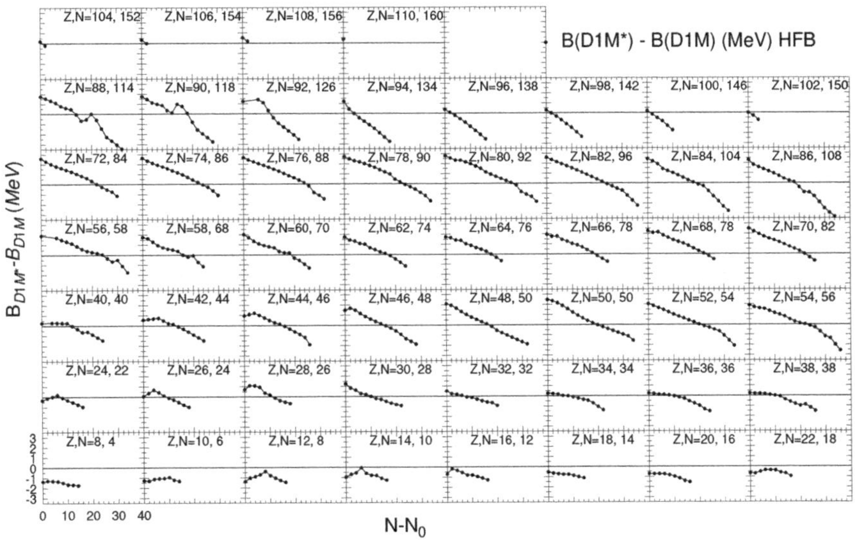

Figure 5. Difference between the binding energies provided by the D1M* and D1M force ΔB (in MeV) plotted as a function of the shifted neutron number N-N_0 for isotopic chains covering the periodic table. The values of the atomic number Z and neutron reference number N_0 are given in each panel. The vertical scale covers from +3.5 MeV to −3.5 MeV, with long ticks every MeV and short ticks every half MeV. The horizontal line in each panel at $\Delta B = 0$ is plotted guide the eye.

3. Neutron Star Crust with Gogny Forces

The outer layer of an NS encircling the homogeneous core is denoted as "crust". It is further subdivided into two or three layers depending on its composition. At the surface of the star, namely the "outer crust", the matter is distributed in a lattice of neutron-rich nuclei immersed in an electron gas. After a certain density $\sim$0.003 fm^{-3} going towards the center of the star, neutrons start to drip from the nuclei forming a background neutron gas but keeping a lattice structure of nuclear clusters. This region is denoted as the NS "inner crust". At a density $\sim$0.08 fm^{-3}, also known as the "crust–core transition density", the inner crust dissolves into an homogeneous core, sometimes with pasta phases in the transition region. As these complicated structures incorporate in-medium many-body effects, a full quantum mechanical treatment of the inner-crust is very difficult and computationally expensive. Nevertheless, there exist some calculations of the EoS in this region of NSs of different degrees of sophistication available in the literature (see for example [21,53–55] for references and a more detailed discussion on this topic). Simplified calculations based on the Thomas–Fermi (TF) approximation or its extended versions are often employed to obtain the EoS of the neutron star crust with different interactions (see [21,53–57] and references quoted therein). Even though global properties like the mass or the radius are not heavily influenced by the crustal properties of the NS, pulsar glitches, quasi-periodic oscillations in soft γ-ray repeaters or thermal relaxations in soft X-ray transients are strongly influenced by the crustal composition of the NS (see for example [54,55,58] and references quoted therein). The crust also might be one of the possible places where the r-process nucleosynthesis occurs during the NS–NS or NS–Black Hole merger events [59–61].

We have organized the description of the crust in this section as follows. In the first subsection we outline the variational Wigner–Kirkwood (VWK) method for describing finite nuclei. After that we describe the restoration of quantum effects like the shell correction with the Strutinsky integral method and the residual pairing correction with state dependent Bardeen–Cooper–Schrieffer (BCS) calculations. In the next subsection we compute with Gogny interactions the structure of the outer crust of a cold, non-accreting star. These calculations are performed within the so-called Wigner–Seitz (WS) approximation, which assumes that the space can be described by non-interacting electrically neutral cells, each one containing a single nuclear cluster embedded in electron (outer crust) or electron and neutron (inner crust) gases. In the inner crust, we restrict ourselves to spherically symmetric nuclear clusters disregarding pasta phases for the sake of simplicity. The results obtained with different Gogny forces are also displayed in the relevant subsections. At the end, we discuss the crust–core transition obtained with different Gogny interactions.

3.1. *Variational Wigner-Kirkwood Method in Finite Nuclei*

Semiclassical estimates of the binding energy of nuclei throughout the whole nuclear chart have been used since the Bethe–Weizsäcker mass formula was proposed [62,63]. The smooth part of the energy can be estimated by considering a Fermi gas-like system of nucleons with different choice of interactions. Further, one can treat quantum shell corrections perturbatively on top of it, using the techniques established by Strutinsky [64]. The residual pairing energy can also be calculated perturbatively using the shell structure corresponding to the average mean-field. The smooth part of the binding energy, i.e., neglecting quantal effects, of a set of non-interacting fermions in an external potential, can easily be obtained using the Wigner–Kirkwood (WK) $\hbar$-expansion of the single particle partition function [65–69]. An important feature of this expansion is that the variational solution of the minimization of WK energy at each $\hbar$-order is simply the WK expansion of the density at the same order. This method of solving a variational equation by sorting order-by-order the $\hbar$-expansion is called the VWK theory, which is discussed in detail in Refs. [68,70,71]. A primary feature of this method is that one needs to calculate one less order in the density expansion to accurately calculate the energy in the next order. For example, a VWK prediction on the energy containing $\hbar^2$-order contribution only needs the information on the $\hbar^0$-order densities, i.e., the bare TF densities.

To calculate the smooth part of the energy with the VWK method using the Gogny interaction (11), we use in this work the extended TF density matrix [72], which allows us to obtain the kinetic and exchange energy densities up to $\hbar^2$ order as a functional of the particle densities of each type of nucleons [53,73]. Therefore we write the VWK energy as

$$\begin{aligned} E_{\text{VWK}} &= \int \mathcal{H} d\mathbf{R} = \int (\mathcal{H}_0 + \mathcal{H}_2) d\mathbf{R} \\ &= \int (\mathcal{H}_{kin,0} + \mathcal{H}_{dir} + \mathcal{H}_{exch,0} + \mathcal{H}_{zr} + \mathcal{H}_{Coul}) d\mathbf{R} \\ &+ \int (\mathcal{H}_{kin,2} + \mathcal{H}_{exch,2} + \mathcal{H}_{SO}) d\mathbf{R}, \end{aligned} \tag{12}$$

where we have decomposed the energy into TF (subindex 0) and $\hbar^2$ (subindex 2) terms. For a detailed derivation of the energy density in (12), the reader is referred to Refs. [53,73].

To find the density profiles, which in turn will allow one to determine the VWK energy, one should solve first the variational TF equations for each type of particles with respect to the TF densities $\rho_q (q = n, p)$,

$$\frac{\delta}{\delta \rho_q} \left[E_{\text{VWK},0} - \mu_q \int \rho_q(\mathbf{R}) d\mathbf{R} \right] = 0, \tag{13}$$

where μ_q are the chemical potentials that ensure the right number of nucleons of each type. Using the solutions of Equation (13) in Equation (12), one can calculate the semiclassical energy up to $\hbar^2$-order in the VWK approach.

Instead of solving the set of Equation (13), we perform a restricted variational calculation by minimizing the TF part of the VWK energy Equation (12) using a trial density of the Fermi type for each type of particles,

$$\rho_q(r) = \frac{\rho_{0,q}}{1 + \exp\left(\frac{r - C_q}{a_q}\right)}, \tag{14}$$

where the radius C_q and the diffuseness parameter a_q of each trial density are the variational parameters and the strengths $\rho_{0,q}$ are fixed by normalizing the neutron and proton numbers. Finally, using these trial densities the $\hbar^2$ part of the VWK energy in Equation (12) is added perturbatively. This restricted minimization of the energy with parametrized neutron and proton densities has been applied successfully in many semiclassical calculations of the energy of finite nuclei using Skyrme interactions [69], the differences with the full variational calculation being very small [74].

3.2. Shell and Pairing Effects

Once the average smooth part of the energy is determined, we add perturbatively the quantum shell energy that is obtained using the so-called Strutinsky integral method [75,76]. In this approximation, the shell correction is estimated as the difference between the quantal energy and its semiclassical counterpart of a set of nucleons moving under the action of an external single-particle Hartree–Fock Hamiltonian (see Refs. [53,73] for more details) generated by the parametrized neutron and proton densities (14). The corresponding Schrödinger equations read,

$$h_q \phi_{i,q} = \left\{ -\nabla \frac{\hbar^2}{2\widetilde{m}_q^*(\mathbf{r})} \nabla + \widetilde{U}_q(\mathbf{r}) - i\widetilde{\mathbf{W}}_q(\mathbf{r})(\nabla \times \sigma) \right\} \phi_i = \widetilde{\epsilon}_{i,q} \phi_{i,q}. \tag{15}$$

It should be noticed that the local particle $\widetilde{\rho}_q$, kinetic energy $\widetilde{\tau}_q$ and spin $\widetilde{J}_q$ densities, which are used to calculate the effective mass $\widetilde{m}_q^*$, the mean-field $\widetilde{U}_q$ and the spin-orbit potential $\widetilde{\mathbf{W}}_q$ appearing in Equation (15), are obtained semi-classically by the restricted variational approach described above.

After the single-particle energies $\widetilde{\epsilon}_{i,q}$ are obtained by solving Equation (15), the shell correction energy for each type of particles is given by

$$E_q^{shell} = \sum_i \widetilde{\epsilon}_{i,q} - \int \left[\frac{\hbar^2}{2\widetilde{m}_q^*} \widetilde{\tau}_q + \widetilde{\rho}_q \widetilde{U}_q + \widetilde{\mathbf{J}}_q \cdot \widetilde{\mathbf{W}}_q \right] d\mathbf{R}. \tag{16}$$

These single-particle energies $\widetilde{\epsilon}_{i,q}$ can be further used to calculate perturbatively the residual neutron and proton pairing energy through a BCS pairing calculation as,

$$E_q^{pair} = -\frac{1}{4} \sum_{k,q} \frac{\Delta_{k,q}^2}{E_{k,q}} , \tag{17}$$

where $E_{k,q}$ and $\Delta_{k,q}$ are the quasiparticle energy and the gap in the state k of the type of particles q, respectively. The quasi-particle energy in the state k reads

$$E_{k,q} = \sqrt{(\widetilde{\epsilon}_{k,q} - \mu_q)^2 + \Delta_{k,q}^2} , \tag{18}$$

which in addition to the state-dependent gap $\Delta_{k,q}$ also depends on the eigenvalue $\widetilde{\epsilon}_{k,q}$ of (15) corresponding to the state k and on the chemical potential μ_q, which is determined by the particle number condition given by

$$N_q = \sum_k \widetilde{n}_{k,q}^2 , \tag{19}$$

where the occupation number $\widetilde{n}_{k,q}^2$ of the state k is given by,

$$\widetilde{n}_{k,q}^2 = \frac{1}{2} \left[1 - \frac{\widetilde{\epsilon}_{k,q} - \mu_q}{E_{k,q}} \right]. \tag{20}$$

For each type of particles the state-dependent gap in a given state i is obtained as the solution of the so-called gap equation

$$\Delta_{i,q} = -\sum_k v_{i\bar{i},k\bar{k}}^{pair} \frac{\Delta_{k,q}}{2E_{q,k}}. \tag{21}$$

Here, the single particle indices denote the usual quantum numbers, $i \equiv nlj$ and $k \equiv n'l'j'$ for each type of particle. We emphasize that the pairing interaction v^{pair} used in (21) is also determined from the same finite range Gogny interaction (11). The sums over k in Equations (17), (19) and (21) run over bound and quasi-bound states. These quasi-bound states of positive energy are retained by the centrifugal (neutrons) or centrifugal plus Coulomb (protons) barriers [77].

Finally, the total binding energy of a nucleus is given by the sum of the smooth part of the energy computed at VWK level (12) plus the quantal shell correction (16) and the pairing energy (17) calculated perturbatively, i.e.,

$$E_B = E_{\text{VWK}} + \sum_q \left[E_q^{shell} + E_q^{pair} \right]. \tag{22}$$

This method of obtaining the binding energy, which we call VWKSP, was applied for ~160 even-even nuclei across the whole nuclear chart using three different Gogny forces of D1 type, including D1M* [53]. For D1M*, the relative deviation from the experimental values or the ones obtained with HFB method were found to be within 1%, with only a few exceptions.

3.3. Outer Crust

As we have mentioned before, the external region of the NS crust consists of a lattice of fully ionized atomic nuclei embedded in a free electron gas. In the outer layers of the outer crust, the nuclei are the ones which are also observed in terrestrial experiments. However, near the inner crust neutron-rich nuclei whose masses have not been measured experimentally start to appear. To determine the composition and EoS of the outer crust, the essential ingredient is the mass table, which is provided by the experimental masses, when they are known, supplemented by the predictions from theoretical models for the unknown masses. In the present calculation of the outer crust we use the experimental masses from the AME2016 atomic mass evaluation [78] and the recently measured masses of the $^{75-79}$Cu isotopes [79]. When the relevant masses are unknown experimentally, we compute them at HFB level [9] using the D1M and D1M* Gogny interactions. D1M was also used in the calculations of the outer crust of Ref. [80], together with the experimental masses known at that moment (our results with D1M may differ a little from those of Ref. [80] for the layers of the outer crust where new experimental masses available in [78,79] were unmeasured when [80] was published).

The energy of the outer crust at a given density ρ_{av} is computed within the WS approximation, where the energy of each cell containing a nucleus with Z protons and A nucleons has primarily three contributions [81]

$$E(A, Z, \rho_{av}) = E_{Nuc} + E_e + E_{lat}, \tag{23}$$

where, E_{Nuc}, E_e and E_{lat} are the nuclear, electronic and lattice contribution to the energy, respectively. The number density of the outer crust ρ_{av} is determined by the volume V of the cell as $\rho_{av} = A/V$. The nuclear contribution essentially comes from the mass as

$$E_{Nuc} = M(A, Z) = (A - Z)m_n + Zm_p - E_B(A, Z). \tag{24}$$

Here, m_n and m_p are the rest masses of the neutron and the proton, respectively. For masses of nuclei which are not measured experimentally, we use the HFB predictions [9] computed with the D1M* interaction. The electronic contribution E_e is determined by the electronic energy density $\mathcal{H}_e$ for a degenerate relativistic free Fermi gas as

$$E_e = \mathcal{H}_e V, \tag{25}$$

where

$$\mathcal{H}_e = \frac{k_{F_e}}{8\pi^2}\left(2k_{F_e}^2 + m_e^2\right)\sqrt{k_{F_e}^2 + m_e^2} - \frac{m_e^4}{8\pi^2}\ln\left[\frac{k_{F_e} + \sqrt{k_{F_e}^2 + m_e^2}}{m_e}\right], \tag{26}$$

with m_e as the rest mass of electron and k_{F_e} the electron Fermi momentum, which is given by $k_{F_e} = (3\pi^2 n_e)^{1/3}$. In (26) $n_e = (Z/A)\rho_{av}$ is the electron number density. The lattice contribution to the energy is given by

$$E_{lat} = -C\frac{Z^2}{A^{1/3}}k_{F_{av}}, \tag{27}$$

where $k_{F_{av}} = (3\pi^2\rho_{av})^{1/3}$ is the average Fermi momentum connected with the electron Fermi momentum as $k_{F_{av}} = (A/Z)^{1/3}k_{F_e}$ due to charge equilibrium. The constant $C = 0.00340665$ for the bcc lattice is taken from Ref. [82].

At zero temperature, the pressure exerted by the outer crust comes completely from the electrons and the lattice while the nuclei produce no pressure. Therefore,

$$P = -\left(\frac{\partial E}{\partial V}\right)_{A,Z} = P_e + P_{lat} = n_e\sqrt{k_{F_e}^2 + m_e^2} - \mathcal{H}_e - \frac{\rho_{av}}{3}C\frac{Z^2}{A^{4/3}}k_{F_{av}}. \tag{28}$$

To obtain the optimal configuration in a WS cell, we proceed as follows. For a given pressure, at zero temperature, the Gibbs free energy G per nucleon is minimized for different nuclei in the nuclear chart,

$$g = \frac{G}{A} = \frac{E(A,Z,\rho_{av})}{A} + \frac{P}{\rho_{av}} = \frac{M(A,Z)}{A} + \frac{Z}{A}\sqrt{k_{F_e}^2 + m_e^2} - \frac{4}{3}C\frac{Z^2}{A^{4/3}}k_{F_{av}}. \quad (29)$$

It is worth mentioning here that recently a new analytical method to evaluate the internal composition of the outer crust has been presented in Ref. [83].

In Figure 6 we plot the composition of the outer crust in terms of the proton number Z and the neutron number N at different average densities ρ_{av}, obtained with the nuclear masses measured experimentally (from AME2016 [78] and from [79] for $^{75-79}$Cu) assisted by theoretical HFB calculations [9] using the D1M and D1M* interactions, where the experimental values are not available. For comparison, we also display in the Figure the composition predicted by the BCPM energy density functional [21]. At very low densities (up to $\rho_{av} \sim 10^{-6}$ fm^{-3}), the primary contribution comes from Ni and Fe isotopes with neutron numbers $N = 30$, 34 and 36. After that the contribution comes from Kr, Se, Ge and Zn isotopes up to $\rho_{av} \sim 5 \times 10^{-5}$ fm^{-3}, with $N = 50$. All three interactions in Figure 6 have the same predictions up to this point because the information primarily comes from the experimental masses. The differences start to appear beyond this density. The elements beyond $\rho_{av} \sim 5 \times 10^{-5}$ fm^{-3} are primarily Ru, Mo, Zr, Sr, Kr or Se isotopes. At these higher densities relevant for the outer crust, the optimal configuration of the WS cell comes from $N = 82$. In the region of the outer crust where the nuclear masses are unknown, the D1M* force predicts the nuclei ^{78}Ni, ^{128}Ru, ^{122}Zr and ^{120}Se, while the calculations performed with the Skyrme interactions BSk19-BSk21 in Ref. [80] and BSk22 and BSk24-BSk26 in Ref. [23] show a somewhat richer composition, as can be seen in Tables I-III of Ref. [80] and Tables 3–6 of Ref. [23], respectively. The composition of the outer crust critically depends on the nuclear masses, which can be slightly different when computed with different models and extrapolated to the region of unknown masses.

3.4. Inner Crust

We resort to the spherical WS approximation for describing the inner crust of NSs. We consider a density range between 0.0004 fm^{-3} and 0.08 fm^{-3} for the inner crust. For the present calculation, we have not considered pasta structures such as cylindrical rods, planar slabs, cylindrical tubes or spherical bubbles, which might be present in between the inner crust and the core of the star. These non-spherical structures may modify the optimal composition of the bottom layers of the inner crust but they do not change the core–crust transition density nor the EoS of the crust in a significant way (see [21] for details). At a given average density of the inner crust, we look for the optimal values of N and Z that satisfy the β-equilibrium condition

$$\mu_n = \mu_p + \mu_e, \quad (30)$$

where μ designates the chemical potential of the corresponding particles in the subindex. Once N and Z are fixed, the size of the WS box is determined. The electrons are treated as a free relativistic Fermi gas, with a constant density throughout the WS box. In practice, we proceed as follows. First, we fix the average density and an integer proton number Z and vary the neutron number N, which in general is not integer, until the β-equilibrium condition (30) is reached. Next, keeping the average density fixed, we repeated the procedure for a wide range of Z values searching for the optimal configuration, which corresponds to the WS cell of minimal energy.

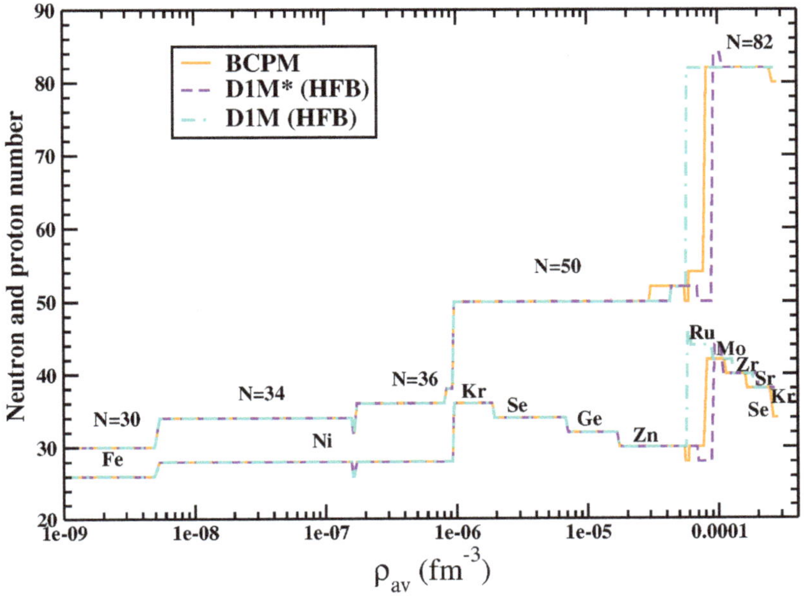

Figure 6. Neutron numbers N and proton numbers Z for the outer crust of NSs with the experimental masses from the AME2016 [78] tabulation plus the recently measured masses of $^{75-79}$Cu [79] aided by theoretical HFB calculations when experimental values are not available, using the D1M and D1M* Gogny forces and the BCPM energy density functional.

For a given N and Z, we calculate the energy of the WS box with the VWKSP method as explained before for finite nuclei in Section 3.1. We have taken a different form of the density profile for the inner crust unlike the finite nuclei, adapted from Refs. [23,76] as

$$\rho_q(r) = \rho_{B,q} + \frac{\rho_{0,q}}{1+\exp\left\{\left(\frac{C_q-R_{WS}}{r-R_{WS}}\right)^2 - 1\right\}\exp\left(\frac{r-C_q}{a_q}\right)} . \tag{31}$$

The first term in the right hand side is well suited to obtain a background density at certain average densities of the inner crust. The first exponential in the denominator of the second term is a damping factor tuned by the size of the WS cell (R_{WS}), which makes sure that the density reaches the background value (or zero) at the edge of the box. It is worthwhile mentioning here that we added the quantum shell and pairing energies only for protons by the reasons pointed out in [84]. A systematic comparison between the predictions of the extended TF plus Strutinsky integral method including pairing correlations and the fully quantal HFB results demonstrates that the perturbative treatment of shell effects and pairing correlations on top of a self-consistent semiclassical calculation provides a very accurate description of the structure of the NS inner crust [85].

In Figure 7 we plot the binding energy per nucleon (E/A) subtracted by the bare nucleon mass m_N for 13 different average densities ρ_{av} in the inner crust, which are indicated in the different panels. For comparison, we provide for each average density ρ_{av} the energy obtained in each of the four steps of the calculation of the energy in a WS cell of the inner crust. The orange line with circles denotes the energy containing only the TF contribution, the blue line with squares additionally contains the $\hbar^2$ contributions. The green line with triangles and the red line with diamonds successively take into account the contribution from the shell correction and the pairing energy, respectively. One can clearly observe that once the shell correction is added, the evolution of $E/A - m_N$ produces some local minima. Further addition of the pairing energy (red) somewhat smoothens this feature out. For all average densities but $\rho_{av} = 0.0789$ fm^{-3} the global minimum appears at Z = 40. At $\rho_{av} = 0.0789$ fm^{-3} it shifts to Z = 92. At $\rho_{av} = 0.0004$ fm^{-3} one can observe shell closures at Z = 20, 28, 40, 50, etc., which are similar to ones found in finite nuclei. With the increase in the average density some of these shell closures like Z = 28 and 50 are washed away (see the panel with $\rho_{av} = 0.07$ fm^{-3}). A systematic study of the inner crust composition performed using the extended TF approach including pairing correlations with a large set of Skyrme forces has been very recently reported [86]. It is shown that the proton content of the WS cells is correlated to the soft or stiff character of the slope of the pure neutron matter EoS for low average densities below 0.05 fm^{-3}. In this region the D1M and D1M* interactions predict a relatively stiff neutron matter EoS, which favors Z = 40 in the minimal energy configuration (see Figure 7 and Table II of [53]) in agreement with the conclusions drawn in [86].

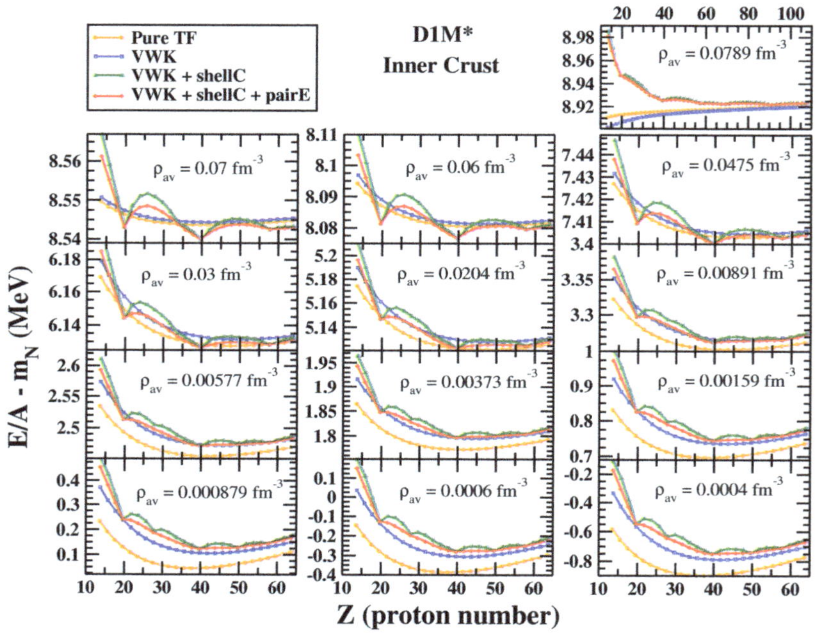

Figure 7. Binding energy per nucleon excluding the bare nucleon mass as a function of proton numbers at different average densities ρ_{av} of the inner crust calculated with D1M* Gogny interaction.

In Figure 8 we show the neutron (red solid line) and proton (blue dashed line) density profiles inside the WS cell at different ρ_{av} in the inner crust calculated with the D1M* interaction. With the increase in the ρ_{av}, the size of the WS cell shrinks significantly and the cells contain more dense neutron gas. With an increase in the density the diffuseness, particularly for protons, increases significantly. However, the central proton density of the cells increase with decrease in ρ_{av}.

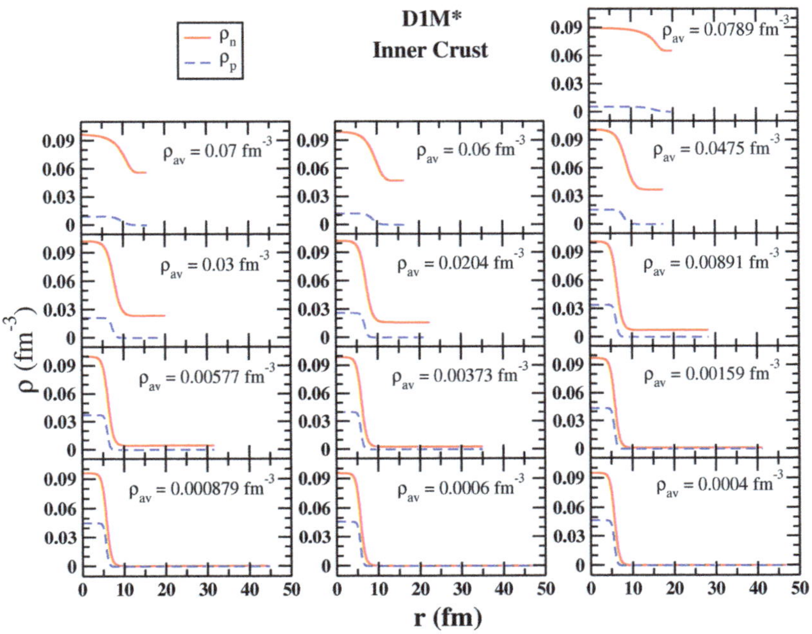

Figure 8. Neutron and proton density distribution inside the Wigner–Seitz cells obtained with variational Wigner–Kirkwood method at different average densities ρ_{av} obtained with D1M* Gogny interaction.

3.5. Core–Crust Transition

From our calculation in the inner crust we observe that the transition from the crust to the core takes place at an average density around $\sim$0.08 fm^{-3}. To find the core–crust transition density within a given model requires, in principle, the computation of the complete EoS of the inner crust, which is not a simple task, as we have seen along this section. However, the search of the crust–core transition density can be considerably simplified by performing the calculation from the core side. In this case one searches for the violation of the stability conditions of the homogeneous core under small amplitude oscillations, which indicate the appearance of nuclear clusters and therefore the transition to the inner crust. There are different ways to determine the core–crust transition from the core side, namely the thermodynamical method (V_{ther}), the dynamical method (V_{dyn}), random phase approximation and the Vlasov equation method (see Refs. [13,20,87] for more details and further references).

In the thermodynamical method the stability of the NS core is discussed in terms of bulk properties only, where the mechanical and chemical stability conditions set the boundaries of the homogeneous core:

$$-\left(\frac{\partial P}{\partial v}\right)_{\mu_{np}} > 0, \quad -\left(\frac{\partial \mu_{np}}{\partial q}\right)_{v} > 0, \tag{32}$$

where P is the total pressure of neutron star matter (1)–(2), μ_{np} is the difference between the neutron and proton chemical potentials, $v = 1/\rho$ is the inverse of the baryon density and q is the charge per baryon. In the low density regime of interest for the core–crust transition the chemical stability is always fulfilled and the mechanical stability condition can be recast through the so-called thermodynamical potential $V_{ther}(\rho)$ [13,20]. The thermodynamical potential is a function of the baryon density only and the transition density corresponds to the value of ρ for which $V_{ther}(\rho)$ changes sign (see [13] and references therein).

The dynamical method, introduced in Ref. [81], assumes that the nuclear energy density can be expressed as the sum of a bulk homogeneous part and an inhomogeneous contribution, which depends on the gradient of the neutron and proton densities as well as on the direct part of the Coulomb potential. The Skyrme forces fit this scheme [88]. However, for finite-range interactions, such as the Gogny forces, the calculation is more involved. Quite often the energy density functional for finite-range forces can be approximated very accurately by a local form using the extended TF density matrix [72] instead of the full HF density matrix. Within this scheme, the energy density can be written as a homogeneous term, provided by the Slater density matrix ($\hbar^0$ term), plus an additional $\hbar^2$ contribution written in terms of the gradients of the neutron and proton densities and of the inverse of the momentum and position dependent effective masses [87]. This inhomogeneous contribution also contains the gradient expansion of the direct nuclear and Coulomb potentials (see [20,87] for more details). Thus,

$$E = E_0 + \frac{1}{2}\sum_{i,j}\int \frac{d\mathbf{k}}{(2\pi)^3}\frac{\delta^2 E}{\delta n_i(\mathbf{k})\delta n_j^*(\mathbf{k})}\delta n_i(\mathbf{k})\delta n_j^*(\mathbf{k}), \tag{33}$$

where E_0 is the unperturbed density and $n_i(\mathbf{k})$ are the momentum distributions (inverse Fourier transform of the density perturbation) for each type of particles. The second variation of the energy defines the so-called curvature matrix, which is the sum of three different terms. The first is the bulk contribution, which defines the stability of uniform NS matter and corresponds to the equilibrium condition in the thermodynamical method. The second term collects the gradient contributions in the energy density functional and is a function of the momentum k. For zero-range Skyrme forces it is a quadratic function [88], but it is a more involved function in the case of finite-range interactions [20,87]. The last contribution is due to the direct Coulomb interactions between protons and electrons. The stability condition requires the curvature matrix to be convex. This allows one to write a dynamical potential $V_{dyn}(\rho, k)$, which is now momentum- and density-dependent. To compute the transition density one first minimizes for each value of the density ρ the dynamical potential respect to k. Next, as in the case of the thermodynamical method, one determines the transition density as the value of the density for which $V_{dyn}(\rho, k(\rho))$ vanishes (see Refs. [20,87] for a detailed description of the dynamical method). Table 3 collects the main core–crust transition properties, namely density, pressure and isospin asymmetry, derived with the thermodynamical and dynamical methods using the D1M, D1M* and D1M** Gogny forces as well as with the BCPM energy density functional, which is used here as a benchmark.

It is known from earlier literature that the core–crust transition density, estimated in the thermodynamical approach, using Skyrme and Relativistic Mean Field (RMF) models, shows a decreasing trend with an increasing value of the slope of the symmetry energy (see [13,87] and references therein). In Refs. [13,87] we have computed the core–crust transition density predicted by finite-range interactions using the thermodynamical and dynamical methods. We find that our results are in harmony with earlier calculations obtained with the Skyrme interactions and RMF parametrizations. This can be seen in Figure 9, where we plot the transition density (left panels) and the transition pressure (right panels) obtained using the thermodynamical (upper panels) and the dynamical (lower panels) methods. We have obtained the transition properties for a large set of Skyrme forces and also for most of Gogny interactions available in the literature. These sets of interactions cover a large range of values of the slope of the symmetry energy L going from around 15 MeV up to 130 MeV. We see that the values of both the transition density and the transition pressure have larger values when they are obtained using the thermodynamical method instead of the dynamical method. The reason behind this is, as we have mentioned, that the dynamical method takes into account the surface and Coulomb contributions that tend to stabilize more the liquid core. Comparing between the transition density and pressure we observe different behaviors. On the one hand the values of the density of the core–crust transition follow a rather linear decreasing trend with respect to the slope of the symmetry energy L for both Skyrme and Gogny forces. On the other hand, the correlation between the transition pressure and L is less obvious, being more visible for Skyrme forces than for the Gogny ones. For example, we can see from Table 3 the decreasing trend of the transition density with the increasing value of L of the different models considered in this Table (see Table 2 in this respect), while the transition pressure is roughly similar computed with the D1M, D1M* and D1M** forces and differs from the prediction of the BCPM energy density functional.

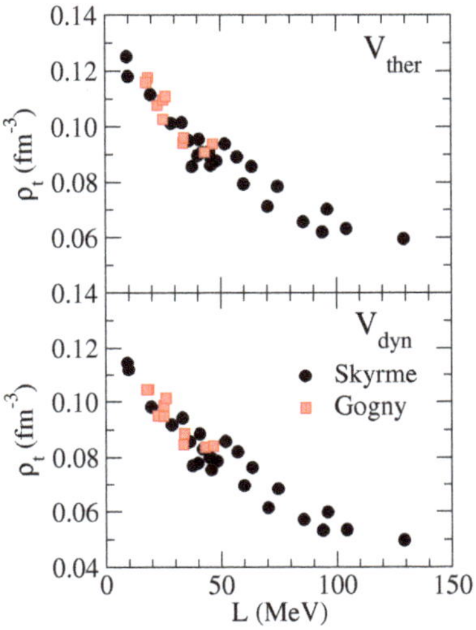

Figure 9. Transition density (left panels) and transition pressure (right panels) against the slope of the symmetry energy computed for some Skyrme and Gogny interactions. The upper panels correspond to the values obtained using the thermodynamical method whereas the lower panels display the results extracted using the dynamical method.

Table 3. Core–crust transition density ρ_t, pressure P_t and and isospin asymmetry δ_t predicted by the D1M, D1M* and D1M** Gogny forces and the BCPM energy density functional.

	ρ_t **(fm^{-3})**	P_t **(MeVfm^{-3})**	δ_t
D1M			
V_{ther}	0.1027	0.3390	0.9241
V_{dyn}	0.0949	0.2839	0.9257
D1M*			
V_{ther}	0.0909	0.3301	0.9275
V_{dyn}	0.0838	0.2702	0.9300
D1M**			
V_{ther}	0.0960	0.3368	0.9257
V_{dyn}	0.0886	0.2786	0.9279
BCPM			
V_{ther}	0.0889	0.5137	0.9339
V_{dyn}	0.0816	0.4132	0.9382

4. Global Properties of Neutron Stars Predicted by Gogny Forces

The unified EoS is obtained from the consistent calculation of the core and the crust, as we have shown in the previous sections. We provide the unified EoSs and the associated stellar matter composition obtained for D1M and D1M* in the Supplementary Material. In addition, for the sake of clarity we display in the left panel of Figure 10 the unified EoS in logarithmic scale computed with these interactions as well as with the BCPM energy density functional. From this panel we see that practically no differences can be observed in the outer crust. In the inner crust the EoSs provided by the different Gogny forces are similar and show some differences with the BCPM predictions. However, in the core region the differences between the original D1M and modified D1M* Gogny forces are more prominent.

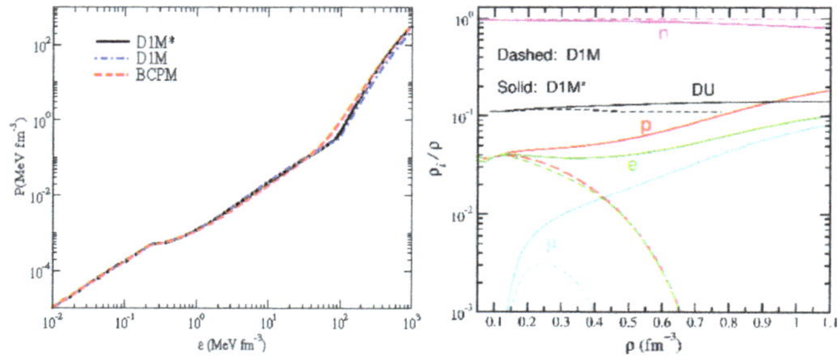

Figure 10. Left: Unified EoS computed with the the D1M and D1M* Gogny force and with the BCPM energy density functional. **Right**: Particle fractions and the proton fraction corresponding to the onset of the direct Urca (DU) process (see text for details) as functions of the nucleonic density from the D1M and D1M* interactions.

In the right panel of Figure 10 we compare the predictions of the D1M and D1M* EoSs for the particle populations in the beta-equilibrated $npe\mu$ matter of the NS core. The impact of the stiffer symmetry energy of the D1M* interaction with respect to D1M, as reflected by the total EoS displayed in the left panel of the Figure, can be clearly seen in the right panel. D1M* predicts a persistent population of protons and leptons in the core of the star with increasing nucleon density. In stark contrast, in the results calculated with D1M we see that matter becomes soon deprotonized and deleptonized when the density increases. This is because in D1M it is much less costly to convert protons into neutrons due to the softer symmetry energy of this interaction. Actually, according to D1M the stellar core would be composed practically of only neutrons after a density $\rho \approx 0.65$ fm^{-3} ($\approx 4\rho_0$), as can be seen from the D1M particle fractions in Figure 10. Notice also that D1M* predicts a growing population of muons with higher density, whereas in D1M the appearance of muons is nominal. According to recent studies in the literature, the presence of muons in NSs may play a significant role in addressing several new physics questions about the interactions and the astrophysical effects of muonphilic dark matter particles, see Ref. [89] and references therein. The proton fraction inside the beta-equilibrated matter also determines whether a proto-neutron star will go through the direct Urca process or not. In $npe\mu$ matter this is attributed to the condition that the proton fraction satisfies $\rho_p/\rho > x_{DU}$, where x_{DU} is defined as [90]

$$x_{DU} = \left[1 + \left\{1 + \left(\frac{\rho_e}{\rho_p}\right)^{1/3}\right\}^3\right]^{-1}. \tag{34}$$

In Figure 10, we plotted this quantity as a function of density, denoted by "DU" (black lines). The density point at which the proton fraction (red) surpasses the quantity x_{DU} indicates the onset of direct Urca. One can see that only D1M* fulfills this condition, though at fairly large densities ($\rho > 0.93$ fm^{-3}). This behavior can be directly attributed to the stiffer symmetry energy for D1M* at suprasaturation densities compared to D1M.

Once the full EoS is obtained, one can look for different global properties of NSs. In this review we will concentrate on three relevant aspects, namely the mass-radius relation in an NS, which provides a detailed information about the structure of the star, the moment of inertia of the NS, and in particular its fraction enclosed by the crust, which may be important for the description of pulsar glitches. Finally, the last aspect to be discussed is the tidal deformability in binary systems of NS. This quantity can be accessed by the detection of gravitational waves (GW), coming for example from the merger of a NS binary as in the GW170817 event recorded recently.

4.1. The Tolman–Oppenheimer–Volkov Equations

In order to study the mass-radius relation of NSs, one has to solve the TOV equations [54,91], which need as an input the full EoS along all of the star. The TOV equations take into account within the general relativity framework the hydrostatic equilibrium in the star between the pressure given by the gravitational field and the pressure coming from the baryons and leptons inside the star. The TOV equations are given by

$$\frac{dP(r)}{dr} = \frac{G}{r^2c^2}[\epsilon(r) + P(r)]\Big[m(r) + 4\pi r^3 P(r)\Big]\left[1 - \frac{2Gm(r)}{rc^2}\right]^{-1} \tag{35}$$

$$\frac{dm(r)}{dr} = 4\pi r^2 \epsilon(r), \tag{36}$$

where $\epsilon(r)$, $P(r)$ and $m(r)$ are, respectively, the energy density (including free nucleon mass), pressure and mass at each radius r inside the NS. Starting with a central energy density $\epsilon(0)$, a central pressure $P(0)$ and a central mass $m(0) = 0$, one integrates outwards the differential equations until reaching the NS surface, where the pressure is zero, $P(R) = 0$. At the same time, the location of the surface of the star determines its total radius R and its total mass $M = m(R)$.

In Figure 11 we plot the mass–radius (MR) relation for the D1M and D1M* Gogny interactions, as well as for the BCPM energy density functional. We stress that all three EoSs used in the calculations are unified EoSs, where the outer crust, the inner crust and the core have been obtained using the same interaction. In the same plot we include constraints coming from different sources. First, we include constraints for the maximum mass obtained from the observation of the highly massive NSs [14,15]. The green vertical constraint comes from cooling tails of type-I X-ray bursts in three low-mass X-ray binaries and a Bayesian analysis [92], and the blue vertical constraint is from five quiescent low-mass X-ray binaries and five photospheric radius expansion X-ray bursters after a Bayesian analysis [93]. The pink–red rectangular constraint at the front is from a Bayesian analysis with the data from the GW170817 detection of gravitational waves from a binary NS merger [94]. Finally, we inserted the constraints coming from the very recent NICER observations for the mass and radius of the pulsars PSR J0030+0451 and PSR J0740+6620 with one-sigma deviation [95,96]. As mentioned in previous sections, we observe that the D1M interaction predicts the NS maximum mass of only $1.74M_\odot$. Moreover, the MR relation obtained from D1M falls outside all considered constraints. If we look at the MR relation obtained using the EoS given by the D1M* interaction, we see that it reaches a maximum NS mass of around $2M_\odot$, similarly to the one given by the BCPM energy density functional, which we included here as a benchmark. The MR relations given by both D1M* and BCPM lie inside most of the constraints for the mass and radius included in the same Figure.

We plot in Figure 12 the mass (left panel) and radius (central panel) enclosed in the NS crust. The values of the crustal mass for the BCPM energy density functional are larger than the ones obtained using Gogny interactions, but are close to the ones computed with D1M* once one approaches the NS maximum mass values. On the other hand, the crustal masses obtained using the D1M interaction are lower than the ones obtained with D1M*. For the crustal radius, or thickness of the crust, we see that the values predicted by D1M* are very similar to the results one achieves for BCPM, at least above $1.4M_\odot$, while the crustal radius computed with D1M is smaller than that for the D1M* interaction or for the BCPM energy density functional.

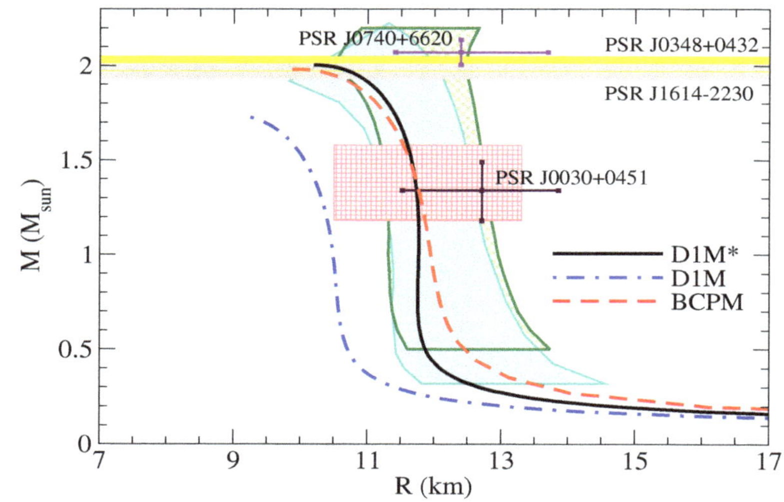

Figure 11. Mass-radius relation obtained using the D1M* and the D1M Gogny forces and the BCPM energy density functional. Constraints from the measurements of $M \approx 2M_{\odot}$ (yellow and grey) [14,15], from cooling tails of type-I X-ray bursts in three low-mass X-ray binaries and a Bayesian analysis (green) [92], from five quiescent low-mass X-ray binaries and five photospheric radius expansion X-ray bursters after a Bayesian analysis (blue) [93] and from a Bayesian analysis with the data from the GW170817 detection of gravitational waves from a binary NS merger (red) [94] are shown. Finally, the very recent constraints coming from the NICER mission are also included [95,96].

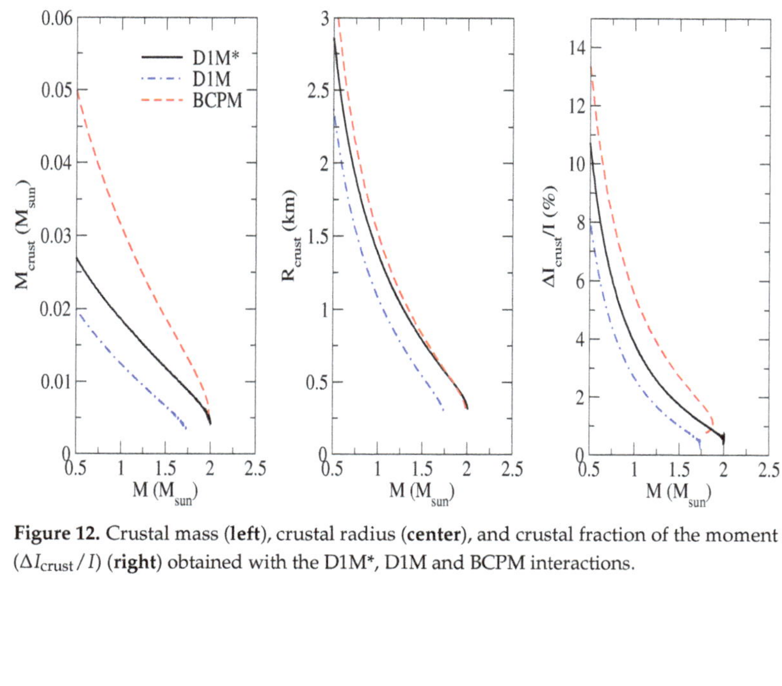

Figure 12. Crustal mass (**left**), crustal radius (**center**), and crustal fraction of the moment of inertia ($\Delta I_{\text{crust}}/I$) (**right**) obtained with the D1M*, D1M and BCPM interactions.

4.2. Moment of Inertia

The moment of inertia of slowly-rotating NSs can be computed from the static mass distribution and the gravitational potentials that one finds when solving the TOV equations [97]. If one studies the slow-rotation limit, the moment of inertia is given by [81,97,98]

$$I \equiv \frac{J}{\Omega} = \frac{8\pi}{3}\int_0^R r^4 e^{-\nu(r)} \frac{\bar{\omega}(r)}{\Omega} \frac{(\epsilon(r)+P(r))}{\sqrt{1-2Gm(r)/rc^2}}dr, \tag{37}$$

where G is the gravitational constant and c the speed of light and one has assumed spherical symmetry. In Equation (37), J is the angular momentum, Ω is the stellar rotational frequency, $\nu(r)$ and $\bar{\omega}$ are radially dependent metric functions and $m(r)$, $\epsilon(r)$ and $P(r)$ are, respectively, the NS mass, energy density and total pressure enclosed in a radius r. The metric function $\nu(r)$ satisfies [98]

$$\nu(r) = \frac{1}{2}\ln\left(1-\frac{2GM}{Rc^2}\right) - \frac{G}{c^2}\int_r^R \frac{(M(x)+4\pi x^3 P(x))}{x^2(1-2GM(x)/xc^2)}dx, \tag{38}$$

and the angular velocity of the fluid measured in a local reference frame is given by the relative frequency $\bar{\omega}(r) \equiv \Omega - \omega(r)$, where $\omega(r)$ is the frequency that appears because of the slow rotation of the star. On the other hand, the relative frequency $\tilde{\omega}(r) \equiv \bar{\omega}(r)/\Omega$ can be obtained by solving the differential equation [98]

$$\frac{d}{dr}\left(r^4 j(r)\frac{d\tilde{\omega}(r)}{dr}\right) + 4r^3\frac{dj(r)}{dr}\tilde{\omega}(r) = 0, \tag{39}$$

with

$$j(r) = \begin{cases} e^{\nu(r)}\sqrt{1-2Gm(r)/rc^2} & \text{if } r \leq R \\ 1 & \text{if } r > R \end{cases}. \tag{40}$$

The relative frequency $\tilde{\omega}(r)$ obtained as a solution of (39) and (40) has to fulfill the following boundary conditions

$$\tilde{\omega}'(0) = 0 \qquad \text{and} \qquad \tilde{\omega}(R) + \frac{R}{3}\tilde{\omega}'(R) = 1. \tag{41}$$

Notice that in the slow-rotation regime the solution of the moment of inertia does not depend on the stellar frequency Ω. Starting from an arbitrary value of $\tilde{\omega}(0)$, one integrates Equation (39) up to the surface. Usually, it will be necessary to re-scale the function $\tilde{\omega}(r)$ and its derivative with an appropriate constant in order to fulfill (41). One can test the accuracy of the final result by checking the condition [98]

$$\tilde{\omega}'(R) = \frac{6GI}{R^4c^2}. \tag{42}$$

The ratio between the fraction of the moment of inertia ΔI_{crust} and the total moment of inertia I is intrinsically connected to pulsar glitches and to the location of the core–crust transition [31,54,99,100]. We plot in the right panel of Figure 12 the ratio between $\Delta I_{crust}/I$ against the total NS mass for the D1M and D1M* interactions and the BCPM energy density functional. Similarly to what happens for the crustal mass and crustal radius, the crustal fraction of the moment of inertia is larger when obtained using the BCPM EoS. On the other hand, the values that one obtains with D1M* fall between the ones of BCPM and the ones given by D1M, which provides the lower values of $\Delta I_{crust}/I$ from these three interactions. As can be seen in the rightmost panel of Figure 12, the values obtained using the D1M* interaction lie between the results predicted by the BCPM and D1M EoSs. Notice that this later provides the lower values of the ratio $\Delta I_{crust}/I$ among all the interactions used in this calculation. To account for the size of the pulsar glitches, the pinning model requires that some amount of angular momentum is carried out by the crust, which can be recast as a constraint on the crustal fraction of the moment of inertia. For example, to explain Vela and another source of glitches, first estimates suggested that $\Delta I_{crust}/I > 1.4\%$ [101], although more recent estimates, which take into account the neutron entrainment in the crust, increases the minimal crustal fraction up to 7% in order to explain the glitching phenomena [102,103]. When the Gogny forces D1M and D1M* are used to evaluate the moment of inertia, the first constraint is fulfilled for NS with masses smaller than 1.4 and $1.7 M_\odot$, respectively, while the second constraint is only fulfilled by very small NS masses, as can be seen in the rightmost panel of Figure 12. If the calculation of the moment of inertia is performed using the BCPM energy density functional instead of the D1M and D1M* forces, the behavior is similar, although the glitching sources have slightly larger masses.

The left panel of Figure 13 encloses the total moment of inertia against the total NS mass for the same interactions as the previous Figure. The values of the moment of inertia obtained with D1M* and BCPM are very similar from low masses up to $1.5 M_\odot$, from where the moment of inertia computed with D1M* is slightly larger than that for BCPM. For these two interactions, the maximum values of the moment of inertia are 1.95×10^{45} g cm^2 and 1.88×10^{45} g cm^2, respectively, which are reached a little bit before the maximum mass configuration. Contrary to these two interactions, the D1M Gogny force gives much smaller values for I, reaching maximum values of only 1.30×10^{45} g cm^2. It is expected that binary pulsar observations can provide new information about the moment of inertia and, therefore, put additional constraints on the EoS of NS [100]. The moment of inertia of the primary component of the pulsar PSR-J0737-3039, which has a mass of $1.338 M_\odot$, has been estimated by Landry and Kumar in the range $I = 1.15^{+0.38}_{-0.24} \times 10^{45}$ g cm^2 [104]. From the left panel of Figure 13 it can be seen that this constraint is fulfilled by the moment of inertia computed using the EoSs based on the D1M and D1M* forces and the BCPM energy density functional (see Ref. [20] for more details). Finally, let us mention that the dimensionless quantity I/MR^2 is found to scale with the NS compactness $\chi = GM/Rc^2$ and to be almost independent of the mass and radius of the NS [99,100,105]. We checked that this is the situation when the moment of inertia is computed using the D1M and D1M* and the BCPM energy density functional on the one hand, and also that the universal relation I/MR^2 vs. χ lies within the region estimated by Lattimer and Schutz [100] and Breu and Rezolla [105] when studied with the same interactions.

4.3. Tidal Deformability

The detection of GW coming from mergers of binary NS systems, and of NS–Black Hole systems, will open new possibilities to study the EoS of highly asymmetric nuclear matter, which one uses to describe the interior of NSs. If we focus on binary NS systems, each of its components induces a gravitational tidal field on its companion. This phenomenon leads to a mass-quadrupole deformation on each member of the binary. To linear order, the tidal deformability Λ describes this tidal deformation of each star in the system, and it is defined as the ratio between the induced quadrupole moment and the external tidal field [106,107].

For each of the stars in the binary, the tidal deformability is given by [106–108]

$$\Lambda = \frac{2}{3} k_2 \left(\frac{Rc^2}{GM} \right)^5, \tag{43}$$

where k_2 is the dimensionless tidal Love number, R is the NS radius, M its total mass. As previously stated in this paper, the solution of the TOV equations provides the values of the mass and radius of a NS, while the Love number k_2 is obtained as

$$
\begin{aligned}
k_2 \;=\; & \frac{8\chi^5}{5}(1-2\chi)^2[2+2\chi(y-1)-y]\times\Big\{2\chi[6-3y+3\chi(5y-8)] \\
+\; & 4\chi^3\Big[13-11y+\chi(3y-2)+2\chi^2(1+y)\Big] \\
+\; & 3(1-2\chi)^2[2-y+2\chi(y-1)]\ln(1-2\chi)\Big\}^{-1},
\end{aligned} \tag{44}
$$

where χ is the compactness of the star and

$$
y = \frac{R\beta(R)}{H(R)}. \tag{45}
$$

In Equations (44) and (45), $\beta(R)$ and $H(R)$ are given by the solution of the following set of coupled differential equations [107,108]:

$$
\frac{dH(r)}{dr} = \beta(r) \tag{46}
$$

$$
\begin{aligned}
\frac{d\beta(r)}{dr} \;=\; & \frac{2G}{c^2}\left(1-\frac{2Gm(r)}{rc^2}\right)^{-1}H(r)\left\{-2\pi\left[5\epsilon+9p+\frac{d\epsilon}{dp}(\epsilon+p)\right]+\frac{3c^2}{r^2G}\right. \\
+\; & \left.\frac{2G}{c^2}\left(1-\frac{2Gm(r)}{rc^2}\right)^{-1}\left(\frac{m(r)}{r^2}+4\pi rp\right)^2\right\} \\
+\; & \frac{2\beta(r)}{r}\left(1-\frac{2Gm(r)}{rc^2}\right)^{-1}\left\{-1+\frac{Gm(r)}{rc^2}+\frac{2\pi r^2G}{c^2}(\epsilon-p)\right\},
\end{aligned} \tag{47}
$$

where $m(r)$ is the mass enclosed inside a radius r, and ϵ and p are the corresponding energy density and pressure. One solves Equations (46) and (47) along with the TOV equations by integrating outwards, with the boundary conditions $H(r) = a_0r^2$ and $\beta(r) = 2a_0r$ as $r \to 0$. The constant a_0 is arbitrary, as it cancels out in the expression for the Love number [108].

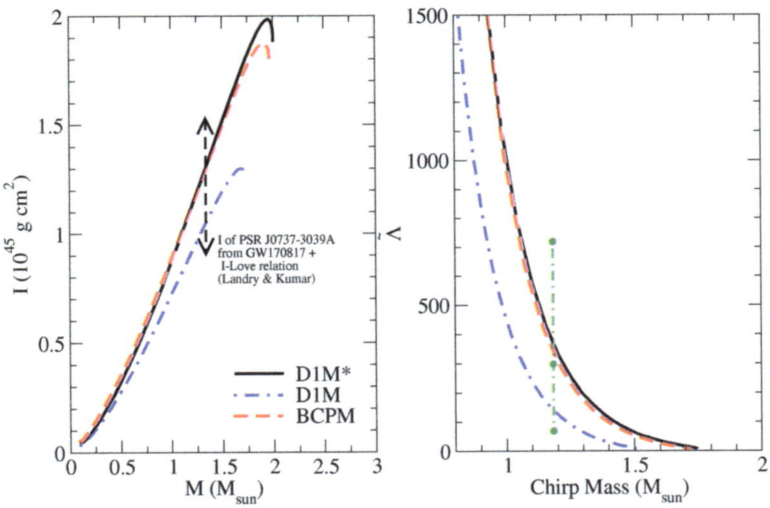

Figure 13. Left: Total moment of inertia against the total mass of neutron stars computed using the D1M* and D1M Gogny forces and the BCPM energy density functional. The constraint proposed in [104] is also displayed. Right: Mass weighted tidal deformability (for symmetric binaries) against the chirp mass of binary neutron star systems obtained using the same interactions as in the left panel. The constraint for $\tilde{\Lambda}$ coming from the GW170817 event is also included [109,110].

When studying the full NS binary system, the mass-weighted tidal deformability $\tilde{\Lambda}$ takes into account the contribution from the tidal effects to the phase evolution of the gravitational wave spectrum of the inspiraling NS binary, and it is defined as

$$\tilde{\Lambda} = \frac{16}{13} \frac{(M_1 + 12M_2)M_1^4\Lambda_1 + (M_2 + 12M_1)M_2^4\Lambda_2}{(M_1 + M_2)^5}, \tag{48}$$

where Λ_1 and Λ_2 are the tidal deformabilities of each NS conforming the system and M_1 and M_2 are their corresponding masses. Notice that the definition (48) fulfills $\tilde{\Lambda} = \Lambda_1 = \Lambda_2$ when $M_1 = M_2$.

The LIGO and Virgo Collaboration have already detected some GW signals coming from the merger of two NSs [109,110], which allow constraining of the mass-weighted tidal deformability $\tilde{\Lambda}$ and the chirp mass of the system $\mathcal{M}$, which is given by

$$\mathcal{M} = \frac{(M_1 M_2)^{3/5}}{(M_1 + M_2)^{1/5}}. \tag{49}$$

In this paper we will use the constraints coming from the GW170817 detection [109,110], as it is at the moment the one that further constrains $\tilde{\Lambda}$ and $\mathcal{M}$, at values of $\tilde{\Lambda} = 300^{+420}_{-230}$, $\mathcal{M} = 1.186^{+0.001}_{-0.001} M_\odot$. Additional constraints for the single NS masses are also given as $M_1 \in (1.36, 1.60) M_\odot$ and $M_2 \in (1.16, 1.36) M_\odot$.

We plot in the right panel of Figure 13 the mass-weighted tidal deformability against the chirp mass obtained using the D1M* and D1M Gogny forces and the BCPM energy density functional. The mass-weighted tidal deformability $\tilde{\Lambda}$ predicted by the BCPM and D1M* EoSs have very similar values, lying well inside the constraint of the GW170817 detection, which is plotted in green in the same Figure. On the other hand, the values obtained with the D1M Gogny interaction are lower than the ones obtained with D1M* and BCPM, even though they also lie inside the GW constraints, but near the lower limit. Finally, let us mention that in Ref. [111] an analysis of the GW170817 constraints has been performed using both Gogny forces and momentum-dependent interactions (MDI). One of the conclusions of this study has been that the successful Gogny and MDI interactions that are compatible with GW170817 restrict the radius of a canonical NS of $1.4 M_\odot$ to within the range of 9.4 km $\leq R_{1.4} \leq$ 13.1 km [111].

5. Conclusions

In this review article we revised and summarized the most relevant aspects of our investigations about the application of effective forces of Gogny type to the NS scenario that have been previously reported in a series of papers. The Gogny interactions were proposed more than forty years ago with the purpose to describe simultaneously the mean field and the pairing field, which usually are disconnected in almost all of the mean field models available in the literature. Although the standard parametrizations of the Gogny force, such as D1S, D1N and D1M, reproduce rather accurately the nuclear masses as well as pairing and deformation properties of finite nuclei, these interactions fail when applied to the NS domain. The basic reason for that is the too soft symmetry energy predicted by these forces at high baryon densities, which are unable to produce heavy enough stellar masses. To cure this limitation, we proposed a reparametrization of the Gogny D1M force in such a way that preserves the accurate description of finite nuclei, the isovector properties of the interaction, in particular the slope of the symmetry energy are modified to obtain a stiffer EoS able to predict maximal NS masses of about $2M_\odot$, in agreement with well-contrasted astronomical observations. Our renormalization procedure has been applied using the D1M force as starting point, because the D1S and D1N interactions are too far from the $2M_\odot$ target. In this way we have built up two new Gogny parametrizations, denoted as D1M* and D1M**, which predict maximal masses of NS of $2M_\odot$ and $1.91M_\odot$, respectively.

Apart from the description of the core of NSs, we also used these new Gogny forces to build up the EoS of the crust of NSs aimed to obtain a unified EoS from the surface to the center of the star. The outermost region of a NS, called outer crust, consists of a lattice of atomic nuclei, which are more neutron rich as the depth increases, embedded in a free electron gas. The basic ingredient to determine the EoS in this region are the nuclear masses, which are taken from the experiment or obtained from a HFB calculation with the same Gogny force when the masses are unknown. After a density around 0.003 fm^{-3}, neutrons cannot be retained by the nuclei and above this density, the matter is arranged still as a lattice structure but now permeated by free neutron and electron gases. The treatment of this region is complicated owing to the presence of the neutron gas. To describe this scenario, called inner crust, we use the Wigner–Seitz approximation and compute the representative nucleus inside each cell using the semiclassical Variational Wigner–Kirkwood approximation, which includes $\hbar^2$ corrections added perturbatively. Moreover, the quantal shell corrections and the pairing correlations for protons are also added perturbatively, using the so-called Strutinsky integral method and the BCS approximation, respectively. At a density roughly around one-half the saturation density the inner crust structure dissolves in a homogeneous core. The precise value of the crust–core transition density is strongly model dependent. To determine the transition point is not an easy task when looking from the crust, as it requires an accurate description of the inner crust. However, it is easier to determine the transition point from the core side searching for the density for which the homogeneous core is unstable against the cluster formation. The simplest approach is the so-called thermodynamical method that only considers the stability of the homogeneous core. A more precise approximation is provided by the dynamical method, which on top of the stability of the homogeneous core, also considers finite-size effects. We have shown that the dynamical method predicts transition densities and pressures in rather good agreement with the estimate obtained from the crust side.

Once the full EoS based on the modified D1M* Gogny force was obtained, we used it to predict different NS properties. In addition to the mass–radius relation, we analyzed the behavior of the moment of inertia and the tidal deformability of the star, which can be related to information extracted from observations in some binary pulsars and from the GW170817 event. We also analyzed some global crustal properties such as the mass and radius of the crust, as well as the crustal fraction of the moment of inertia, which can be relevant for the description of the glitches. We find that these global properties obtained with the Gogny-based EoS are in good agreement with the predictions of other well contrasted EoS as the ones based on the SLy4 Skyrme force [22] or the microscopic BCPM energy density functional [21], which is used as a benchmark in this work. Although a detailed study of some other nuclear structure phenomena, such as the description of odd nuclei, fission phenomena or giant resonances computed with the new D1M* and D1M** Gogny forces is still pending, we conclude that these new interactions are promising alternatives to describe simultaneously finite nuclei and neutron stars providing results in harmony with the experimental data and astronomical observations.

Supplementary Materials: The following are available online at https://www.mdpi.com/article/10.3390/sym13091613/s1. This supplementary material consists of the D1M and D1M* EoS from the outer crust to the core, files D1M_complete_EOS.dat and D1MSTAR_complete_EOS.dat, respectively, as well as a README.txt with details about how the these EoS files are written.

Author Contributions: All the authors contributed equally to the writing of the review with X.V. leading the coordination of all different tasks. All authors have read and agreed to the published version of the manuscript.

Funding: C.G., M.C. and X.V. were partially supported by Grants No. FIS2017-87534-P, No. PID2020-118758GB-I00 (DOI 10.13039/501100011033) and No. CEX2019-000918-M (through the "Unit of Excellence María de Maeztu 2020–2023" award to ICCUB) from the State Agency for Research (AEI) of the Spanish Ministry of Science and Innovation (MICINN). The work of C.M. was partially supported by the IN2P3 Master Project "NewMAC". The work of L.M.R. was partly supported by the Spanish MINECO Grant No. PGC2018-094583-B-I00.

Institutional Review Board Statement: This manuscript has gone through VIRGO document review system as C.M. is obliged to make sure any infringement of the VIRGO data has not taken place, being part of the VIRGO collaboration through Caen-Meudon group.

Informed Consent Statement: Not applicable.

Data Availability Statement: See Supplemental Material accompanying this paper, further inquiries can be directed to the corresponding author.

Acknowledgments: The authors are indebted to J.N. De for a careful reading of the manuscript.

Conflicts of Interest: The authors declare no conflict of interest.

References

1. Dechargé, J.; Gogny, D. Hartree-Fock-Bogoliubov calculations with the *D*1 effective interaction on spherical nuclei. *Phys. Rev. C* **1980**, *21*, 1568. [CrossRef]
2. Berger, J.-F.; Girod, M.; Gogny, D. Time-dependent quantum collective dynamics applied to nuclear fission. *Comp. Phys. Comm.* **1991**, *63*, 365. [CrossRef]
3. Hilaire, S.; Girod, M. AMEDEE database, International Conference on Nuclear Data for Science and Technology. *Nucl. Struct. Decay Data* **2008**. [CrossRef]
4. Pillet, N.; Hilaire, S. Towards an extended Gogny force. *Eur. Phys. J. A* **2017**, *53*, 193. [CrossRef]
5. Chappert, F.; Girod, M.; Hilaire, S. Towards a new Gogny force parametrization: Impact of the neutron matter equation of state. *Phys. Lett. B* **2008**, *668*, 420. [CrossRef]
6. Goriely, S.; Hilaire, S.; Girod, M.; Péru, S. First Gogny-Hartree-Fock-Bogoliubov Nuclear Mass Model. *Phys. Rev. Lett.* **2009**, *102*, 242501. [CrossRef]
7. Audi, G.; Waspra, A.H.; Thibault, C. The AME2003 atomic mass evaluation (II). Tables, graphs and references. *Nucl. Phys. A* **2003**, *729*, 337. [CrossRef]
8. Friedman, B.; Pandharipande, V.R. Hot and cold, nuclear and neutron matter. *Nucl. Phys. A* **1981**, *361*, 502. [CrossRef]
9. Robledo, L.M. HFBaxial Computer Code. 2002; unpublsihed.
10. Robledo, L.M.; Bertsch, G.F. Application of the gradient method to Hartree-Fock-Bogoliubov theory. *Phys. Rev. C* **2011**, *84*, 014312. [CrossRef]
11. Wang, M.; Audi, G.; Waspra, A.H.; Kondev, F.G.; MacCormick, M.; Xu, X.; Pfeifer, B. The AME2012 atomic mass evaluation (II). Tables, graphs and references. *Chin. Phys. C* **2012**, *36*, 1603. [CrossRef]
12. Sellahewa, R.; Rios, A. Isovector properties of the Gogny interaction. *Phys. Rev. C* **2014**, *90*, 054327. [CrossRef]
13. Gonzalez-Boquera, C.; Centelles, M.; Viñas, X.; Rios, A. Higher-order symmetry energy and neutron star core-crust transition with Gogny forces. *Phys. Rev. C* **2017**, *96*, 065806. [CrossRef]
14. Demorest, P.B.; Pennucci, T.; Ramson, S.M.; Roberts, M.S.E.; Hessels, J.W.T. A two-solar mass neutron star measured using the Shapiro delay. *Nature* **2010**, *467*, 1081. [CrossRef]
15. Antoniadis, J.; Freire, P.C.; Wex, N.; Tauris, T.M.; Lynch, R.S.; Van Kerkwijk, M.H.; Kramer, M.; Bassa, C.; Dhillon, V.S.; Driebe, T.; et al. A Massive Pulsar in a Compact Relativistic Binary. *Science* **2013**, *340*, 448. [CrossRef] [PubMed]
16. Fonseca, E.; Pennucci, T.T.; Ellis, J.A.; Stairs, I.H.; Nice, D.J.; Ransom, S.M. The NANOGrav nine-year data set: Mass and geometric measurements of binary millisecond pulsars. *Astrophys. J.* **2016**, *832*, 167. [CrossRef]
17. Arzoumanian, Z.; Brazier, A.; Burke-Spolaor, S.; Chamberlin, S.; Chatterjee, S. The NANOGrav 11-year data set: High precision timing of 45 millisecond pulsars. *Astrophys. J. Suppl. Ser.* **2018**, *235*, 37. [CrossRef]
18. Chappert, F.; Pillet, N.; Girod, M.; Berger, J.-F. Gogny force with a finite-range density dependence. *Phys. Rev. C* **2015**, *91*, 034312. [CrossRef]
19. Gonzalez-Boquera, C.; Centelles, M.; Viñas, X.; Robledo, L.M. New Gogny interaction suitable for astrophysical applications. *Phys. Lett. B* **2018**, *779*, 195. [CrossRef]
20. Gonzalez-Boquera, C. Neutron-rich matter in atomic nuclei and neutron stars. *arXiv* **2019**, arXiv:2003.00490.
21. Sharma, B.K.; Centelles, M.; Viñas, X.; Baldo, M.; Burgio, G.F. Unified equation of state for neutron stars on a microscopic basis. *Astron. Astrophys.* **2015**, *A584*, 103. [CrossRef]
22. Douchin, F.; Haensel, P. A unified equation of state of dense matter and neutron star structure. *Astron. Astrophys.* **2001**, *380*, 151. [CrossRef]
23. Pearson, J.M.; Chamel, N.; Potekhin, A.Y.; Fantina, A.F.; Ducoin, C.; Dutta, A.K.; Goriely, S. Unified equations of state for cold non-accreting neutron stars with Brussels-Montreal functionals–I. Role of symmetry energy. *Mon. Not. R. Astron. Soc.* **2018**, *481*, 2994. [CrossRef]
24. Perot, L.; Chamel, N.; Sourie, A. Role of the symmetry energy and the neutron-matter stiffness on the tidal deformability of a neutron star with unified equations of state. *Phys. Rev. C* **2019**, *100*, 035801. [CrossRef]
25. Danielewicz, P.; Lacey, R.; Lynch, W.G. Determination of the Equation of State of Dense Matter. *Science* **2002**, *298*, 1592. [CrossRef]
26. Danielewicz, P.; Lee, J. Symmetry energy II: Isobaric analog states. *Nucl. Phys. A* **1992**, *922*, 1. [CrossRef]

27. Zhang, Z.; Chen, K.-W. Electric dipole polarizability in ^{208}Pb as a probe of the symmetry energy and neutron matter around $\rho_0/3$. *Phys. Rev. C* **2015**, *92*, 031301. [CrossRef]
28. Tsang, M.B.; Zhang, Y.; Danielewicz, P.; Famiano, M.; Li, Z.; Lynch, W.G.; Steiner, A.W. Constraints on the density dependence of the symmetry energy. *Phys. Rev. Lett.* **2009**, *102*, 122701. [CrossRef] [PubMed]
29. Tsang, M.B.; Stone, J.R.; Camera, F.; Danielewicz, P.G.; olfi, S.; Hebeler, K.; Horowitz, C.J.; Lee, J.; Lynch, W.G.; Kohley, Z.; et al. Constraints on the symmetry energy and neutron skins from experiments and theory. *Phys. Rev. C* **2012**, *86*, 015803. [CrossRef]
30. Lattimer, J.M.; Lim, Y. Constraining the symmetry parameters of the nuclear interaction. *Astrophys. J.* **2013**, *771*, 51. [CrossRef]
31. Lattimer, J.M.; Prakash, M. The equation of state of hot, dense matter and neutron stars. *Phys. Rep.* **2016**, *621*, 127. [CrossRef]
32. Centelles, M.; Roca-Maza, X.; Viñas, X.; Warda, M. Nuclear Symmetry Energy Probed by Neutron Skin Thickness of Nuclei. *Phys. Rev. Lett.* **2009**, *102*, 122502. [CrossRef]
33. Roca-Maza, X.; Centelles, M.; Viñas, X.; Warda, M. Neutron skin of ^{208}Pb, Nuclear Symmetry Energy, and the Parity Radius Experiment. *Phys. Rev. Lett.* **2011**, *106*, 252501. [CrossRef]
34. Viñas, X.; Centelles, M.; Roca-Maza, X.; Warda, M. Density dependence of the symmetry energy from neutron skin thickness in finite nuclei. *Eur. Phys. J. A* **2014**, *50*, 27. [CrossRef]
35. Agrawal, B.K.; De, J.N.; Samaddar, S.K. Determining the density content of symmetry energy and neutron skin: An empirical approach. *Phys. Rev. Lett.* **2012**, *109*, 262501. [CrossRef] [PubMed]
36. Li, B.-A.; Han, X. Constraining the neutron-proton effective mass splitting using empirical constraints on the density dependence of nuclear symmetry energy around normal density. *Phys. Lett. B* **2013**, *727*, 276. [CrossRef]
37. Oertel, M.; Hempel, M.; Klähn, T.; Typel, S. Equations of state for supernovae and compact stars. *Rev. Mod. Phys.* **2017**, *89*, 015007. [CrossRef]
38. Birkhan, J.; Miorelli, M.; Bacca, S.; Bassauer, S.; Bertulani, C.A.; Hagen, G.; Matsubara, H.; von Neumann-Cosel, P.; Papenbrock, T.; Pietralla, N.; et al. Electric Dipole Polarizability of ^{48}Ca and implications for neutron stars. *Phys. Rev. Lett.* **2017**, *118*, 252501. [CrossRef]
39. Holt, J.W.; Kaiser, N. Equation of state of nuclear and neutron matter at third order in perturbation theory from chiral effective field theory. *Phys. Rev. C* **2017**, *95*, 034326. [CrossRef]
40. Drischler, C.; Hebeler, K.; Schwenk, N. Chiral interactions up to next-to-next-to-next-to-leading order and nuclear saturation. *Phys. Rev. Lett.* **2019**, *122*, 042501. [CrossRef] [PubMed]
41. Roca-Maza, X.; Viñas, X.; Centelles, M.; Agrawal, B.K.; Colò, G.; Paar, N.; Piekarewicz, J.; Vretenar, D. Neutron skin thickness from the measured electric dipole polarizability in ^{68}Ni, ^{120}Sn, and ^{208}Pb. *Phys. Rev. C* **2015**, *92*, 064304. [CrossRef]
42. Hagen, G.; Ekström, A.; Forssén, C.; Jansen, G.R.; Nazarewicz, W.; Papenbrock, T.; Wendt, K.A.; Bacca, S.; Barnea, N.; Carlsson, B.; et al. Neutron and weak-charge distributions of the ^{48}Ca nucleus. *Nat. Phys.* **2015**, *12*, 186. [CrossRef]
43. Viñas, X.; Gonzalez-Boquera, C.; Centelles, M.; Robledo, L.M.; Mondal, C. Gogny forces in the astrophysical context. *Bulg. J. Phys.* **2018**, *37*, 68.
44. Viñas, X.; Gonzalez-Boquera, C.; Centelles, M.; Mondal, C. The modified D1M interaction: New Gogny forces adapted for neutron star calculations. *Acta Phys. Pol. B Proc. Suppl.* **2019**, *12*, 705. [CrossRef]
45. Roca-Maza, X.; Brenna, M.; Agrawal, B.K.; Borington, P.F.; Colò, G.; Cao, L.G.; Paar, N.; Vretenar, D. Giant quadrupole resonances in ^{208}Pb, the nuclear symmetry energy, and the neutron skin thickness. *Phys. Rev. C* **2013**, *87*, 034301. [CrossRef]
46. Agrawal, B.K.; Shlomo, S.; Au, V.K. Determination of the parameters of a Skyrme type effective interaction using the simulated annealing approach. *Phys. Rev. C* **2005**, *72*, 014310. [CrossRef]
47. Piekarewicz, J. Pygmy resonances and neutron skins. *Phys. Rev. C* **2011**, *83*, 034319. [CrossRef]
48. Fattoyev, F.J.; Piekarewicz, J. Has a thick neutron skin in ^{208}Pb been ruled out? *Phys. Rev. Lett.* **2013**, *111*, 162501. [CrossRef]
49. Horowitz, C.J.; Piekarewicz, J. Neutron star structure and the neutron radius of ^{208}Pb. *Phys. Rev. Lett.* **2001**, *86*, 5647. [CrossRef]
50. Hellemans, V.; Pastore, A.; Duguet, T.; Bennaceur, K.; Davesne, D.; Meyer, J.; Bender, M.; Heenen, P.-H. Spurious finite-size instabilities in nuclear energy density functionals. *Phys. Rev. C* **2013**, *88*, 064323. [CrossRef]
51. Martini, M.; De Pace, A.; Bennaceur, K. Spurious finite-size instabilities with Gogny-type interactions. *Eur. Phys. J. A* **2019**, *55*, 150. [CrossRef]
52. Gonzalez-Boquera, C.; Centelles, M.; Viñas, X.; Robledo, L.M. Finite-size instabilities in finite-range forces. *Phys. Rev. C* **2021**, *103*, 064314. [CrossRef]
53. Mondal, C.; Viñas, X.; Centelles, M.; De, J.N. Structure and composition of the inner crust of neutron stars from Gogny interactions. *Phys. Rev. C* **2020**, *102*, 015802. [CrossRef]
54. Haensel, P.; Potekhin, A.Y.; Yakovlev, D.G. *Neutron Stars 1: Equation of State and Structure*; Springer-Verlag: New York, NY, USA, 2007. [CrossRef]
55. Chamel, N.; Haensel, P. Physics of neutron star crusts. *Liv. Rev. Relat.* **2008**, *11*, 10. [CrossRef]
56. Grill, F.; Providência, C.; Avancini, S.S. Neutron star inner crust and symmetry energy. *Phys. Rev. C* **2012**, *85*, 055808. [CrossRef]
57. Avancini, S.S.; Bertolino, B.P. Cold pasta phase in the extended Thomas-Fermi approximation. *Int. J. Mod. Phys. E* **2015**, *24*, 1550078. [CrossRef]
58. Newton, W.G.; Gearheart, M.; Li, B.-A. A survey of the parameter space of the compressible liquid drop model as applied to the neutron star inner crust. *Astrophys. J. Suppl. Ser.* **2012**, *204*, 9. [CrossRef]
59. Freiburghaus, C.; Rosswog, S.; Thielemann, F.K. r-process in neutron star mergers. *Astrophys. J.* **1999**, *525*, L121. [CrossRef]

60. Goriely, S.; Demetriou, P.; Janka, H.-T.; Pearson, J.M.; Samyn, M. The r-process nucleosynthesis: A continued challenge for nuclear physics and astrophysics. *Nucl. Phys. A* **2005**, *758*, 587. [CrossRef]
61. Baiotti, L.; Rezzolla, L. Binary neutron star mergers: A review of Einstein's richest laboratory. *Rep. Prog. Phys.* **2017**, *80*, 096901. [CrossRef] [PubMed]
62. von Weizsäcker, C.F. Zur theorie der kernmassen. *Z. Phys.* **1935**, *96*, 431. [CrossRef]
63. Bethe, H.A.; Bacher, R.F. Nuclear Physics A. Stationary states of nuclei. *Rev. Mod. Phys.* **1936**, *8*, 82. [CrossRef]
64. Strutinsky, V.M. Shell effects in nuclear masses and deformation energies. *Nucl. Phys. A* **1967**, *95*, 420. [CrossRef]
65. Wigner, E. On the quantum correction for thermodynamic equilibrium. *Phys. Rev.* **1932**, *40*, 749. [CrossRef]
66. Kirkwood, J.G. Quantum statistics of almost classical assemblies. *Phys. Rev.* **1933**, *44*, 31. [CrossRef]
67. Ring, P.; Schuck, P. *The Nuclear Many-Body Problem*; Springer: Berlin/Heidelberg, Germany, 1980.
68. Schuck, P.; Viñas, X. A variational Wigner-Kirkwood theory of finite nuclei. *Phys. Lett. B* **1993**, *302*, 1. [CrossRef]
69. Brack, M.; Bhaduri, P.K. *Semiclassical Physics*; Adisson-Wesley: London, UK, 1997.
70. Centelles, M.; Viñas, X.; Durand, M.; Schuck, P.; Von-Eiff, D. Variational Wigner-Kirkwood $\hbar$ expansion. *Ann. Phys.* **1998**, *266*, 207. [CrossRef]
71. Centelles, M.; Schuck, P.; Viñas, X. Thomas-Fermi theory for atomic nuclei revisited. *Ann. Phys.* **2007**, *322*, 363. [CrossRef]
72. Soubbotin, V.B.; Viñas, X. Extended Thomas-Fermi approximation to the one-body density matrix. *Nucl. Phys. A* **2000**, *665*, 291. [CrossRef]
73. Bhagwat, A.; Centelles, M.; Viñas, X.; Schuck, P. Woods-Saxon type of mean-field potentials with effective mass derived from the D1S Gogny force. *Phys. Rev. C* **2021**, *103*, 024320. [CrossRef]
74. Centelles, M.; Pi, M.; Viñas, X.; Garcias, F.; Barranco, M. Self-consistent extended Thomas-Fermi calculations in nuclei. *Nucl. Phys. A* **1990**, *510*, 397. [CrossRef]
75. Chu, Y.H.; Jennings, B.K.; Brack, M. Nuclear binding energies and liquid drop parameters in the Extended. Thomas-Fermi approximation *Phys. Lett. B* **1977**, *68*, 407.
76. Pearson, J.M.; Chamel, N.; Goriely, S.; Ducoin, C. Inner crust of neutron stars with mass-fitted Skyrme functionals. *Phys. Rev. C* **2012**, *85*, 065803. [CrossRef]
77. Del Estal, M.; Centelles, M.; Viñas, X.; Patra, S.K. Pairing properties in relativistic mean field models obtained from effective field theory. *Phys. Rev. C* **2001**, *63*, 044321. [CrossRef]
78. Wang, M.; Audi, G.; Kondev, F.G.; Huang, W.J.; Naimi, S.; Xu, X. The AME2016 atomic mass evaluation (II). Tables, graphs and references. *Chin. Phys. C* **2017**, *41*, 030003. [CrossRef]
79. Welker, A.; Althubiti, N.A.S.; Atanasov, D.; Blaum, K.; Cocolios, T.E.; Herfurth, F.; Kreim, S.; Lunney, D.; Manea, V.; Mougeot, M.; et al. Binding energy of ^{79}Cu: Probing the structure of the doubly magic ^{78}Ni from only one proton away. *Phys. Rev. Lett.* **2017**, *119*, 192502. [CrossRef] [PubMed]
80. Pearson, J.M.; Goriely, S.; Chamel, N. Properties of the outer crust of neutron stars from Hartree-Fock-Bogoliubov mass models. *Phys. Rev. C* **2011**, *83*, 065810. [CrossRef]
81. Baym, G.; Pethick, C.; Sutherland, P. The ground state of matter at high densities: Equation of state and stellar models. *Astrophys. J.* **1971**, *170*, 299. [CrossRef]
82. Roca-Maza, X.; Piekarewicz, J. Impact of the symmetry energy on the outer crust of non accreting neutron stars. *Phys. Rev. C* **2008**, *78*, 025807. [CrossRef]
83. Chamel, N. Analytical determination of the structure of the outer crust of a cold nonaccreted neutron star. *Phys. Rev. C* **2020**, *101*, 032801. [CrossRef]
84. Pearson, J.M.; Chamel, N.; Pastore, A.; Goriely, S. Role of proton pairing in a semimicroscopic treatment of the inner crust of neutron stars. *Phys. Rev. C* **2015**, *91*, 018801. [CrossRef]
85. Shelley, M.; Pastore, A. Comparison between the Thomas-Fermi and Hartree-Fock-Bogolibov Methods in the Inner Crust of a Neutron Star: The role of Pairing Correlations. *Universe* **2020**, *6*, 206. [CrossRef]
86. Shelley, M.; Pastore, A. Systematic analysis of the inner crust composition using the extended Thomas-Fermi approximation with pairing correlations. *Phys. Rev. C* **2021**, *103*, 035807. [CrossRef]
87. Gonzalez-Boquera, C.; Centelles, M.; Viñas, X.; Routray, T.R. Core-crust transition in neutron stars with finite-range interactions: The dynamical method. *Phys. Rev. C* **2019**, *100*, 015806. [CrossRef]
88. Xu, J.; Chen, L.-W.; Li, B.-A.; Ma, H.-R. Nuclear constraints on properties of neutron star crusts. *Astrophys. J.* **2009**, *697*, 1549. [CrossRef]
89. Zhang, N.-B.; Li, B.-A. Constraints on the muon fraction and density profile in neutron stars. *Astrophys. J.* **2020**, *893*, 61. [CrossRef]
90. Klähn, T.; Blaschke, D.; Typel, S.; van Dalen, E.N.E.; Faessler, A.; Fuchs, C.; Gaitanos, T.; Grigorian, H.; Ho, A.; Kolomeitsev, E.E.; et al. Constraints on the high-density nuclear equation of state from the phenomenology of compact stars and heavy-ion collisions. *Phys. Rev. C* **2006**, *74*, 035802. [CrossRef]
91. Shapiro, S.L.; Teukolsky, S.A. *Black Holes, White Dwarfs, and Neutron Stars: The Physics of Compact Objects*; John Wiley & Sons: Hoboken, NJ, USA, 1983.
92. Nättilä, J.; Steiner, A.W.; Kajava, J.J.E.; Suleimanov, V.F.; Poutanen, J. Equation of state constraints for the cold dense matter inside neutron stars using the cooling tail method. *Astron. Astrophys.* **2016**, *591*, A25. [CrossRef]

93. Lattimer, J.M.; Steiner, A.W. Constraints on the symmetry energy using the mass-radius relation of neutron stars. *Eur. Phys. J. A* **2014**, *50*, 40. [CrossRef]
94. Abbott, B.; Jawahar, S.; Lockerbie, N.; Tokmakov, K. LIGO Scientific Collaboration and Virgo Collaboration. GW170817: Measurements of neutron star radii and equation of state. *Phys. Rev. Lett.* **2018**, *121*, 161101. [CrossRef]
95. Riley, T.E.; Watts, A.L.; Bogdanov, S.; Ray, P.S.; Ludlam, R.M.; Guillot, S.; Arzoumanian, Z.; Baker, C.L.; Bilous, A.V.; Chakrabarty, D.; et al. A NICER view of PSR J0030+0451: Millisecond pulsar parameter estimation. *Astrophys. J. Lett.* **2019**, *887*, L21. [CrossRef]
96. Riley, T.E.; Watts, A.L.; Ray, P.S.; Bogdanov, S.; Guillot, S.; Morsink, S.M.; Bilous, A.V.; Arzoumanian, Z.; Choudhury, D.; Deneva, J.S.; et al. A NICER view of the massive pulsar PSR J0740+6620 informed by radio timing and XMM-Newton spectroscopy. *arXiv* **2021**, arXiv:2105.06980.
97. Hartle, J.B. Slowly rotating relativistic stars. I. equations of structure. *Astrophys. J.* **1967**, *150*, 1005. [CrossRef]
98. Fattoyev, F.J.; Piekarewicz, J. Sensitivity of the moment of inertia of neutron stars to the equation of state of neutron-rich matter. *Phys. Rev. C* **2010**, *82*, 025810. [CrossRef]
99. Ravenhall, D.G.; Pethick, C.J. Neutron star moments of inertia. *Astrophys. J.* **1994**, *424*, 846. [CrossRef]
100. Lattimer, J.M.; Schutz, B.F. Constraining the equation of state with moment of inertia measurements. *Astrophys. J.* **2005**, *629*, 979. [CrossRef]
101. Link, B.; Epstein, R.I.; Lattimer, J.M. Pulsar constraints on neutron stars structure and equation of state. *Phys. Rev. Lett.* **1999**, *83*, 3362. [CrossRef]
102. Anderson, N.; Glampedakis, K.; Ho, W.C.G.; Espinoza, C.M. Pulsar glitches: The crust is not enough. *Phys. Rev. Lett.* **2012**, *109*, 241103. [CrossRef]
103. Chamel, N. Crustal Entrainment and Pulsar Glitches. *Phys. Rev. Lett.* **2013**, *110*, 011101. [CrossRef]
104. Landry, P.; Kumar, B. Constraints on the moment of inertia of PSR J037-3039A from GW170817. *Astrophys. J.* **2018**, *868*, L22. [CrossRef]
105. Breu, C.; Rezzolla, L. Maximum mass, moment of inertia and compactness of relativistic stars. *Mon. Not. R. Astron. Soc.* **2016**, *459*, 646. [CrossRef]
106. Flanagan, É.É.; Hinderer, T. Constraining neutron-star tidal Love numbers with gravitational-wave detectors. *Phys. Rev. D* **2008**, *77*, 021502. [CrossRef]
107. Hinderer, T. Tidal Love Numbers of Neutron Stars. *Astrophys. J.* **2008**, *677*, 1216. [CrossRef]
108. Hinderer, T.; Lackey, B.D.; Lang, R.N.; Read, J.S. Tidal deformability of neutron stars with realistic equations of state and their gravitational wave signatures in binary inspiral. *Phys. Rev. D* **2010**, *81*, 123016. [CrossRef]
109. Abbott, B.; Jawahar, S.; Lockerbie, N.; Tokmakov, K. LIGO scientific collaboration and Virgo collaboration. GW170817: Observation of gravitational waves from a binary neutron star inspiral. *Phys. Rev. Lett.* **2017**, *119*, 161101. [CrossRef]
110. Abbott, B.; Jawahar, S.; Lockerbie, N.; Tokmakov, K. LIGO scientific collaboration and Virgo collaboration. Properties of the binary neutron star merger GW170817. *Phys. Rev. X* **2019**, *9*, 011001.
111. Lourenço, O.; Bhuyan, M.; Lenzi, C.H.; Dutra, M.; Gonzalez-Boquera, C.; Centelles, M.; Viñas, X. GW170817 constraints analyzed with Gogny forces and momentum-dependent interactions. *Phys. Lett. B* **2020**, *803*, 135306. [CrossRef]

symmetry

Article

Uniqueness of the Inflationary Higgs Scalar for Neutron Stars and Failure of Non-Inflationary Approximations

Vasilis K. Oikonomou [1,2,3]

[1] Department of Physics, Aristotle University of Thessaloniki, 54124 Thessaloniki, Greece; v.k.oikonomou1979@gmail.com or voikonomou@auth.gr or voiko@physics.auth.gr
[2] Laboratory of Theoretical Cosmology, Tomsk State University of Control Systems and Radioelectronics (TUSUR), 634050 Tomsk, Russia
[3] Department of Physics, Tomsk State Pedagogical University, 634061 Tomsk, Russia

Abstract: Neutron stars are perfect candidates to investigate the effects of a modified gravity theory, since the curvature effects are significant and more importantly, potentially testable. In most cases studied in the literature in the context of massive scalar-tensor theories, inflationary models were examined. The most important of scalar-tensor models is the Higgs model, which, depending on the values of the scalar field, can be approximated by different scalar potentials, one of which is the inflationary. Since it is not certain how large the values of the scalar field will be at the near vicinity and inside a neutron star, in this work we will answer the question, which potential form of the Higgs model is more appropriate in order for it to describe consistently a static neutron star. As we will show numerically, the non-inflationary Higgs potential, which is valid for certain values of the scalar field in the Jordan frame, leads to extremely large maximum neutron star masses; however, the model is not self-consistent, because the scalar field approximation used for the derivation of the potential, is violated both at the center and at the surface of the star. These results shows the uniqueness of the inflationary Higgs potential, since it is the only approximation for the Higgs model, that provides self-consistent results.

Keywords: neutron stars; scalar-tensor gravity; Higgs inflationary model

Citation: Oikonomou, V.K. Uniqueness of the Inflationary Higgs Scalar for Neutron Stars and Failure of Non-Inflationary Approximations. *Symmetry* **2022**, *14*, 32. https://doi.org/10.3390/sym14010032

Academic Editor: Kazuharu Bamba

Received: 21 November 2021
Accepted: 17 December 2021
Published: 28 December 2021

Publisher's Note: MDPI stays neutral with regard to jurisdictional claims in published maps and institutional affiliations.

1. Introduction

The next two decades will possibly bring sensational observational results to the cosmology, theoretical physics and theoretical astrophysics community. All of these observations are related to gravitational wave detections, either stochastic inflationary gravitational waves, such as the LISA [1,2] and DECIGO [3,4], or ordinary astrophysical originating gravitational waves. With regard to astrophysical sources of gravitational waves, neutron stars are in the epicenter of current theoretical and experimental research. This is because neutron stars (NSs) [5–9] are superstars among stars, a wide range of physics research areas must be used to describe these accurately, such as nuclear and high energy physics [10–24], modified gravity can also describe NSs [25–37], and theoretical astrophysics [38–49]. Four decades passed since the first observation of a NS, and to date serious questions remain regarding the inner structure and physics of NSs. The equation of state (EoS) of nuclear matter is still a mystery in addition to the fundamental question whether general relativity (GR) or modified gravity [50–57] controls the physics of the star. A particularly appealing form of modified gravity is scalar-tensor gravity, and many works on NSs in the context of scalar-tensor gravity already exist in the literature [58–71]. Also scalar-tensor gravity is popular in cosmological contexts too [72–84], where viable inflationary models can be realized. The model with the highest importance in scalar-tensor gravity is the Higgs model, since the Higgs boson is the first fundamental (elementary) scalar elementary particle that has ever been observed [85]. The Higgs inflationary potential is capable of producing a viable inflationary era [76] and this occurs for a specific range of values of the scalar field and the non-minimal coupling constant to the Ricci scalar, usually denoted as ξ. In a previous work we studied NSs in the context of scalar-tensor theories, using the inflationary Higgs potential [34]. In this work we extend our work to account for different limiting values of the scalar field and the combined non-minimal coupling of the form $\sim \xi\phi^2$. We shall be interested in values $\xi\phi^2 \ll 1$ in Geometrized units. In this approximation, we shall derive the Einstein frame potential and the relevant conformal transformation function $A(\phi)$. Accordingly, we shall derive the corresponding Tolman–Oppenheimer–Volkoff (TOV) equations in the Einstein frame, for static NSs, and we shall solve these numerically, assuming piecewise polytropic EoSs [86,87]. We shall find the $M-R$ relations for static NSs. Our results indicate an important fact, that the only correct description of the Higgs potential for static NSs is the one we developed in Ref. [34]. The results of the current article indicate that the maximum masses of NSs exceed the 3 $M_{\odot}$ limit, but the approximation $\xi\phi^2 \ll 1$ fails to hold true at the center and at the surface of the NSs. This result indicates how unique is the inflationary Higgs potential, for providing a self-consistent neutron star phenomenology.

2. Non-Inflationary Higgs Scalar-Tensor Gravity in the Einstein Frame and Static NSs Phenomenology

We are interested in extracting the Einstein frame counterpart theory of the Jordan frame Higgs theory, and we shall do so by using a conformal transformation, see [58,79,88–91] for details on conformal transformations. The Jordan frame action of the Higgs model as it appears in cosmological contexts [76], in Geometrized units ($G=1$) is the following,

$$\mathcal{S} = \int d^4x \frac{\sqrt{-g}}{16\pi}\Big[f(\phi)\mathcal{R} - \frac{1}{2}g^{\mu\nu}\partial_\mu\phi\partial_\nu\phi - U(\phi)\Big] + S_m(\psi_m, g_{\mu\nu}), \tag{1}$$

where $f(\phi)$ is the non-minimal coupling function and $U(\phi)$ is the potential, defined as follows,

$$f(\phi) = 1 + \xi\phi^2, \;\; U(\phi) = \lambda\phi^4, \tag{2}$$

where ϕ denotes the Jordan frame scalar field. In addition, $g^{\mu\nu}$, $S_m(\psi_m, g_{\mu\nu})$, g and $\mathcal{R}$ denote the metric tensor, the action for the matter fluids, the determinant of the metric tensor and the Ricci scalar in the Jordan frame.

Performing the conformal transformation $\tilde{g}_{\mu\nu} = A^{-2}g_{\mu\nu}$, where the function $A(\phi)$ is defined as,

$$A(\phi) = f^{-1/2}(\phi), \tag{3}$$

and the Einstein frame action in terms of the canonical scalar field φ reads,

$$\mathcal{S} = \int d^4x\sqrt{-\tilde{g}}\Big(\frac{\tilde{\mathcal{R}}}{16\pi} - \frac{1}{2}\tilde{g}_{\mu\nu}\partial^\mu\varphi\partial^\nu\varphi - \frac{V(\varphi)}{16\pi}\Big) + S_m(\psi_m, A^2(\varphi)g_{\mu\nu}), \tag{4}$$

where the "tilde" denotes quantities evaluated in the Einstein frame. Specifically, $\tilde{g}^{\mu\nu}$, $S_m(\psi_m, A^2(\varphi)g_{\mu\nu})$, $\tilde{g}$ and $\tilde{\mathcal{R}}$ denote the metric tensor, the action for the matter fluids, the determinant of the metric tensor and the Ricci scalar in the Einstein frame.

Recall that $A(\phi)$ enters in the conformal transformation $\tilde{g}_{\mu\nu} = A^{-2}g_{\mu\nu}$ and by using Equations (2) and (3) we have,

$$A(\phi) = \Big(1 + \xi\phi^2\Big)^{-1/2}. \tag{5}$$

Also the Einstein frame potential is,

$$V(\phi) = \frac{U(\phi)}{f^2(\phi)}, \tag{6}$$

and when expressed in terms of ϕ this is written as,

$$V(\phi) = \frac{\lambda\phi^4}{(1+\xi\phi^2)^2}, \tag{7}$$

Using relation between the Einstein frame canonical scalar field φ and the Jordan frame scalar field ϕ,

$$\frac{d\varphi}{d\phi} = \frac{1}{\sqrt{4\pi}}\sqrt{\left(\frac{3}{4}\frac{1}{f^2}\left(\frac{df}{d\phi}\right)^2 + \frac{1}{4f}\right)}, \tag{8}$$

and combined with Equation (2), we get,

$$\frac{d\varphi}{d\phi} = \frac{1}{\sqrt{16\pi}}\frac{\sqrt{1+\xi\phi^2+12\xi^2\phi^2}}{1+\xi\phi^2}. \tag{9}$$

For the Higgs inflationary potential, the field values approximations are the following

$$\xi^2\phi^2 \gg 1,\ \xi^2\phi^2 \gg \xi\phi^2, \tag{10}$$

however, we shall use another approximation relevant in Higgs potential physics, namely [76],

$$\frac{1}{\sqrt{12\xi}} \ll \phi \ll \frac{1}{\sqrt{\xi}}, \tag{11}$$

which is equivalent to the following two approximations,

$$\xi\phi^2 \ll 1, \tag{12}$$

$$12\xi^2\phi^2 \gg 1. \tag{13}$$

In view of the approximations (12) and (13) we get approximately at leading order,

$$\frac{d\varphi}{d\phi} \simeq \frac{\sqrt{12}}{\sqrt{16\pi}}\xi\phi. \tag{14}$$

Thus, integrating the above we get the final relation between φ and ϕ,

$$\varphi = \frac{\sqrt{12}}{2\sqrt{16\pi}}\xi\phi^2. \tag{15}$$

At leading order the function $A(\varphi)$ reads,

$$A(\varphi) = 1 - \frac{2\sqrt{12}}{\sqrt{16\pi}}\varphi, \tag{16}$$

thus, at leading order $\alpha(\varphi) = \frac{\mathrm{d}\ln \mathrm{A}}{\mathrm{d}'} = -2\sqrt{\frac{16\pi}{12}}\left(1+2\sqrt{\frac{16\pi}{12}}\varphi\right)$. Moreover, the potential as function of φ is,

$$V(\varphi) \simeq \lambda\left(\frac{2\sqrt{16\pi}}{\sqrt{12}}\right)^2\frac{\varphi^2}{\xi^2}. \tag{17}$$

For phenomenological reasoning [76], we shall choose $\xi \sim 11.455 \times 10^4$ with $\lambda = 0.1$. With regard to the EoS, we shall use a piecewise polytropic EoS, the details of which can be found in [34].

For $\xi \sim 11.455 \times 10^4$, the requirement (13) can in principle be satisfied, but the constraint of Equation (12) is not necessarily satisfied. As we will show, this is the case for the non-inflationary Higgs potential, and we shall verify this numerically. For the study we shall consider static NSs, which are described by a spherically symmetric static spacetime of the form,

$$ds^2 = -e^{\nu(r)}dt^2 + \frac{dr^2}{1-\frac{2m}{r}} + r^2(d\theta^2 + \sin^2\theta d\phi^2)\,, \tag{18}$$

where $m(r)$ denotes the gravitational mass of the stellar object confined inside a radius r.

For Geometrized units ($c = G = 1$), the TOV equations for the spherically symmetric spacetime are,

$$\frac{dm}{dr} = 4\pi r^2 A^4(\varphi)\epsilon + 2\pi r(r-2m)\omega^2 + 4\pi r^2 V(\varphi)\,, \tag{19}$$

$$\frac{d\nu}{dr} = 4\pi r\omega^2 + \frac{2}{r(r-2m)}\left[4\pi A^4(\varphi)r^3 P - 4\pi V(\varphi)r^3\right] + \frac{2m}{r(r-2m)}\,, \tag{20}$$

$$\frac{d\omega}{dr} = \frac{rA^4(\varphi)}{r-2m}\Big(\alpha(\varphi)(\epsilon-3P) + 4\pi r\omega(\epsilon-P)\Big) - \frac{2\omega(r-m)}{r(r-2m)} + \frac{8\pi\omega r^2 V(\varphi) + r\frac{dV(\varphi)}{d\varphi}}{r-2m}\,, \tag{21}$$

$$\frac{dP}{dr} = -(\epsilon+P)\left[\alpha(\varphi)\omega + 2\pi r\omega^2 + \frac{m - 4\pi r^3(-A^4P + V)}{r(r-2m)}\right], \tag{22}$$

$$\frac{d\varphi}{dr} = \omega\,, \tag{23}$$

The TOV equations must be solved numerically subject to the following initial conditions,

$$P(0) = P_c,\;\; m(0) = 0,\;\; \nu(0), = -\nu_c,\;\; \varphi(0) = \varphi_c,\;\; \omega(0) = 0\,, \tag{24}$$

where P_c, ν_c, ϕ_c are the pressure of the NS, the value of the function $\nu(r)$ and the value of the scalar field at the center of the NS. The values of nu_c and φ_c at the center of the star, shall be obtained using a double shooting method, in order for the optimal values of them to be obtained. The requirement for obtaining the optimal values is the scalar field values to vanish at numerical infinity, which proves to be the same numerically as in the inflationary Higgs potential, namely $r \sim 67.94378528694695$ km in the Einstein frame, see [34]. In addition, for the derivation of the $M-R$ gravity we need to consider the ADM Jordan frame mass and the Jordan frame radius. Denoting with r_E the Einstein frame radius at large distances, and $\frac{d\varphi}{dr} = \frac{d\varphi}{dr}\Big|_{r=r_E}$, the Jordan frame mass $M_J \equiv M$ is related to the Einstein frame mass as follows,

$$M_J = A(\varphi(r_E))\left(M_E - \frac{r_E^2}{2}\alpha(\varphi(r_E))\frac{d\varphi}{dr}\left(2 + \alpha(\varphi(r_E))r_E\frac{d\varphi}{dr}\right)\left(1 - \frac{2M_E}{r_E}\right)\right), \tag{25}$$

with $\frac{d\varphi}{dr} = \frac{d\varphi}{dr}\Big|_{r=r_E}$ and $r_J = Ar_E$, and r_J is the Jordan frame radius. The Einstein frame radius R_s of the star can be obtained by the numerical code by using the condition $P(R_s) = 0$, so it is basically determined by the condition that the pressure of the star vanishes at the surface of the star. Accordingly, by finding R_s we can obtain the Jordan frame radius R using the relation $R = A(\varphi(R_s))\,R_s$, where $\varphi(R_s)$ is the value of the scalar field at the surface of the star. Finally and important note is to verify numerically the validity of the approximation (12) in the Jordan frame. For the numerical analysis, we shall use a freely available PYTHON code pyTOV-STT [92], and we shall derive the solutions for both the interior and the exterior of the NS, using the "LSODA" numerical method. The EoSs we shall use are the WFF1 [93], the SLy [94], and the APR EoS [95]. Let us proceed to the results of our analysis, and we start off with the $M-R$ graphs for all the EoSs which we present in Figure 1. The purple curve corresponds to the WFF1 EoS, while the red and blue to the SLy and APR EoSs respectively. From the graphs it is apparent that for the non-inflationary Higgs model, the maximum masses are comparably higher with regard to the GR ones. In addition, in Table 1 we present all the maximum masses for all the EoSs corresponding to the alternative Higgs model. As it can be seen in Table 1, the maximum masses for the alternative Higgs model are quite elevated compared to the GR ones. In addition, the GW170817 constraint which indicates that the radius corresponding to the maximum NS mass must be larger than is satisfied $R = 9.6^{+0.14}_{-0.03}$ km.

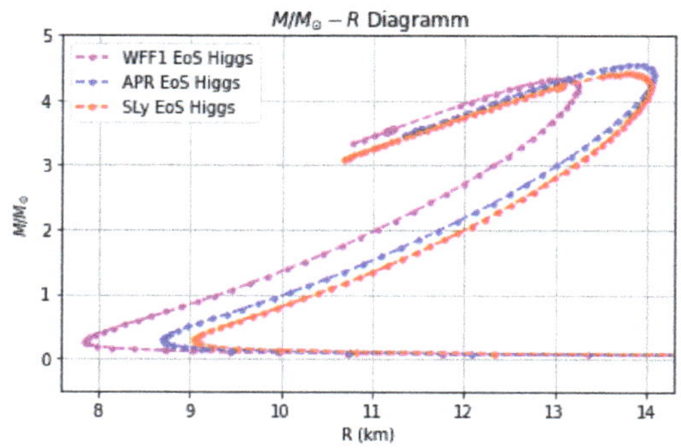

Figure 1. $M - R$ graphs for the alternative non-inflationary Higgs model for the WFF1 EoS (purple curve), the APR EoS (blue curve), and the SLy EoS (red curve). The y-axis is expressed in $M/M_\odot$ units, with M denoting the Jordan frame ADM mass, and the x-axis is the circumferential radius.

Table 1. Maximum Masses and the of Static NS for the non-Inflationary Higgs Model and for GR.

Model	APR EoS	SLy EoS	WFF1 EoS
GR	$M_{max} = 2.18739372\,M_\odot$	$M_{max} = 2.04785291\,M_\odot$	$M_{max} = 2.12603999\,M_\odot$
Alternative Higgs $\xi \sim 10^4$	$M_{max} = 4.55374471\,M_\odot$	$M_{max} = 4.41766131\,M_\odot$	$M_{max} = 4.33460622\,M_\odot$

The results are deemed quite interesting; however, the non-inflationary Higgs model has inherent issues with the approximation (12) as we proved numerically. Particularly it is not satisfied neither at the center nor at the surface of the star. This feature can be clearly seen in Figure 2, where we present the values of $\xi\phi^2$ in the Jordan frame for all the EoS for the surface scalar field values. The same applies for the values of the scalar field in the center of the star. Therefore, to our original question whether inflationary scalar potentials or other approximations must be used for static NSs phenomenology, the answer seems to be that only inflationary potentials provide consistent results.

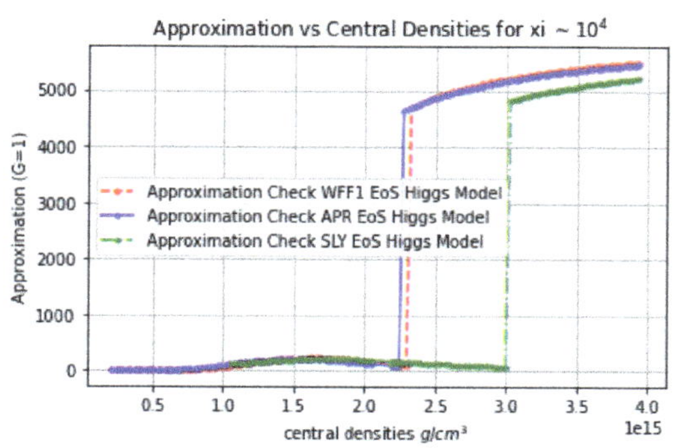

Figure 2. The quantity $\xi\phi^2$ (y-axis) in Geometrized units, versus the central densities in CGS units, for $\xi \sim 11.455 \times 10^4$, for the WFF1 (red curve), APR (blue curve) and Sly (green curve) EoSs. As it can be seen the constraint (12) is not satisfied.

3. Concluding Remarks

In the field of cosmology, there exist several massive scalar field theories which can potentially play an important role for describing NSs phenomenology. From these theories, the most important is the Higgs inflationary theory in its various forms. Specifically, depending on the scalar field values, the Higgs potential can take various forms, each of which may describe a different era in the cosmological theory. Thus, the question is which approximate Higgs potential can describe in a viable and consistent static NS phenomenology. In this paper we addressed this question for the most fundamental of all the scalar field cosmologies, the Higgs inflationary theory. We considered the theory in the Jordan frame and upon conformally transforming it, we derived the Einstein frame theory. Accordingly, assuming a specific range for for the scalar field values, we derived the appropriate quantities which are relevant for studying static NSs in the Einstein frame. For a static spherically symmetric spacetime we derived the TOV equations and we numerically solved them using a double shooting method for optimizing the results. The numerical analysis yielded the Einstein frame masses and radii of the static NS, and also the Einstein frame values of the scalar field, from which we found the corresponding Jordan frame quantities. We constructed the $M - R$ graphs and we investigated the validity of the approximations holding true for the non-inflationary Higgs model. As we showed, the maximum masses for the alternative Higgs model are quite elevated, compared with the GR case; however, for all the EoSs studies, the approximation we assumed for deriving the theory break down. Thus, although the theory provides interesting result, the inherent structure of it is not correct and consistent. This indicates strongly the suitability of inflationary potentials for studying NSs phenomenology, regardless how well motivated other forms of potentials might be. Moreover, it seems that the approximations for the scalar field values used for deriving the inflationary potentials, are well respected on the surface, center and at numerical infinity of the NS. Hence, in conclusion the Higgs potential that is used for inflationary phenomenology is the only suitable for describing consistently NSs.

We need to note with regard to the EoSs we used, that we used the PAR, and more importantly the WFF1, known as FPS EoS, and the SLy, which both are known to provide a unified description of the crust and core of NSs. However, all these EoSs are to date rather old (nearly 20 years old); thus, it is compelling to incorporate to the analysis more timely and to date EoSs, such as the BSk24 [96,97].

Funding: This research received no external funding.

Institutional Review Board Statement: Not applicable.

Informed Consent Statement: Not applicable.

Data Availability Statement: Not applicable.

Acknowledgments: I am grateful to the referee 3 of the manuscript who pointed out the timely BSk24, which I intend to incorporate in my code for the piecewise polytropic EoS.

Conflicts of Interest: The author declares no conflict of interest.

References

1. Baker, J.; Bellovary, J.; Bender, P.L.; Berti, E.; Caldwell, R.; Camp, J.; Conklin, J.W.; Cornish, N.; Cutler, C.; DeRosa, R.; et al. The Laser Interferometer Space Antenna: Unveiling the Millihertz Gravitational Wave Sky. *arXiv* **2019**, arXiv:1907.06482.
2. Smith, T.L.; Caldwell, R. LISA for Cosmologists: Calculating the Signal-to-Noise Ratio for Stochastic and Deterministic Sources. *Phys. Rev. D* **2019**, *100*, 104055. [CrossRef]
3. Seto, N.; Kawamura, S.; Nakamura, T. Possibility of direct measurement of the acceleration of the universe using 0.1-Hz band laser interferometer gravitational wave antenna in space. *Phys. Rev. Lett.* **2001**, *87*, 221103. [CrossRef]
4. Kawamura, S.; Ando, M.; Seto, N.; Sato, S.; Musha, M.; Kawano, I.; Yokoyama, J.; Tanaka, T.; Ioka, K.; Akutsu, T.; et al. Current status of space gravitational wave antenna DECIGO and B-DECIGO. *Prog. Theor. Exp. Phys.* **2021**, *2021*, 05A105. [CrossRef]
5. Haensel, P.; Potekhin, A.Y.; Yakovlev, D.G. Neutron stars 1: Equation of state and structure, Astrophys. *Space Sci. Libr.* **2007**, *326*, 1–619.

6. Friedman, J.L.; Stergioulas, N. *Rotating Relativistic Stars*; Cambridge University Press: Cambridge, UK, 2013. [CrossRef]
7. Baym, G.; Hatsuda, T.; Kojo, T.; Powell, P.D.; Song, Y.; Takatsuka, T. From hadrons to quarks in neutron stars: A review. *Rept. Prog. Phys.* **2018**, *81*, 056902. [CrossRef] [PubMed]
8. Lattimer, J.M.; Prakash, M. The physics of neutron stars. *Science* **2004**, *304*, 536–542. [CrossRef]
9. Olmo, G.J.; Rubiera-Garcia, D.; Wojnar, A. Stellar structure models in modified theories of gravity: Lessons and challenges. *Phys. Rept.* **2020**, *876*, 1–75. [CrossRef]
10. Lattimer, J.M. The nuclear equation of state and neutron star masses. *Ann. Rev. Nucl. Part. Sci.* **2012**, *62*, 485–515. [CrossRef]
11. Steiner, A.W.; Gandolfi, S. Connecting Neutron Star Observations to Three-Body Forces in Neutron Matter and to the Nuclear Symmetry Energy. *Phys. Rev. Lett.* **2012**, *108*, 081102. [CrossRef]
12. Horowitz, C.J.; Perez-Garcia, M.A.; Berry, D.K.; Piekarewicz, J. Dynamical response of the nuclear 'pasta' in neutron star crusts, *Phys. Rev. C* **2005**, *72*, 035801.
13. Watanabe, G.; Iida, K.; Sato, K. Thermodynamic properties of nuclear 'pasta' in neutron star crusts. *Nucl. Phys. A* **2000**, *676*, 455–473; Erratum in *Nucl. Phys. A* **2003**, *726*, 357–365. [CrossRef]
14. Shen, H.; Toki, H.; Oyamatsu, K.; Sumiyoshi, K. Relativistic equation of state of nuclear matter for supernova and neutron star. *Nucl. Phys. A* **1998**, *637*, 435–450. [CrossRef]
15. Xu, J.; Chen, L.W.; Li, B.A.; Ma, H.R. Nuclear constraints on properties of neutron star crusts. *Astrophys. J.* **2009**, *697*, 1549–1568 [CrossRef]
16. Hebeler, K.; Lattimer, J.M.; Pethick, C.J.; Schwenk, A. Equation of state and neutron star properties constrained by nuclear physics and observation. *Astrophys. J.* **2013**, *773*, 11. [CrossRef]
17. de Jesús Mendoza-Temis, J.; Wu, M.R.; Martínez-Pinedo, G.; Langanke, K.; Bauswein, A.; Janka, H.T. Nuclear robustness of the r process in neutron-star mergers. *Phys. Rev. C* **2015**, *92*, 055805. [CrossRef]
18. Ho, W.C.G.; Elshamouty, K.G.; Heinke, C.O.; Potekhin, A.Y. Tests of the nuclear equation of state and superfluid and superconducting gaps using the Cassiopeia A neutron star. *Phys. Rev. C* **2015**, *91*, 015806. [CrossRef]
19. Kanakis-Pegios, A.; Koliogiannis, P.S.; Moustakidis, C.C. Probing the nuclear equation of state from the existence of a ~2.6 $M_\odot$ neutron star: The GW190814 puzzle. *Symmetry* **2021**, *13*, 183. [CrossRef]
20. Buschmann, M.; Co, R.T.; Dessert, C.; Safdi, B.R. X-ray Search for Axions from Nearby Isolated Neutron Stars. *Phys. Rev. Lett.* **2021**, *126*, 021102. [CrossRef]
21. Safdi, B.R.; Sun, Z.; Chen, A.Y. Detecting Axion Dark Matter with Radio Lines from Neutron Star Populations. *Phys. Rev. D* **2019**, *99*, 123021. [CrossRef]
22. Hook, A.; Kahn, Y.; Safdi, B.R.; Sun, Z. Radio Signals from Axion Dark Matter Conversion in Neutron Star Magnetospheres. *Phys. Rev. Lett.* **2018**, *121*, 241102. [CrossRef]
23. Edwards, T.D.P.; Kavanagh, B.J.; Visinelli, L.; Weniger, C. Transient Radio Signatures from Neutron Star Encounters with QCD Axion Miniclusters. *Phys. Rev. Lett.* **2021**, *127*, 131103. [CrossRef] [PubMed]
24. Nurmi, S.; Schiappacasse, E.D.; Yanagida, T.T. Radio signatures from encounters between Neutron Stars and QCD-Axion Minihalos around Primordial Black Holes. *arXiv* **2021**, arXiv:2102.05680.
25. Astashenok, A.V.; Capozziello, S.; Odintsov, S.D.; Oikonomou, V.K. Extended Gravity Description for the GW190814 Supermassive Neutron Star. *Phys. Lett. B* **2020**, *811*, 135910. [CrossRef]
26. Capozziello, S.; De Laurentis, M.; Farinelli, R.; Odintsov, S.D. Mass-radius relation for neutron stars in f(R) gravity. *Phys. Rev. D* **2016**, *93*, 023501. [CrossRef]
27. Astashenok, A.V.; Capozziello, S.; Odintsov, S.D. Extreme neutron stars from Extended Theories of Gravity. *JCAP* **2015**, *2015*, 001. [CrossRef]
28. Astashenok, A.V.; Capozziello, S.; Odintsov, S.D. Maximal neutron star mass and the resolution of the hyperon puzzle in modified gravity. *Phys. Rev. D* **2014**, *89*, 103509. [CrossRef]
29. Astashenok, A.V.; Capozziello, S.; Odintsov, S.D. Further stable neutron star models from f(R) gravity. *JCAP* **2013**, *2013*, 040. [CrossRef]
30. Arapoglu, A.S.; Deliduman, C.; Eksi, K.Y. Constraints on Perturbative f(R) Gravity via Neutron Stars. *JCAP* **2011**, *2011*, 020. [CrossRef]
31. Astashenok, A.V.; Odintsov, S.D. Rotating Neutron Stars in F(R) Gravity with Axions. *Mon. Not. Roy. Astron. Soc.* **2020**, *498*, 3616. [CrossRef]
32. Lobato, R.V.; Carvalho, G.A.; Bertulani, C.A. Neutron stars in $f(\mathtt{R}, \mathtt{L_m})$ gravity with realistic equations of state: Joint-constrains with GW170817, massive pulsars, and the PSR J0030+0451 mass-radius from *NICER* data. *Eur. Phys. J. C* **2021**, *81*, 1013. [CrossRef]
33. Oikonomou, V.K. Universal inflationary attractors implications on static neutron stars. *Class. Quant. Grav.* **2021**, *38*, 175005. [CrossRef]
34. Odintsov, S.D.; Oikonomou, V.K. Neutron Stars in Scalar-tensor Gravity with Higgs Scalar Potential. *arXiv* **2021**, arXiv:2104.01982.
35. Odintsov, S.D.; Oikonomou, V.K. Neutron stars phenomenology with scalar–tensor inflationary attractors. *Phys. Dark Univ.* **2021**, *32*, 100805. [CrossRef]
36. Astashenok, A.V.; Capozziello, S.; Odintsov, S.D.; Oikonomou, V.K. Causal limit of neutron star maximum mass in $f(R)$ gravity in view of GW190814. *Phys. Lett. B* **2021**, *816*, 136222. [CrossRef]

37. Astashenok, A.V.; Capozziello, S.; Odintsov, S.D.; Oikonomou, V.K. Novel stellar astrophysics from extended gravity. *EPL* **2021**, *134*, 59001. [CrossRef]
38. Sedrakian, A. Axion cooling of neutron stars. *Phys. Rev. D* **2016**, *93*, 065044. [CrossRef]
39. Khadkikar, S.; Raduta, A.R.; Oertel, M.; Sedrakian, A. Maximum mass of compact stars from gravitational wave events with finite-temperature equations of state. *Phys. Rev. C* **2021**, *103*, 055811. [CrossRef]
40. Sedrakian, D.M.; Hayrapetyan, M.V.; Shahabasyan, M.K. Gravitational radiation of slowly rotating neutron stars. *Astrophysics* **2006**, *49*, 194–200. [CrossRef]
41. Sedrakian, A. Axion cooling of neutron stars. II. Beyond hadronic axions. *Phys. Rev. D* **2019**, *99*, 043011. [CrossRef]
42. Bauswein, A.; Guo, G.; Lien, J.H.; Lin, Y.H.; Wu, M.R. Compact Dark Objects in Neutron Star Mergers. *arXiv* **2020**, arXiv:2012.11908.
43. Vretinaris, S.; Stergioulas, N.; Bauswein, A. Empirical relations for gravitational-wave asteroseismology of binary neutron star mergers. *Phys. Rev. D* **2020**, *101*, 084039. [CrossRef]
44. Bauswein, A.; Blacker, S.; Vijayan, V.; Stergioulas, N.; Chatziioannou, K.; Clark, J.A.; Bastian, N.U.F.; Blaschke, D.B.; Cierniak, M.; Fischer, T. Equation of state constraints from the threshold binary mass for prompt collapse of neutron star mergers. *Phys. Rev. Lett.* **2020**, *125*, 141103. [CrossRef] [PubMed]
45. Bauswein, A.; Just, O.; Janka, H.T.; Stergioulas, N. Neutron-star radius constraints from GW170817 and future detections. *Astrophys. J. Lett.* **2017**, *850*, L34. [CrossRef]
46. Most, E.R.; Weih, L.R.; Rezzolla, L.; Schaffner-Bielich, J. New constraints on radii and tidal deformabilities of neutron stars from GW170817. *Phys. Rev. Lett.* **2018**, *120*, 261103. [CrossRef]
47. Rezzolla, L.; Most, E.R.; Weih, L.R. Using gravitational-wave observations and quasi-universal relations to constrain the maximum mass of neutron stars. *Astrophys. J. Lett.* **2018**, *852*, L25. [CrossRef]
48. Nathanail, A.; Most, E.R.; Rezzolla, L. GW170817 and GW190814: Tension on the maximum mass. *Astrophys. J. Lett.* **2021**, *908*, L28. [CrossRef]
49. Köppel, S.; Bovard, L.; Rezzolla, L. A General-relativistic Determination of the Threshold Mass to Prompt Collapse in Binary Neutron Star Mergers. *Astrophys. J. Lett.* **2019**, *872*, L16. [CrossRef]
50. Nojiri, S.; Odintsov, S.D.; Oikonomou, V.K. Modified Gravity Theories on a Nutshell: Inflation, Bounce and Late-time Evolution. *Phys. Rept.* **2017**, *692*, 1. [CrossRef]
51. Capozziello, S.; De Laurentis, M. Extended Theories of Gravity. *Phys. Rept.* **2011**, *509*, 167. [CrossRef]
52. Faraoni, V.; Capozziello, S. *Beyond Einstein Gravity: A Survey of Gravitational Theories for Cosmology and Astrophysics*; Springer: Dordrecht, The Netherland, 2010; p. 170.
53. Nojiri, S.; Odintsov, S.D. Introduction to modified gravity and gravitational alternative for dark energy . *Int. J. Geom. Meth. Mod. Phys.* **2007**, *4*, 115. [CrossRef]
54. Nojiri, S.; Odintsov, S.D. Unified cosmic history in modified gravity: From F(R) theory to Lorentz non-invariant models. *Phys. Rept.* **2011**, *505*, 59. [CrossRef]
55. de la Cruz-Dombriz, A.; Saez-Gomez, D. Black holes, cosmological solutions, future singularities, and their thermodynamical properties in modified gravity theories. *Entropy* **2012**, *14*, 1717. [CrossRef]
56. Olmo, G.J. Palatini Approach to Modified Gravity: f(R) Theories and Beyond. *Int. J. Mod. Phys. D* **2011**, *20*, 413. [CrossRef]
57. Dimopoulos, K. *Introduction to Cosmic Inflation and Dark Energy*; CRC Press: Boca Raton, FL, USA, 2021.
58. Pani, P.; Berti, E. Slowly rotating neutron stars in scalar-tensor theories. *Phys. Rev. D* **2014**, *90*, 024025. [CrossRef]
59. Staykov, K.V.; Doneva, D.D.; Yazadjiev, S.S.; Kokkotas, K.D. Slowly rotating neutron and strange stars in R^2 gravity. *JCAP* **2014**, *2014*, 006. [CrossRef]
60. Horbatsch, M.; Silva, H.O.; Gerosa, D.; Pani, P.; Berti, E.; Gualtieri, L.; Sperhake, U. Tensor-multi-scalar theories: Relativistic stars and 3 + 1 decomposition. *Class. Quant. Grav.* **2015**, *32*, 204001. [CrossRef]
61. Silva, H.O.; Macedo, C.F.B.; Berti, E.; Crispino, L.C.B. Slowly rotating anisotropic neutron stars in general relativity and scalar–tensor theory. *Class. Quant. Grav.* **2015**, *32*, 145008. [CrossRef]
62. Doneva, D.D.; Yazadjiev, S.S.; Stergioulas, N.; Kokkotas, K.D. Rapidly rotating neutron stars in scalar-tensor theories of gravity. *Phys. Rev. D* **2013**, *88*, 084060. [CrossRef]
63. Xu, R.; Gao, Y.; Shao, L. Strong-field effects in massive scalar-tensor gravity for slowly spinning neutron stars and application to X-ray pulsar pulse profiles. *Phys. Rev. D* **2020**, *102*, 064057. [CrossRef]
64. Salgado, M.; Sudarsky, D.; Nucamendi, U. On spontaneous scalarization. *Phys. Rev. D* **1998**, *58*, 124003. [CrossRef]
65. Shibata, M.; Taniguchi, K.; Okawa, H.; Buonanno, A. Coalescence of binary neutron stars in a scalar-tensor theory of gravity. *Phys. Rev. D* **2014**, *89*, 084005. [CrossRef]
66. Savaş Arapoğlu, A.; Yavuz Ekşi, K.; Emrah Yükselci, A. Neutron star structure in the presence of nonminimally coupled scalar fields. *Phys. Rev. D* **2019**, *99*, 064055. [CrossRef]
67. Ramazanoğlu, F.M.; Pretorius, F. Spontaneous Scalarization with Massive Fields. *Phys. Rev. D* **2016**, *93*, 064005. [CrossRef]
68. Altaha Motahar, Z.; Blázquez-Salcedo, J.L.; Doneva, D.D.; Kunz, J.; Yazadjiev, S.S. Axial quasinormal modes of scalarized neutron stars with massive self-interacting scalar field. *Phys. Rev. D* **2019**, *99*, 104006. [CrossRef]
69. Chew, X.Y.; Dzhunushaliev, V.; Folomeev, V.; Kleihaus, B.; Kunz, J. Rotating wormhole solutions with a complex phantom scalar field. *Phys. Rev. D* **2019**, *100*, 044019. [CrossRef]

70. Blázquez-Salcedo, J.L.; Scen Khoo, F.; Kunz, J. Ultra-long-lived quasi-normal modes of neutron stars in massive scalar-tensor gravity. *EPL* **2020**, *130*, 50002. [CrossRef]
71. Altaha Motahar, Z.; Blázquez-Salcedo, J.L.; Kleihaus, B.; Kunz, J. Scalarization of neutron stars with realistic equations of state. *Phys. Rev. D* **2017**, *96*, 064046. [CrossRef]
72. Bezrukov, F.; Shaposhnikov, M. Higgs inflation at the critical point. *Phys. Lett. B* **2014**, *734*, 249–254. [CrossRef]
73. Garcia-Bellido, J.; Rubio, J.; Shaposhnikov, M.; Zenhausern, D. Higgs-Dilaton Cosmology: From the Early to the Late Universe. *Phys. Rev. D* **2011**, *84*, 123504. [CrossRef]
74. Bezrukov, F.; Magnin, A.; Shaposhnikov, M.; Sibiryakov, S. Higgs inflation: Consistency and generalisations. *JHEP* **2011**, *2011*, 016. [CrossRef]
75. Bezrukov, F.L.; Shaposhnikov, M. The Standard Model Higgs boson as the inflaton. *Phys. Lett. B* **2008**, *659*, 703–706. [CrossRef]
76. Mishra, S.S.; Sahni, V.; Toporensky, A.V. Initial conditions for Inflation in an FRW Universe. *Phys. Rev. D* **2018**, *98*, 083538. [CrossRef]
77. Steinwachs, C.F.; Kamenshchik, A.Y. Non-minimal Higgs Inflation and Frame Dependence in Cosmology. *AIP Conf. Proc.* **2013**, *1514*, 161–164.
78. Rubio, J. Higgs inflation. *Front. Astron. Space Sci.* **2019**, *5*, 50. [CrossRef]
79. Kaiser, D.I. Primordial spectral indices from generalized Einstein theories. *Phys. Rev. D* **1995**, *52*, 4295–4306. [CrossRef] [PubMed]
80. Gundhi, A.; Steinwachs, C.F. Scalaron-Higgs inflation. *Nucl. Phys. B* **2020**, *954*, 114989. [CrossRef]
81. Cervantes-Cota, J.L.; Dehnen, H. Induced gravity inflation in the standard model of particle physics. *Nucl. Phys. B* **1995**, *442*, 391–412. [CrossRef]
82. Kamada, K.; Kobayashi, T.; Takahashi, T.; Yamaguchi, M.; Yokoyama, J. Generalized Higgs inflation. *Phys. Rev. D* **2012**, *86*, 023504. [CrossRef]
83. Schlogel, S.; Rinaldi, M.; Staelens, F.; Fuzfa, A. Particlelike solutions in modified gravity: The Higgs monopole. *Phys. Rev. D* **2014**, *90*, 044056. [CrossRef]
84. Füzfa, A.; Rinaldi, M.; Schlögel, S. Particlelike distributions of the Higgs field nonminimally coupled to gravity. *Phys. Rev. Lett.* **2013**, *111*, 121103. [CrossRef]
85. ATLAS Collaboration; Aad, G.; Abajyan, T.; Abbott, B.; Abdallah, J.; Khalek, S.A.; Abdelalim, A.A.; Abdinov, O.; Aben, R.; Abi, B. Observation of a new particle in the search for the Standard Model Higgs boson with the ATLAS detector at the LHC. *Phys. Lett. B* **2012**, *716*, 1–29. [CrossRef]
86. Read, J.S.; Lackey, B.D.; Owen, B.J.; Friedman, J.L. Constraints on a phenomenologically parameterized neutron-star equation of state. *Phys. Rev. D* **2009**, *79*, 124032. [CrossRef]
87. Read, J.S.; Markakis, C.; Shibata, M.; Uryu, K.; Creighton, J.D.E.; Friedman, J.L. Measuring the neutron star equation of state with gravitational wave observations. *Phys. Rev. D* **2009**, *79*, 124033. [CrossRef]
88. Faraoni, V. *Cosmology in Scalar-Tensor Gravity*; Springer: Berlin/Heidelberg, Germany, 2004.
89. Faraoni, V. Conformally coupled inflation. *Galaxies* **2013**, *1*, 96–106. [CrossRef]
90. Buck, M.; Fairbairn, M.; Sakellariadou, M. Inflation in models with Conformally Coupled Scalar fields: An application to the Noncommutative Spectral Action. *Phys. Rev. D* **2010**, *82*, 043509. [CrossRef]
91. Faraoni, V.; Gunzig, E.; Nardone, P. Conformal transformations in classical gravitational theories and in cosmology. *Fund. Cosmic Phys.* **1999**, *20*, 121.
92. Nikolaos Stergioulas. Available online: https://github.com/niksterg (accessed on 14 January 2021).
93. Wiringa, R.B.; Fiks, V.; Fabrocini, A. Equation of state for dense nucleon matter. *Phys. Rev. C* **1988**, *38*, 1010–1037. [CrossRef]
94. Douchin, F.; Haensel, P. A unified equation of state of dense matter and neutron star structure. *Astron. Astrophys.* **2001**, *380*, 151. [CrossRef]
95. Akmal, A.; Pandharipande, V.R.; Ravenhall, D.G. The Equation of state of nucleon matter and neutron star structure. *Phys. Rev. C* **1998**, *58*, 1804–1828. [CrossRef]
96. Pearson, J.M.; Chamel, N.; Potekhin, A.Y.; Fantina, A.F.; Ducoin, C.; Dutta, A.K.; Goriely, S. Unified equations of state for cold non-accreting neutron stars with Brussels–Montreal functionals—I. Role of symmetry energy. *Mon. Not. Roy. Astron. Soc.* **2018**, *481*, 2994–3026; Erratum in *Mon. Not. Roy. Astron. Soc.* **2019**, *486*, 768. [CrossRef]
97. Pearson, J.M.; Chamel, N.; Potekhin, A.Y. Unified equations of state for cold nonaccreting neutron stars with Brussels-Montreal functionals. II. Pasta phases in semiclassical approximation. *Phys. Rev. C* **2020**, *101*, 015802. [CrossRef]

symmetry

MDPI

Article

Relativistic Magnetized Astrophysical Plasma Outflows in Black-Hole Microquasars

Theodora Papavasileiou [1], Odysseas Kosmas [2,*] and Ioannis Sinatkas [1]

[1] Department of Informatics, University of Western Macedonia, GR-52100 Kastoria, Greece; th.papavasileiou@uowm.gr (T.P.); isinatkas@uowm.gr (I.S.)

[2] Department of MACE, University of Manchester, George Begg Building, Manchester M1 3BB, UK

* Correspondence: odysseas.kosmas@manchester.ac.uk

Abstract: In this work, we deal with collimated outflows of magnetized astrophysical plasma known as astrophysical jets, which have been observed to emerge from a wide variety of astrophysical compact objects. The latter systems can be considered as either hydrodynamic (HD) or magnetohydrodynamic (MHD) in nature, which means that they are governed by non-linear partial differential equations. In some of these systems, the velocity of the jet is very high and they require relativistic MHD (RMHD) treatment. We mainly focus on the appropriate numerical solutions of the MHD (and/or RMHD) equations as well as the transfer equation inside the jet and simulate multi-messenger emissions from specific astrophysical compact objects. We use a steady state axisymmetric model assuming relativistic magnetohydrodynamic descriptions for the jets (astrophysical plasma outflows) and perform numerical simulations for neutrino, gamma-ray and secondary particle emissions. By adopting the existence of such jets in black hole microquasars (and also in AGNs), the spherical symmetry of emissions is no longer valid, i.e., it is broken, and the system needs to be studied accordingly. One of the main goals is to estimate particle collision rates and particle energy distributions inside the jet, from black-hole microquasars. As concrete examples, we choose the Galactic Cygnus X-1 and the extragalactic LMC X-1 systems.

Keywords: XRBs; astrophysical jets; photo-pion production; extragalactic; γ-ray emission; lepto-hadronic model

Citation: Papavasileiou, T.; Kosmas, O.; Sinatkas, I. Relativistic Magnetized Astrophysical Plasma Outflows in Black-Hole Microquasars. *Symmetry* **2022**, *14*, 485. https://doi.org/10.3390/sym14030485

Academic Editor: Charalampos Moustakidis

Received: 12 November 2021
Accepted: 17 December 2021
Published: 27 February 2022

1. Introduction

The last decades, collimated plasma outflows have been observed to emerge from a wide variety of astrophysical objects. These include the proto-planetary nebulae, the compact objects (like galactic black holes or microquasars and X-ray binary stars), as well as the nuclei of active galaxies (AGNs) [1,2]. Despite their different scales concerning the size, the velocity, the amount of energy transported, etc., these cosmic structures have strong similarities. The observations of multi-wavelength and multi-particle emissions from the black-hole X-ray binaries (BHXRBs) [3–5] and the AGNs have shown that they are mainly due to the mass accretion onto the compact object (black hole or neutron star). The accreted matter comes from a companion (donor) star which is often a nearby main sequence star [6,7]. Thus, the inflow (from the companion star) into the accretion disc and the outflow in the jet motivate the investigation of the disc–jet connection [8] in the latter systems. They are governed by non-linear partial differential equations that must be solved through the use of advanced numerical techniques [9–11].

Recently, the X-ray binary systems (XRBs) and the microquasars (MQ) along with their astrophysical magnetohydrodynamical outflows have aroused great interest among the researches dealing with the mechanisms of production and emission of high-energy particles (mostly neutrinos) and/or photons (X-rays, gamma-rays, etc.) from such cosmic structures. This is not restricted only in the analysis of the very large amount of observations [3–5,12] compared to emissions of smaller energy ranges but also in their advanced theoretical modeling [9–11,13,14]. Up to now, multidimensional numerical simulations and theoretical calculations have attempted to shed light on the nature of the interaction mechanisms as well as on the dynamics involved in the associated emissions [7,11]. They offer good support for experimental efforts of astrophysical particles and radiation detection (for more details the reader is referred to References [13,15]).

From an astronomy and astrophysics viewpoint, XRBs are binary systems consisted of a compact object (mostly stellar black hole or neutron star) and a donor star in rotational trajectory around the central star. Due to the presence of very high gravitational field around the compact object, mass is absorbed out of the companion star. The result is the creation of a rotating accretion disc of very high temperature matter and gas along the equatorial plane of the compact object. Subsequently, the magnetic field lines created by the rotating charged matter of the disc, collimate the ejected plasma. This way two oppositely directed astrophysical jets are formed [16]. From a magnetohydrodynamical point of view, the jets (astrophysical plasma outflows) are considered as fluid flows emanating from the region of the compact object. Then, they may be strongly accelerated within a cone of radius $r(z)$ dependent on its half-opening angle ξ ($r = ztan\xi$). It is worth noting that, in some systems (e.g., the M87 and some AGNs) the consideration of a parabolic jet is geometrically more realistic. For example, in describing the jet acceleration, the parabolic shape favors the region near the jet's base. In the conical geometry, however, assumed in this work the acceleration region is put at greater distances from the base.

The plasma ejection is closely connected with the accretion disc formation and its thickness as well as the jet creation mechanisms. Ideally, spherical symmetry would be well suited to describe thin accretion-disc plasma outflows. Initially, prominent theoretical models have been developed by assuming isotropic emissions from the jets (in such models solution through analytic calculations are possible). Realistically (even under slow black-hole rotation and small-scale magnetic fields) there are rather strong deviations from a perfectly symmetrical geometry. In addition, the formation of two oppositely directed jets destroys the spherical symmetry of many systems which become mostly axisymmetric (around the z-axis, the jet-ejection axis). That is why in various types of microquasars (and AGNs) the statistical analysis was made on the basis of the jet's orientation. Moreover, astrophysical jets are often observed to be one-sided and associated with a Doppler factor that confirms the existence of bulk relativistic motion inside the jet. Recently, with the development of advanced efficient computational tools, the employment of more realistic, anisotropic emission (non-symmetric) models became possible. A prominent example would be the relativistic hydrocode PLUTO implemented in References [9–11,17].

For the purposes of our present study, we adopt a lepto-hadronic model for neutrino and gamma-ray production [18–20]. Therefore, we consider that the jet's matter consists mainly of hadrons and electrons (their portion is determined by defining the ratio α of protons to leptons) that are strongly collimated by the system's magnetic field. Furthermore, we assume that a portion of the main jet's content (electrons and protons) [21] is accelerated to rather relativistic velocities through shock-waves [22–25]. A power-law is best suited to describe the energy distribution of the fast protons $N'(E')$, which in the jet's rest frame, is given by the expression:

$$N'(E') = K_0 E'^{-2}\, GeV^{-1} cm^{-3}, \tag{1}$$

where K_0 is a normalization constant. The accelerated fast protons scatter with the cold protons of the jet or the protons of the stellar wind [26–28]. They can also scatter with the radiation fields (gamma-rays) emanating from sources inside or outside the jet. The result is high-energy secondary particles (pions, muons, neutrinos, etc.) production as well as secondary gamma-ray photons emission through various reaction chains [18,19,29].

In this work, initially, we discuss some of the possible interactions resulting in secondary particle and radiation production. In our previous works, we have addressed the proton-proton (p-p) mechanism and estimated the respective energy distributions as well as the emissivities of neutrino and gamma-ray emissions [7,30,31]. In the current study, we concentrate on the proton-photon (p-γ) interaction mechanism and its dependence on geometric characteristics of the binary systems of our interest. These include the Galactic Cygnus X-1 microquasar [32] and the extragalactic LMC X-1 system [33] located at the neighboring galaxy of the Large Magellanic Cloud. For the detection of such emissions, extremely sensitive detector facilities, like the IceCube (at the South pole), the ANTARES and KM3NeT (at the Mediterranean Sea), etc., are in operation to record the relevant signals reaching the Earth [34,35].

2. Radiation Field Density and Transport Equation in Microquasar Jets

The radiation fields that interact with the accelerated (fast) protons of the jet, may consist of soft X-ray photons emanating from the system's accretion disc. They could also include synchrotron radiation emitted by the charged particles (accelerated by the magnetic field inside the jet), or ultra violet (UV) photons originated from the corona region [18,19,36]. In the context of this study, we will take into account the first two cases that are involved in p-γ collisions leading to a reaction chain producing neutrinos and gamma-rays of high energies. We note that UV photons do not contribute significantly to the energy range of our interest.

2.1. *Microquasar Jet Mechanisms Leading to Neutrino and Gamma-Ray Production*

In our previous works [7,9], we have adopted the proton-proton (p-p) interaction mechanism taking place inside the relativistic astrophysical outflows of microquasar jets. This also leads to the particle (pions, muons, neutrinos, etc.) production and radiation emission (gamma-rays, etc.) (see Ref. [7]). Such emissions present axial symmetry around the jet's ejection z-axis. In Refs. [10,11], the employed PLUTO hydro-code permits emission calculations without the assumption of axial symmetry which are considered more realistic jet emission simulations. In this paper, however, we focus on the proton-photon (p-γ) interactions described below.

The p-γ mechanism reflects the collisions of the relativistic protons with the photons of the radiation fields discussed before. This results in the known photo-pion production shortly represented by the following scattering and/or decay reactions:

$$\begin{aligned} & p+\gamma \to p+\pi^0 \\ & p+\gamma \to p+\pi^0+\pi^+ +\pi^- \\ & p+\gamma \to n+\pi^+ . \end{aligned} \tag{2}$$

The above pions (π^0, π^+, π^-) decay to gamma-ray photons, charged muons, neutrinos, etc., according to the following reactions:

$$\begin{aligned} & \pi^0 \to \gamma+\gamma \\ & \pi^+ \to \mu^+ + \nu_\mu \to e^+ + \bar{\nu}_\mu + \nu_e + \nu_\mu \\ & \pi^- \to \mu^- + \bar{\nu}_\mu \to e^- + \bar{\nu}_e + \nu_\mu + \bar{\nu}_\mu . \end{aligned} \tag{3}$$

In the above reaction chains (2) and (3), gamma-ray photons are emitted through the neutral pions decay. Furthermore, through the subsequent decay of the secondary muons which create electrons e^- (or positrons e^+), neutrinos or anti-neutrinos are produced too. By implementing the p-γ mechanism, the above reactions constitute the main processes feeding the neutrino and gamma-ray production channels in the lepto-hadronic model employed [18,19].

In general, the radiation density that interacts with the non-thermal (relativistic) protons, is due to mainly two contributing factors. The first is the synchrotron emission, by accelerated charged particles, that create a photon distribution, n_{phS} (produced by relativistic electrons as well as protons). It is given by [18,19]:

$$n_{phS}(\epsilon, z) \approx \frac{\varepsilon_{syn} r(z)}{\epsilon c}, \tag{4}$$

where ε_{syn} corresponds to the total power radiated by electron and proton distributions and is given in Appendix A.1. The second contributing factor is an X-ray distribution, n_{phX} (for 2 keV < ϵ < 100 keV), originated from the corona that surrounds the inner accretion disc. For the latter distribution, we have [19,36]:

$$n_{phX}(\epsilon, z) = \frac{L_X e^{-\epsilon/kT_e}}{4\pi c z^2 \epsilon^2}, \tag{5}$$

where X-ray luminosity is $L_X = 10^{36}$ ergs^{-1} and $kT_e \simeq 30$ keV.

Furthermore, within our approximation it holds $n_{ph} = n_{phX} + n_{phS}$. It should be noted that, the primary particles (protons, electrons) as well as the secondary particles take part also in energy loss interactions and processes. These include the energy losses due to jet adiabatic expansion, the losses due to inelastic collisions with the cold protons and losses due to the emission of synchrotron radiation [30,31,37].

Concerning the fast (non-thermal or relativistic) proton distribution, its shape in the one-zone approximation, resembles to a power-law type (defined by a normalization constant K_0) [38] assumed in our model (see Equation (1)).

2.2. *Solution of the Transfer Equation*

In the lepto-hadronic model employed in the present work, the acceleration mechanism (it is not included in the transport equation) is used to fix the injection function of the primary electrons and protons. Then, it determines the maximum energies that can be acquired by the relativistic particles inside the jet. To this aim, the particle transfer (transport) equation must be solved which, assuming a steady-state model, is written as [9–11]:

$$\frac{\partial N(E,z) b(E,z)}{\partial E} + t^{-1} N(E,z) = Q(E,z)\,. \tag{6}$$

$N(E,z)$ represents the particle density per unit of energy (cm^{-3}GeV^{-1}) and $Q(E,z)$ denotes the particle source function (in units of cm^{-3}GeV^{-1}s^{-1}). Obviously, this is not a spherical but an axisymmetric model, hence the particle distributions and injection functions depend on z (i.e., the distance to the central object on the ejection-axis of the jet). In the latter equation, $b(E)$ stands for the total energy loss rate given by:

$$b(E) = \frac{dE}{dt} = -E t_{loss}^{-1}, \tag{7}$$

while t^{-1} represents the particles' reduction rate as:

$$t^{-1} = t_{esc}^{-1} + t_{dec}^{-1}\,. \tag{8}$$

Here, t_{dec}^{-1} is the decay rate (in the case of pions and muons) and t_{esc}^{-1} the escape rate of the particles from the jet's region. The latter rate is given by:

$$t_{esc}^{-1} = \frac{c}{z_{max} - z_0}, \tag{9}$$

with $(z_{max} - z_0)$ being the length of the acceleration zone inside the jet.

The solution of the differential Equation (6) is written as:

$$N(E,z) = \frac{1}{\mid b(E) \mid} \int_{E}^{E_{max}} Q(E', z) e^{-\tau(E,E')} dE', \tag{10}$$

where the maximum proton energy is approximated by $E_p^{max} \simeq 10^7$ GeV, while:

$$\tau(E,E') = \int_E^{E'} \frac{dE'' t^{-1}}{\mid b(E'') \mid} \,. \tag{11}$$

The source function $Q(E,z)$, for the relativistic particles (in the jet's rest frame) is given by [19]:

$$Q(E',z) = Q_0 \left(\frac{z_0}{z}\right)^3 E'^{-2} \,. \tag{12}$$

In the observer's reference frame, $Q(E,z)$ takes the form given in the Appendix (see Equation (A8) of Appendix A.2) where the normalization constant Q_0 of Equation (12) is also given in the Appendix A.

3. Interaction Frequency and Particle Emission through p-γ Mechanism

The injection function for the pions produced through the p-γ interaction mechanism is the following:

$$Q_\pi^{(p\gamma)}(E,z) = 5N_p(5E,z)\omega_{p\gamma}(5E,z)\bar{N}_\pi^{(p\gamma)}(5E) \,, \tag{13}$$

where $N_p(E,z)$ denotes the proton energy distribution. Furthermore, $\omega_{p\gamma}$ is the p-γ collision frequency that results in pion production. The mean number $\bar{N}_\pi^{(p\gamma)}$ of positive or negative pions produced per p-γ collision is given by:

$$\bar{N}_\pi^{(p\gamma)} = p_{p\to n} p_1 + 2p_2 \,, \tag{14}$$

where the parameter $p_{p\to n} \simeq 0.5$ is to express the probability of conversion of a primary proton to a neutron. Furthermore, p_1 and p_2 are defined as:

$$p_1 = \frac{K_2 - \bar{K}_{p\gamma}}{K_2 - K_1} \,, \tag{15}$$

and $p_2 = 1 - p_1$. Moreover, we have $K_1 = 0.2$ and $K_2 = 0.6$. In the latter equation, for the mean inelasticity parameter $\bar{K}_{p\gamma}$ it holds:

$$\bar{K}_{p\gamma} = \frac{t_{p\gamma}^{-1}}{\omega_{p\gamma}} \,. \tag{16}$$

The p-γ collision frequency $\omega_{p\gamma}$ in Equations (13) and (16) is given by [19,39]:

$$\omega_{p\gamma}(E_p,z) = \frac{c}{2\gamma_p^2} \int_{\epsilon_{th}/2\gamma_p}^{\infty} \frac{n_{ph}(\epsilon,z)}{\epsilon^2} d\epsilon \int_{\epsilon_{th}}^{2\epsilon\gamma_p} \sigma_{p\gamma}^{(\pi)}(\epsilon')\epsilon' d\epsilon' \,, \tag{17}$$

where $\gamma_p = E_p/m_p c^2$, $n_{ph}(\epsilon,z)$ is the radiation field density that was discussed in Section 2. The threshold energy ϵ_{th} is assumed to be $\epsilon_{th} = 0.15$ GeV [39].

After the above definitions, the respective cross-section $\sigma_{p\gamma}^{(\pi)}$ is given by [39,40]:

$$\sigma_{p\gamma}^{(\pi)} = [3.4\Theta(\epsilon' - 0.2)\Theta(0.5 - \epsilon') + 1.2\Theta(\epsilon' - 0.5)] \times 10^{-28} cm^2 \,, \tag{18}$$

where $\Theta(x)$ denotes the well-known step function. In the latter relationships, the proton-photon collision rate $t_{p\gamma}^{-1}$ was obtained from the expression [39]:

$$t_{p\gamma}^{-1} = \frac{c}{2\gamma_p^2} \int_{\epsilon_{th}/2\gamma_p}^{\infty} \frac{n_{ph}(\epsilon,z)}{\epsilon^2} d\epsilon \int_{\epsilon_{th}}^{2\epsilon\gamma_p} \sigma_{p\gamma}^{(\pi)}(\epsilon') K_{p\gamma}^{(\pi)}(\epsilon')\epsilon' d\epsilon' \,. \tag{19}$$

In the above expression, $K^{(\pi)}_{p\gamma}(\epsilon')$ is the respective inelasticity function [39,40] which is equal to:

$$K^{(\pi)}_{p\gamma}(\epsilon') = 0.2\Theta(\epsilon' - 0.2)\Theta(0.5 - \epsilon') + 0.6\Theta(\epsilon' - 0.5)\,. \tag{20}$$

4. Results and Discussion

In the context of the model chosen in this study, at first the p-γ collision rate $t^{-1}_{p\gamma}$ of Equation (19) was calculated for the X-ray binary systems Cygnus X-1 and LMC X-1. The geometric properties and other parameters of these systems are listed in Table 1. The results obtained for the $t^{-1}_{p\gamma}$ in the latter microquasars are presented in Figure 1. In the upper two sub-figures, we illustrate the interaction rate as a function of the proton energy E for three different distances of the studied point up to the system's central object (z). The range of z covers the length of the acceleration zone (from z_0 to $z_{max} = 5z_0$). It is supported by theoretical calculations that, for greater distances, the p-γ collision rate decreases. The reason is the dependence of X-ray photon density (see Equation (5)) as well as the fast particle density on z. It is reasonable that as the jet expands (e.g., its radius increases), the mean-free path of the particles and photons involved in the collisions increases as well. That leads to reduced collision rates.

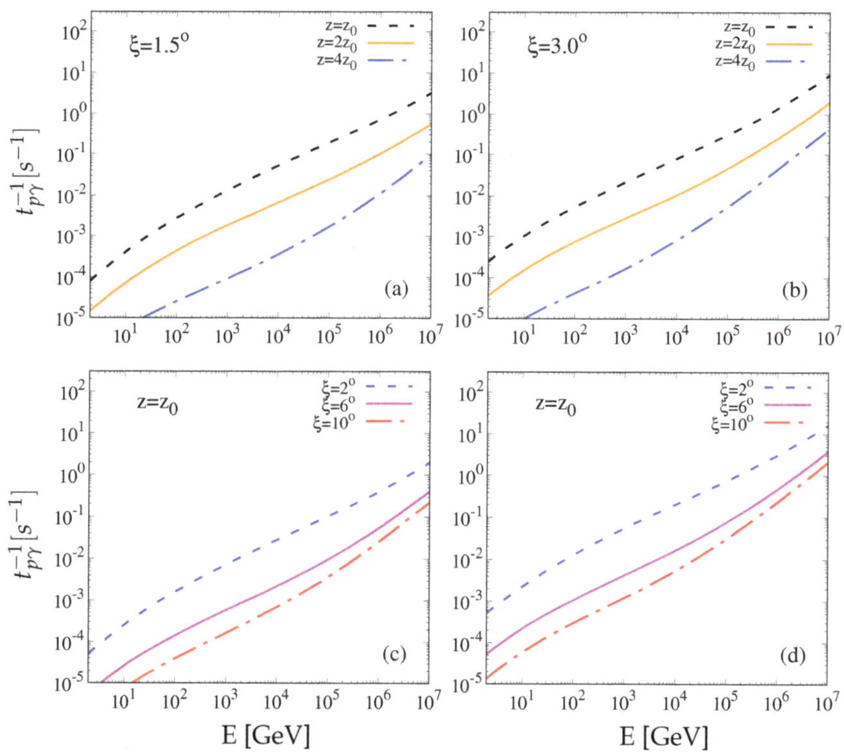

Figure 1. Proton-photon (p-γ) interaction rate $t^{-1}_{p\gamma}$ as a function of proton energy E for the binary systems Cygnus X-1 (left column) and LMC X-1 (right column). In the upper two sub-figures (**a**,**b**), we show the p-γ interaction rate for three different values of the distance z (in factors of z_0) inside the jet from the compact object. In the lower two sub-figures (**c**,**d**), we plot the $t^{-1}_{p\gamma}$ for three different values of the jet's half-opening angle ξ.

Table 1. Model parameters describing geometric characteristics of the Galactic Cygnus X-1 binary system and the extragalactic LMC X-1 located at the Large Magellanic Cloud.

Description	Parameter	Cygnus X-1	LMC X-1
Jet's base	z_0 (cm)	$191R_{Sch}$	$95R_{Sch}$
Acceleration zone limit	z_{max} (cm)	$956R_{Sch}$	$477R_{Sch}$
Mass of compact object	M_{BH}	$14.8M_{\odot}$ [41]	$10.91M_{\odot}$ [42]
Angle to the line-of-sight	θ (°)	27.1 [41]	36.38 [42]
Jet's half-opening angle	ξ (°)	1.5 [43]	3 *
Jet's bulk velocity	v_b	0.6c [43]	0.92c *

* Indicative values that we consider for our calculations.

In the lower two sub-figures of Figure 1, we illustrate the p-γ collision rate for three different values of the half-opening angle ξ of the jet. This angle strongly depends on the magnetic field strength and characterizes the jet's collimation. The values of ξ considered are representative and cover the assumed range extended up to 10°. We notice that the collision rate decreases as the jet expands perpendicularly to its ejection axis (i.e., for greater angles). This is validated by the corresponding theoretical expressions if we consider the decrease of the synchrotron emission which is due to the magnetic field that controls the jet's collimation. In addition, the mean-free path of the proton-photon collisions increases causing the reduction of the respective rate.

We have, also, calculated the pion energy distributions for different typical values of the angle (θ) between the jet axis and the direction of the line of sight as well as the bulk velocity of the jet's matter (u_b) for the Cygnus X-1 binary system, see Figure 2. It is evident that the particle production is higher for smaller values of θ as demonstrated in Equation (A8) in the Appendix A.2. Furthermore, the particle production increases for greater bulk velocities u_b (i.e., it is u_b=$\beta_j c$ where c corresponds to the speed of light constant) as shown in sub-figure (b) of Figure 2. This result is justified by considering the relation of bulk velocity to the jet's collimation and the proton-photon interaction rate which increases for larger energies. It is hoped that, our present results would be helpful for future relevant experimental and observational measurements.

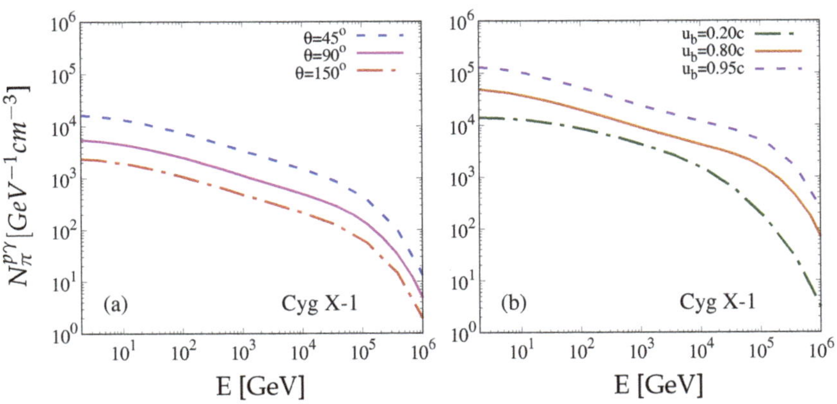

Figure 2. For Cygnus X-1, we present pion distributions for different values of the system's angle to the line of sight (**a**) and bulk velocity of the jet's ejected matter (**b**). The pions in this case have not been subjected to energy loss mechanisms.

In this work, we, furthermore study the pion distributions produced by p-γ interactions for three different values of the parameter α defined as:

$$\alpha = \frac{L_h}{L_e}. \tag{21}$$

L_h and L_e denote the hadronic and leptonic luminosity, respectively. These values correspond to: (i) a leptonic model ($\alpha = 0.001$), (ii) a hadronic model ($\alpha = 100$) and (iii) an extreme-hadronic model ($\alpha = 1000$). The results are shown in Figure 3. Our main purpose, here, is the comparison between the particle (pions) production in these well-implemented models (leptonic and hadronic). For that reason, we have not considered the possible energy losses that the aforementioned particles are subjected through the various mechanisms [30,31]. In Figure 3, we notice the reasonable decrease in pion production in the leptonic case as the protons are reduced compared to electrons inside the jet. However, as can be seen, there is no essential difference between the hadronic and the extreme-hadronic model, therefore, a ratio of $\alpha \simeq 100$ is deemed sufficient enough.

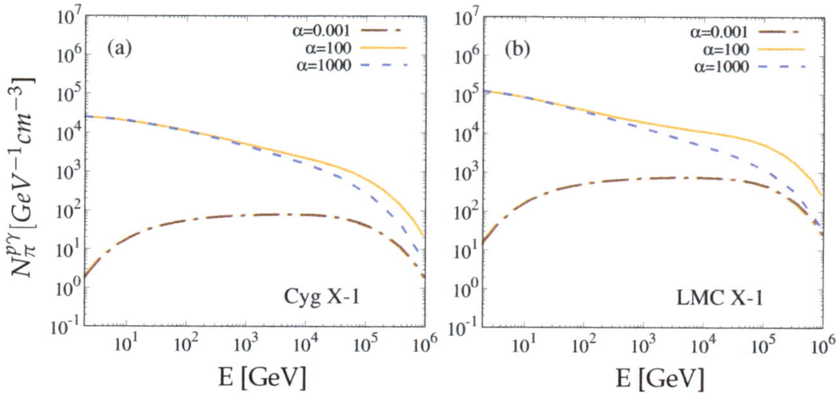

Figure 3. Pion energy distributions from p-γ collisions in the jets of binary systems Cygnus X-1 (**a**) and LMC X-1 (**b**) for three different values of hadron-to-lepton ratio α.

5. Summary and Conclusions

Astrophysical binary systems, consisted of a high-mass compact object (black hole or neutron star) that absorbs mass out of a companion star (forming a rotating accretion disc) are studied extensively in recent years. More specifically, their magnetohydrodynamical astrophysical flow ejection and high-energy radiation as well as particle emissions have been the research subject of many authors working in this field. In our present work, we go beyond the spherically symmetric models of neutrino and gamma-ray emissions and adopt conical jets, i.e., axially symmetric plasma outflows from microquasar jets.

We, furthermore, assume that strong shock-waves accelerate a portion of the jets' charged particles (mostly protons but also electrons) to rather relativistic energies obeying a power-law energy-dependent distribution. Under the above circumstances, these particles interact with thermal protons of the jet (or of stellar winds) or with the radiation fields originating from energy-exchanging mechanisms (i.e., between electrons and low-energy photons) inside or outside the jet's region. The outcomes of these interactions are secondary particles and photons such as pions, muons, neutrinos, gamma-ray photons, etc. The multi-messenger signals created this way are detectable by the terrestrial extremely sensitive detectors like the IceCube, ANTARES, KM3NeT, etc.

In our study, we mainly focus on the p-γ interaction mechanism (which was not taken explicitly into account in our previous works) and the parameters (e.g., geometric characteristics of the binary systems, the jet's matter composition of leptons or hadrons, etc.) which affect the associated emissions of particle and radiation. In particular, we have selected two concrete examples, the Galactic Cygnus X-1 and the extragalactic LMC X-1 binary systems, to study with the model chosen and present numerical calculations for the collision rates and energy distributions of particles that come out (pions). Our results show that the particle density production strongly increases in more collimated flow ejections which is highly dependent on the prevailing magnetic field. In addition, we find that the smaller the distance z from the central region the larger the production of particles and radiation is emitted.

Author Contributions: Conceptualization, O.K. and I.S.; methodology, O.K.; software, T.P.; validation, O.K. and I.S.; formal analysis, T.P.; investigation, T.P. and O.K.; writing—original draft preparation, T.P. and O.K.; writing—review and editing, O.K. and I.S.; visualization, T.P. and O.K.; supervision, O.K. and I.S.; project administration, O.K. and I.S. All authors have read and agreed to the published version of the manuscript.

Funding: This research was funded in the context of the PhD dissertation of T.P. by the Department of Informatics, School of Sciences of University of Western Macedonia.

Institutional Review Board Statement: Not applicable.

Informed Consent Statement: Not applicable.

Data Availability Statement: There is no data used in this paper.

Conflicts of Interest: The authors declare no conflict of interest.

Appendix A

Appendix A.1. Synchrotron Power Radiation by Particle Distributions

When a charged particle of energy E is being accelerated by a magnetic field with pitch angle a, it emits synchrotron radiation. The power of the radiation per units of the emitted photons' energy and pitch angle a is given by [37]

$$P_{syn}(\epsilon, E, z, \alpha) = \frac{\sqrt{3}e^3 B(z) sin(a)}{hmc^2} \frac{\epsilon}{E_{cr}} \int_{\frac{\epsilon}{E_{cr}}}^{\infty} K_{5/3}(\zeta) d\zeta \,, \tag{A1}$$

where $B(z)$ is the binary system's magnetic field responsible for the jet's collimation. e is the electric charge, h denotes the Planck constant and E_{cr} corresponds to the critical frequency of the emitted radiation which is written as

$$E_{cr} = \frac{3heB(z)sin(a)}{4\pi mc} \gamma^2 \,. \tag{A2}$$

In Equation (A1), we integrate over the modified Bessel function of order 5/3 which we calculate through the following relationships

$$K_{\frac{1}{3}}(\zeta) = \sqrt{3} \int_0^{\infty} exp\left[-\zeta \left(1 + \frac{4x^2}{3} \right) \sqrt{1 + \frac{x^2}{3}} \right] dx \tag{A3}$$

$$K_{\frac{2}{3}}(\zeta) = \frac{1}{\sqrt{3}} \int_0^{\infty} \frac{3 + 2x^2}{\sqrt{1 + \frac{x^2}{3}}} exp\left[-\zeta \left(1 + \frac{4x^2}{3} \right) \sqrt{1 + \frac{x^2}{3}} \right] dx \tag{A4}$$

$$K_{\frac{5}{3}}(\zeta) = K_{\frac{1}{3}}(\zeta) + \frac{4}{3\zeta} K_{\frac{2}{3}}(\zeta) \,. \tag{A5}$$

After calculating P_{syn}, we integrate over the pitch angle a as well as the particle energy distribution in order to obtain the total power radiated per unit energy by a particle (electron or proton) distribution such as those we discussed before. The result for electron or proton distributions is given below

$$\varepsilon_{syn}^{(e,p)}(\epsilon) = \int d\Omega_{\alpha} \int_{E_{e,p}^{(min)}}^{E_{e,p}^{(max)}} P_{syn} N_{e,p}(E, z) dE \,, \tag{A6}$$

where $E_p^{min} = 1.2$ GeV and $E_e^{min} = 0.001$ GeV are the minimum proton and electron energies, respectively. For the maximum energies, we have $E_p^{max} \simeq 10^7$ GeV and $E_e^{max} \simeq 7$ GeV.

The total power radiated by both electrons and protons is given by

$$\varepsilon_{syn}(\epsilon) = \varepsilon_{syn}^{(e)}(\epsilon) + \varepsilon_{syn}^{(p)}(\epsilon). \tag{A7}$$

Appendix A.2. Injection Function in Observer's Reference Frame

The transformation of the proton injection function to the observer's reference frame is given by [10,11,44]

$$Q(E,z) = Q_0\left(\frac{z_0}{z}\right)^3 \frac{\Gamma_b^{-1}(E - \beta_b cos\theta\sqrt{E^2 - m^2c^4})^{-2}}{\sqrt{sin^2\theta + \Gamma_b^2\left(cos\theta - \frac{\beta_b E}{\sqrt{E^2 - m^2c^4}}\right)^2}}. \tag{A8}$$

Here, Γ_b is the Lorentz factor responding to the jet's bulk velocity ($u_b = \beta_b c$ and $\Gamma_b = (1 - \beta_b^2)^{-\frac{1}{2}}$) and Q_0 is given by

$$Q_0 = \frac{8 q_r L_k}{z_0 r_0^2 ln(E_p^{max}/E_p^{min})}, \tag{A9}$$

where r_0 is the jet radius that corresponds to the distance z_0 from the central object. Moreover, the kinetic luminosity that is transferred in the jet L_k is considered to be 10% of the central object's Eddington luminosity [8]. We, also, adopt the value $q_r = 0.1$ for the portion of relativistic protons and electrons inside the jet.

References

1. Heinz, S.; Sunyaev, R. Cosmic rays from microquasars: A narrow component to the CR spectrum? *Astron. Astrophys.* **2002**, *390*, 751–766. [CrossRef]
2. Fabrika, S. The jets and supercritical accretion disk in SS433. *Astrophys. Space Phys. Rev.* **2004**, *12*, 1–152.
3. Aharonian, F.; Akhperjanian, A.G.; Aye, K.M.; Bazer-Bachi, A.R.; Beilicke, M.; Benbow, W.; Berge, D.; Berghaus, P.; Bernlöhr, K.; Boisson, C.; et al. Discovery of Very High Energy Gamma Rays Associated with an X-ray Binary. *Science* **2005**, *309*, 746–749. [CrossRef]
4. Albert, J.; Aliu, E.; Anderhub, H.; Antoranz, P.; Armada, A.; Asensio, M.; Baixeras, C.; Barrio, J.A.; Bartelt, M.; Bartko, H.; et al. Variable Very-High-Energy Gamma-Ray Emission from the Microquasar LS I +61 303. *Science* **2006**, *312*, 1771–1773. [CrossRef] [PubMed]
5. Albert, J.; Aliu, E.; Anderhub, H.; Antoranz, P.; Armada, A.; Baixeras, C.; Barrio, J.A.; Bartko, H.; Bastieri, D.; Becker, J.K.; et al. Very High Energy Gamma-Ray Radiation from the Stellar Mass Black Hole Binary Cygnus X–1. *Astrophys. J.* **2007**, *665*, L51–L54. [CrossRef]
6. Falcke, H.; Biermann, P. The jet-disk symbiosis. 1. Radio to X-ray emission models for quasars. *Astron. Astrophys.* **1994**, *293*, 665–682.
7. Papavasileiou, T.V.; Kosmas, O.T.; Sinatkas, J. Simulations of Neutrino and Gamma-Ray Production from Relativistic Black-Hole Microquasar Jets. *Galaxies* **2021**, *9*, 67. [CrossRef]
8. Körding, E.G.; Fender, R.P.; Migliari, S. Jet-dominated advective systems: Radio and X-ray luminosity dependence on the accretion rate. *Mon. Not. R. Astron. Soc.* **2006**, *369*, 1451–1458. [CrossRef]
9. Smponias, T.; Kosmas, O. High Energy Neutrino Emission from Astrophysical Jets in the Galaxy. *Adv. High Energy Phys.* **2015**, 921757. [CrossRef]
10. Smponias, T.; Kosmas, O. Neutrino Emission from Magnetized Microquasar Jets. *Adv. High Energy Phys.* **2017**, *2017*, 496274. [CrossRef]
11. Kosmas, O.T.; Smponias, T. Simulations of Gamma-Ray Emission from Magnetized Microquasar Jets. *Adv. High Energy Phys.* **2018**, *2018*, 960296. [CrossRef]
12. Aartsen, M.; Ackermann, M.; Adams, J.; Aguilar, J.A.; Ahlers, M.; Ahrens, M.; Al Samarai, I.; Altmann, D.; Andeen, K.; Anderson, T.; et al. Neutrino emission from the direction of the blazar TXS 0506+056 prior to the IceCube-170922A alert. *Science* **2018**, *361*, 147–151.
13. Romero, G.E.; Boettcher, M.; Markoff, S.; Tavecchio, F. Relativistic Jets in Active Galactic Nuclei and Microquasars. *Space Sci. Rev.* **2017**, *207*, 5–61. [CrossRef]
14. Vieyro, F.L.; Romero, G.E. Particle transport in magnetized media around black holes and associated radiation. *Astron. Astrophys.* **2012**, *542*, A7. [CrossRef]
15. Romero, G.E.; Vila, G.S. The proton low-mass microquasar: High-energy emission. *Astron. Astrophys.* **2008**, *485*, 623–631. [CrossRef]

16. Mirabel, I.F.; Rodríguez, L.F. Sources of Relativistic Jets in the Galaxy. *Annu. Rev. Astron. Astrophys.* **1999**, *37*, 409–443. [CrossRef]
17. Beall, J.H. A Review of Astrophysical Jets. *PoS Proc. Sci.* **2015**, *246*, 58–70. [CrossRef]
18. Reynoso, M.M.; Romero, G.E.; Christiansen, H.R. Production of gamma rays and neutrinos in the dark jets of the microquasar SS433. *Mon. Not. R. Astron. Soc.* **2008**, *387*, 1745–1754. [CrossRef]
19. Reynoso, M.M.; Romero, G.E. Magnetic field effects on neutrino production in microquasars. *Astron. Astrophys.* **2009**, *493*, 1–11. [CrossRef]
20. Bosch-Ramon, V.; Romero, G.E.; Paredes, J.M. A broadband leptonic model for gamma-ray emitting microquasars. *Astron. Astrophys.* **2006**, *447*, 263–276. [CrossRef]
21. Begelman, M.C.; Rudak, B.; Sikora, M. Consequences of Relativistic Proton Injection in Active Galactic Nuclei. *Astrophys. J.* **1990**, *362*, 38. [CrossRef]
22. Berezhko, E.G.; Ellison, D.C. A Simple Model of Nonlinear Diffusive Shock Acceleration. *Astrophys. J.* **1999**, *526*, 385–399. [CrossRef]
23. Gallant, Y.A.; Achterberg, A. Ultra-high-energy cosmic ray acceleration by relativistic blast waves. *Mon. Not. R. Astron. Soc.* **1999**, *305*, L6–L10. [CrossRef]
24. Kirk, J.G.; Guthmann, A.W.; Gallant, Y.A.; Achterberg, A. Particle Acceleration at Ultrarelativistic Shocks: An Eigenfunction Method. *Astrophys. J.* **2000**, *542*, 235–242. [CrossRef]
25. Achterberg, A.; Gallant, Y.A.; Kirk, J.G.; Guthmann, A.W. Particle acceleration by ultrarelativistic shocks: Theory and simulations. *Mon. Not. R. Astron. Soc.* **2001**, *328*, 393–408. [CrossRef]
26. Romero, G.E.; Torres, D.F.; Kaufman Bernadó, M.M.; Mirabel, I.F. Hadronic gamma-ray emission from windy microquasars. *Astron. Astrophys.* **2003**, *410*, L1–L4. [CrossRef]
27. Romero, G.E.; Okazaki, A.T.; Orellana, M.; Owocki, S.P. Accretion vs. colliding wind models for the gamma-ray binary LS I +61 303: An assessment. *Astron. Astrophys.* **2007**, *474*, 15–22. [CrossRef]
28. Friend, D.B.; Castor, J.I. Radiation driven winds in X-ray binaries. *Astrophys. J.* **1982**, *261*, 293–300. [CrossRef]
29. Kelner, S.R.; Aharonian, F.A.; Bugayov, V.V. Energy spectra of gamma rays, electrons, and neutrinos produced at proton-proton interactions in the very high energy regime. *Phys. Rev. D* **2006**, *74*, 034018. [CrossRef]
30. Papavasileiou, T.V.; Papadopoulos, D.A.; Kosmas, T.S. Astrophysical magnetohydrodynamical outflows in the extragalactic binary system LMC X-1. *J. Phys. Conf. Ser.* **2020**, *1730*, 012138. [CrossRef]
31. Papadopoulos, D.A.; Papavasileiou, T.V.; Kosmas, T.S. High energy neutrino and gamma-ray emissions from the jets of M33 X-7 microquasar. *J. Phys. Conf. Ser.* **2021**, *1730*, 012137. [CrossRef]
32. Ahnen, M.L.; Ansoldi, S.; Antonelli, L.A.; Arcaro, C.; Babić, A.; Banerjee, B.; Bangale, P.; de Almeida, U.B.; Barrio, J.A.; González, J.B.; et al. Search for very high-energy gamma-ray emission from the microquasar Cygnus X-1 with the MAGIC telescopes. *Mon. Not. R. Astron. Soc.* **2017**, *472*, 3474–3485.
33. Hyde, E.A.; Russell, D.M.; Ritter, A.; Filipovic, M.D.; Kaper, L.; Grieve, K.; O'Brien, A.N. LMC X-1: A New Spectral Analysis of the O-star in the Binary and Surrounding Nebula. *Astron. Soc. Pac.* **2017**, *129*, 094201. [CrossRef]
34. Aartsen, M.G.; Abraham, K.; Ackermann, M.; Adams, J.; Aguilar, J.A.; Ahlers, M.; Ahrens, M.; Altmann, D.; Andeen, K.; Anderson, T.; et al. Searches for Sterile Neutrinos with the IceCube Detector. *Phys. Rev. Lett.* **2016**, *117*, 071801. [CrossRef] [PubMed]
35. Adrián-Martínez, S.; Ageron, M.; Aharonian, F.; Aiello, S.; Albert, A.; Ameli, F.; Anassontzis, E.; Andre, M.; Androulakis, G.; Anghinolfi, M.; et al. Letter of intent for KM3NeT 2.0. *J. Phys. G Nucl. Part Phys.* **2016**, *43*, 084001. [CrossRef]
36. Cherepashchuk, A.M.; Sunyaev, R.A.; Fabrika, S.N.; Postnov, K.A.; Molkov, S.V.; Barsukova, E.A.; Antokhina, E.A.; Irsmambetova, T.R.; Panchenko, I.E.; Seifina, E.V.; et al. INTEGRAL observations of SS433: Results of a coordinated campaign. *Astron. Astrophys.* **2005**, *437*, 561–573. [CrossRef]
37. Blumenthal, G.R.; Gould, R.J. Bremsstrahlung, Synchrotron Radiation, and Compton Scattering of High-Energy Electrons Traversing Dilute Gases. *Rev. Mod. Phys.* **1970**, *42*, 237–271. [CrossRef]
38. Khangulyan, D.; Hnatic, S.; Aharonian, F.; Bogovalov, S. TeV light curve of PSR B1259–63/SS2883. *Mon. Not. R. Astron. Soc.* **2007**, *380*, 320–330. [CrossRef]
39. Atoyan, A.M.; Dermer, C.D. Neutral Beams from Blazar Jets. *Astrophys. J.* **2003**, *586*, 79. [CrossRef]
40. Kelner, S.R.; Aharonian, F.A. Energy spectra of gamma rays, electrons, and neutrinos produced at interactions of relativistic protons with low energy radiation *Phys. Rev. D* **2008**, *78*, 034013. [CrossRef]
41. Orosz, J.A.; McClintock, J.E.; Aufdenberg, J.P.; Remillard, R.A.; Reid, M.J.; Narayan, R.; Gou, L. The mass of the black hole in cygnus X-1. *Astrophys. J.* **2011**, *742*, 84. [CrossRef]
42. Orosz, J.A.; Steeghs, D.; McClintock, J.E.; Torres, M.A.P.; Bochkov, I.; Gou, L.; Narayan, R.; Blaschak, M.; Levine, A.M.; Remillard, R.A.; et al. A new dynamical model for the black hole binary lmc X-1. *Astrophys. J.* **2009**, *697*, 573–591. [CrossRef]
43. Stirling, A.; Spencer, R.; de La Force, C.; Garrett, M.; Fender, R.; Ogley, R. A relativistic jet from Cygnus X-1 in the low/hard X-ray state. *Mon. Not. R. Astron. Soc.* **2001**, *327*, 1273–1278. [CrossRef]
44. Torres, D.F.; Reimer, A. Hadronic beam models for quasars and microquasars. *Astron. Astrophys.* **2011**, *528*, L2. [CrossRef]

symmetry

MDPI

Article

Mapping Topology of Skyrmions and Fractional Quantum Hall Droplets to Nuclear EFT for Ultra-Dense Baryonic Matter

Mannque Rho

Institut de Physique Théorique, Université Paris-Saclay, CNRS, CEA, CEDEX, 91191 Gif-sur-Yvette, France; mannque.rho@ipht.fr

Abstract: We describe the mapping at high density of topological structure of baryonic matter to a nuclear effective field theory that implements hidden symmetries emergent from strong nuclear correlations. The theory constructed is found to be consistent with no conflicts with the presently available observations in both normal nuclear matter and compact-star matter. The hidden symmetries involved are "local flavor symmetry" of the vector mesons identified to be (Seiberg-)dual to the gluons of QCD and hidden "quantum scale symmetry" with an IR fixed point with a "genuine dilaton (GD)" characterized by non-vanishing pion and dilaton decay constants. Both the skyrmion topology for $N_f \geq 2$ baryons and the fractional quantum Hall (FQH) droplet topology for $N_f = 1$ baryons are unified in the "homogeneous/hidden" Wess–Zumino term in the hidden local symmetry (HLS) Lagrangian. The possible indispensable role of the FQH droplets in going beyond the density regime of compact stars approaching scale-chiral restoration is explored by moving toward the limit where both the dilaton and the pion go massless.

Keywords: topology for baryons; hidden symmetries; ultra dense matter; compact stars

Citation: Rho, M. Mapping Topology of Skyrmions and Fractional Quantum Hall Droplets to Nuclear EFT for Ultra-Dense Baryonic Matter. *Symmetry* **2022**, *14*, 994. https://doi.org/10.3390/sym14050994

Academic Editors: Kazuharu Bamba and Sergei D. Odintsov

Received: 17 March 2022
Accepted: 30 April 2022
Published: 12 May 2022

1. Introduction

1.1. The Problem

In nuclear processes going from low density near-normal nuclear matter to high density relevant to massive compact stars, two or possibly more density regimes are involved. They are most likely delineated by changes of degrees of freedom (DoF). At low densities up to slightly above the equilibrium nuclear matter density ($n_0 \simeq 0.16$ fm^{-3}), say, to $n \lesssim 2n_0$, the relevant degrees of freedom are the nucleons N and pions π figuring in what is referred to as "standard chiral effective field theory" (given the acronym *s*ChEFT) with a cut-off scale set, typically, at $\Lambda_{sChEFT} \sim (400 - 500)$ MeV. When treated to N^{m}LO for $m = (3 - 4)$ in systematic chiral power expansion, ab initio calculations in *s*ChEFT have been seen to work highly satisfactorily for nuclear structure in finite nuclei, as well as for properties of nuclear matter. This impressive success in nuclear physics could be taken as a convincing proof of Weinberg's "Folk Theorem for EFT" as applied to QCD [1]. This *s*ChEFT is expected to be extendable to ∼2 times n_0, but it is to break down by the premise of EFT at a high density as the relevant degrees of freedom are no longer just the nucleons and pions but more massive hadronic degrees of freedom, and ultimately the QCD degrees of freedom, i.e., quarks and gluons, must figure as density increases. Thus, there must be one or more changes of DoFs from hadronic to quarkonic.

Due to the paucity of trustful theoretical tools for guidance in the absence of lattice approach to QCD in dense medium—in contrast to thermal matter—there is no clear indication how many and in what form(s) these changeovers of DoFs could take place, in the density regime relevant to the center of massive compact stars, $\sim(5 - 7)n_0$. This presents a totally uncharted domain that could very well encompass several different fields, say, condensed matter, nuclear, and particle, in addition to astrophysics.

There are several treatments in the literature that invoke quarks in various different forms, perturbative or nonperturbative, to explore this uncharted domain. We describe a possible strategy that exploits the topological structure of baryonic matter—without explicit QCD variables—as density increases beyond $\sim 2n_0$ to access the putative baryon-quark continuity. In this approach, no explicit quark-gluon DoFs are invoked, but fractional (baryon-charged) quasiparticles induced from topology change(s) seem to figure.

Among various topological structures of baryons, we adopt the skyrmion topology for the number of flavor $N_f \geq 2$ and the fractional quantum Hall (FQH) topology for $N_f = 1$. The former is intricately relevant to QCD—i.e., large N_c limit—in nuclear physics at low density involving the u(p) and d(own) quarks and the latter will be relevant at a high density. The s(trange) quark does enter essentially in the formulation both in the way the scalar dilaton and η' figure, but we will focus first on the $N_f = 2$ systems and then come to the case where η' could enter in FQH topology.

Ultimately this feat should be feasible in terms of skyrmions—ignoring FQH topology—with a Lagrangian modeling QCD with all relevant degrees of freedom and work out the full topological structure of baryons going beyond the Skyrme model [2] with pions only (denoted as skyrmion$_\pi$), but this is currently out of reach. There is an on-going effort with the skyrmion$_\pi$ at suitable chiral derivative orders [3], but it is mathematically daunting to do realistic calculations even with the truncated model to confront Nature. Our strategy is to map what are considered—though not firmly confirmed—as *robust* topological inputs, independent of the details of the Lagrangian, to an EFT going beyond *s*ChEFT—that we shall give the acronym G*n*EFT standing for "generalized nuclear EFT"—and treat it in terms of a Landau Fermi-liquid approach to baryonic matter which we put in the class of the density functional ("DF") à la Hohenberg–Kohn theorem.

This type of approach is highly unconventional and admittedly incomplete in various aspects. Nonetheless, it is found not only to successfully post-dict the overall properties of nuclear matter at density $\sim n_0$ in quality more or less comparable to high chiral-order (i.e., $N^{\gtrsim}3LO$) *s*ChEFT but also account for the properties of massive compact stars at $n \gg n_0$ in fair consistency with what has been established in astrophysical observations. There is, however, a surprising new prediction—which is extremely simple—that follows from the topology-change exploited in the approach, namely, the "pseudo-conformal" sound velocity of the star and its impact on the structure of the core of compact stars. Our claim is that it exposes certain emergent symmetries hidden in QCD at low density.

1.2. The Motivation

The basic idea figuring here was largely inspired and motivated by the predominant role topology plays in quantum critical phenomena in condensed matter systems, stunningly exemplified, among others, by the fractional quantized Hall (FQH) effects [4]. This idea is germinated by the observation that many-body interactions in strongly correlated condensed matter systems and in nuclear many-body systems, although the basic interactions are different, i.e., QED vs. QCD, intricately share certain common features. In the former, there have been remarkable breakthroughs by formulating correlated electron problems in terms of topology, thereby mapping many-electron interactions to topological field theories and also the other way around. The strategy we describe here is prompted to do something analogous to what has been done in the physics of the quantum Hall effects (QHEs), with certain arguments—and ideas—borrowed therefrom. This effort is currently largely unrecognized in the nuclear community. Of course they are necessarily of different nature given that we are dealing with strong interactions (QCD) with inherently more complex dynamics than in QED. Furthermore strong interactions have much less direct access to experiments than in condensed matter systems.

1.3. The Objective

Let us first state briefly what the problem is and what the objective of this article is.

The main stream of current activities in nuclear/astro-physics community motivated by what is heralded as "first-principles" approaches to nuclear physics *and* gravity-wave signals of merging neutron stars is anchored on ab initio treatments of the *s*ChEFT at higher chiral orders and its related density-functional-type theories *defined and valid* at the density regime $\lesssim 2n_0$. They are then extrapolated, resorting to various sophisticated "meta-modelings", relying on the Bayesian inference, high-order "uncertainty analysis", etc., to higher densities relevant for massive compact stars. In doing this, one confronts the inherent obstacles due to the paucity of trustful theoretical tools, given the inaccessibility of lattice QCD to high density.

The spirit of this article is drastically different from the majority of approaches pursued in the field.

The aim of the approach adopted is to construct as simple and economical an effective field theory as possible, implementing what are deemed to be necessary to meet the requirement for the "Folk Theorem for EFT" appropriate for normal as well as compact-star matter. The philosophy is then to see how far one can go forward with this extremely simple theory—never mind the nitty-gritty "error uncertainties"—before being hit by "torpedoes". That is, In the spirit of Farragut's famous uttering "Damn the torpedoes. Full speed ahead!"

In a nutshell, this article attempts to explain why the approach developed by us—which is admittedly (over)simplified and more intuitive than of rigor—seems to work surprisingly well for nuclear dynamics ranging from low density at $n \sim n_0$ – where the *s*ChEFT is *believed* to be applicable – to high density $n \sim (5-7)n_0$ where it is *suspected* to break down. The objective of this article is to see to what extent we can offer justifications and what need to be improved upon for the results obtained up to date.

Although the approach presented here predates the arrival of *s*ChEFT in the 1990s, the core idea was the principal theme of the five year "World Class University Project" (WCUP) established at Hanyang University in Seoul in 2007 funded by the Korean Government. This WCUP was in some sense in anticipation of the upcoming ambitious Institute of Basic Science (IBS) with the purpose to put Korean basic science on the world's frontier. We will base our discussion on what was initiated in 2007 and continued after the termination of the WCUP/Hanyang up to today.

The results that we will refer to are mostly available in the literature. Their up-to-date status will be discussed in an accompanying contribution by Yong-Liang Ma [5] to which we will refer for quantitative details. The development up to 2018 was summarized in [6], written in tribute to Gerry Brown who had made invaluable contributions to the development of theoretical nuclear physics in Korea. More recent developments, which make the story more exciting, are found in [7,8]. We must say that this development remains more or less unrecognized in the field. We hope that this note makes the basic idea involved better understood.

1.4. The Strategy

The best way to motivate the reliance on topology to go from nucleons and pions at low density to quarks (and gluons) at high density is to think of a possible parallel of the approach adopted to how the physics of quantum Hall effects is formulated in terms of a topological field theory. What we have in mind in particular is the mapping of the fractional quantum Hall effect (FQHE) given in Chern–Simons topological field theory to the Kohn–Sham density functional theory (DFT) [9,10]. The parallel is, of course, far from direct, given the totally different physics involved, but what figures in both the fractional quantum Hall effect and compact-star matter involves mapping between the *microscopic description*, DFT, and the *macroscopic description*, Chern–Simons field theory. The possible presence of such a parallel, although present from the very beginning of the WCU/Hanyang program, was only very recently recognized by us thanks to the on-going works of string-theory-oriented theorists. It is suggested that both the KS-DFT-type microscopic approach and the Chern–Simons field theory-type macroscopic approach figured conceptually in the development made at the WCU/Hanyang and since then.

Briefly, the parallel that we see is as follows.

In [9,10], the system of strongly interacting electrons in the FQHE regime is formulated in terms of composite fermions of electrons bound with even number of quantum vortices involving an $U(1)$ gauge field *emergent* from strong correlations of the electrons. The complex effects of many-electron interactions are cast in a single-particle formalism in Kohn–Sham (KS) density functional theory incorporating the emergent $U(1)$ gauge interactions between *weakly interacting* composite fermions (CFs), i.e., quasiparticles, induced by the quantum mechanical vortices. The gist of the approach then is that *the topology of Chern–Simons field theory is translated into the effective field theory, DFT*. What is remarkable in this approach is that their DFT does "faithfully capture the topological characteristics" of the FQHE.

The approach that we will follow is inspired from the analogy of accessing the strongly correlated strong interactions in the density regime $n \gtrsim (2-4)n_0$—where *s*ChEFT is presumably broken down—to the mapping of the FQHE to the problem of nearly non-interacting composite electrons in the KS-DFT subject to an emerging "magnetic field". This analogy is not totally unfamiliar in its generic form in nuclear dynamics where the effective field theories of QCD, e.g., *s*ChEFT at low density, are extended to higher density regime with the cut-off set higher than Λ_{sChEFT} in the form of "relativistic mean-field theory" with heavier meson DoFs included. In fact one could consider these EFT approaches—including the G*n*EFT discussed in this note—to generically belong to the class of Kohn's DFTs (including KS-DFT) applied in nuclear physics. In the literature, there is a huge variety of "mean-field theories", both relativistic and non-relativistic, for treating both finite nuclei and infinite nuclear matter. Some of them are capable of explaining the ensemble of available terrestrial and astrophysical observables with success. However, they remain mostly phenomenological, having little if any to do with the fundamental theory QCD, at the densities relevant to massive compact stars.

We define our principal strategy as follows: Incorporate into an EFT—called "G*n*EFT" from here on to be distinguished from the standard ChEFT—what are deemed to be "robust" properties of the dense skyrmion matter built with the DoFs heavier than the pion. This EFT is to capture as "faithfully" as feasible the topological characteristics of the skyrmion matter. The basic assumption made is that, at high density and in the large N_c limit, the skyrmion matter is a crystal with a *negligible* contribution from the kinetic energy term [11].

The ω meson figuring as the $U(1)$ component of the hidden local symmetry (HLS) can be identified with the $U(1)$ Chern–Simons field playing a role in the fractional quantum Hall droplet structure for the $N_f = 1$ baryon associated with the η' singularity. We will return to the η' singularity because it is currently argued to be crucial for chiral restoration involving topology at high density, which is most likely outside of the range of densities relevant to compact stars, an issue that has only very recently been raised.

In short, what comes out from the approach treated in this note, simple in concept—and albeit unorthodox—turns out to work well ranging from nuclear matter density to high density relevant to massive compact stars. Up to date we see no serious tension with Nature as reviewed in [5] for this Special Issue of MDPI.

2. Topology in Baryonic Matter

2.1. Change of DoFs: Hidden Symmetries

Limiting our considerations for the moment to the density regime relevant to stars of mass $\sim 2M_\odot$, we will ignore the role that the FQH droplets may play in strong correlations involved. We will return to that matter later. To simulate the change of DoFs in terms of topology as density goes up from below to above the putative baryon-quark continuity density denoted as $n_{\rm BQC}$, two hidden symmetries invisible in QCD in the vacuum are found to be *absolutely* essential. One is hidden local symmetry (HLS) [12] and the other is hidden scale symmetry (HSS). The cut-off scale involved for the G*n*EFT should be greater than the cut-off effective for *s*ChEFT. The precise value of the relevant cut-off scale is not needed for what follows, but to be specific, one can take the HLS scale to that given by the vector meson ($V = (\rho, \omega)$) mass m_V~700 MeV. The scale symmetry is associated with the possible dilaton scalar $f_0(500)$ which will be later considered as a pseudo-Nambu–Goldstone boson of broken scale symmetry.

As for the baryon-quark crossover density, we will be considering $n_{\rm BQC} \approx (2-4)n_0$ which will be identified later with the topology change density $n_{1/2}$.

Defining precisely what the hidden symmetries to be incorporated are requires the details which have been given in review articles, e.g., [7]. The basic ideas can however be explained rather simply without losing physical content. Here, we will summarize the key ingredients that enter in the G*n*EFT.

One of the two symmetries that play a crucial role in the G*n*EFT is the hidden local symmetry (HLS) first formulated in [12] and made powerfully applicable to nuclear physics, as comprehensively reviewed in [13]. An important property of the HLS concerned is that it is a gauge symmetry *dynamically generated* giving rise to "composite" gauge field of pions. The existence of such a composite gauge boson is proven to be "inevitable" if such a symmetry is implicit in the dynamics [14] as assumed in our approach. It implies that the vacuum could be tweaked under extreme conditions, say, by high temperature or high density, such that

$$m_V \propto g_V \to 0 \tag{1}$$

where g_V is the gauge coupling for the vector mesons V. The point (1) is called the "vector manifestation (VM) fixed point" [13]. How to expose such a symmetry at high temperature, as in heavy-ion dilepton experiments or at high density as in compact stars, is an extremely subtle issue. An important point to note in this connection is the modern realization that there is a possible duality à la Seiberg (referred to as "Seiberg-(type) duality") between the vector mesons of HLS and the gluons of QCD. This involves a conjecture, but our study points to its validity as we will try to argue.

The other hidden symmetry that figures equally importantly is the scale symmetry. There is a long history with a still on-going controversy on how the scale symmetry is manifested in gauge theories, e.g., in strong-interaction physics, and in going beyond the Standard Model (BSM). In the literature are found strong arguments that "dilatons do not exist in QCD for $N_f \sim 3$". We differ from such arguments and eschew going into that highly controversial issue for which we refer to [7,8] viewed vis-à-vis with nuclear physics. As argued there, what is relevant to G*n*EFT is the "genuine dilaton (GD)" scenario of [15,16]. (Note added in proof: While this article was being drafted, an article appeared in which a very similar IR fixed-point structure with a "conformal dilaton" was arrived at [17]. In our view, at a high density, the genuine dilaton will coincide with the conformal dilaton when the IR fixed point is approached. This matter will be discussed elsewhere.). The GD scenario posits that there is an infrared (IR) fixed point with $\beta(\alpha_{s\rm IR}) = 0$ in the chiral limit (with u(p), d(own) and s(trange) quark masses equal to 0) and that the $f_0(500)$ is the scalar pseudo-Nambu–Goldstone (pNG) boson of spontaneously broken scale symmetry which is also explicitly broken by quantum (scale) anomaly. One of the most distinctive characteristics of this scenario is that the IR fixed point, which is most likely non-perturbative, is realized in the *NG mode* with non-zero dilaton condensate (or decay constant) and non-zero pion condensate (or decay constant). (As a side remark, we note that this scenario differs basically from the scenario popular in the Beyond the Standard Model circle working with $N_f \sim 8$ in the conformal window of the IR fixed point realized in the Wigner mode.) Given that the dilaton χ is a pNG boson as the pions π are and both satisfy soft theorems, one can make a systematic power counting expansion in chiral-scale-symmetric theory as in chiral-symmetric theory. The power counting in chiral expansion is well established. Scale symmetry brings an additional power counting in terms of the expansion of the β function. Expanding the gluon stress-tensor β function in the QCD coupling α_s near the IR fixed point,

$$\beta(\alpha_s) = \epsilon\beta'(\alpha_{s\rm IR}) + O(\epsilon^2) \tag{2}$$

where $\epsilon = \alpha_s - \alpha_{s\text{IR}}$ and $\beta' > 0$. The non-zero mass of f_0 is attributed to $|\beta'(\alpha_{sIR})\epsilon|$. Thus, the power counting in the scale symmetry is

$$\epsilon \sim O(p^2) \sim O(\partial^2). \tag{3}$$

One can, in principle, make a systematic chiral-scale counting comparable to chiral symmetry [15,18].

The matter-free-space mass of the dilaton ∼500 MeV is comparable to that of kaons, so the dilaton is put on the same mass scale as the octet pions. In principle, one has an $SU(3)$ chiral Lagrangian coupled to the dilaton. In the background of nuclear medium, however, as is well known from nuclear phenomena, the lowest-mass meson of the scalar quantum number is expected to undergo significant mass-drop whereas kaons do not appreciably. Therefore, in medium, identifying the scalar meson to be the dilaton, one can ignore the s quark in chiral-scale dynamics at high density. Unless otherwise stated, this will be what we will do.

We write the dilaton field as

$$\chi = f_\chi e^{\sigma/f_\chi} \tag{4}$$

which is referred, in the literature, to as "conformal compensator" field that figures in the notion of "quantum scale invariance (QSI)". As written, χ transforms linearly under scale transformation whereas σ transforms nonlinearly. One can use either. In what follows, we find it more convenient to employ the linearly transforming field χ. The Lagrangian that combines the (octet) pion field π, the HLS fields $V = (\rho, \omega)$ and the dilaton field χ, suitably written in a chiral-scale invariant way with the scale-symmetry breaking term put in the dilaton potential $V(\chi)$, will be denoted $\mathcal{L}_{\chi\text{HLS}}$ with χ standing for the dilaton. It is of the form

$$\begin{aligned} \mathcal{L}_{\chi\text{HLS}} &= \left(\frac{\chi}{f_\chi}\right)^2 \left(f_\pi^2 \text{Tr}\left[\hat{\alpha}_{\perp\mu}\hat{\alpha}_{\perp}^{\mu}\right] + a f_\pi^2 \text{Tr}\left[\hat{\alpha}_{\parallel\mu}\hat{\alpha}_{\parallel}^{\mu}\right]\right) \\ &- \frac{1}{2g^2}\text{Tr}\left[V_{\mu\nu}V^{\mu\nu}\right] + \cdots \\ &+ \mathcal{L}_{\text{hWZ}} + \frac{1}{2}\partial_\mu\chi\partial^\mu\chi + V(\chi) \end{aligned} \tag{5}$$

where $\hat{\alpha}_{\parallel\mu}$ and $\hat{\alpha}_{\perp\mu}$ are Maurer–Cartan 1-forms and $V(\chi)$ is the dilaton potential. $\mathcal{L}_{\text{hWZ}}$ is what is known as "homogeneous" (or "hidden") Wess–Zumino term which is scale-invariant that will figure later. For $N_f > 2$, there is the 5-D topological Wess–Zumino term which we have left out as it does not figure in our discussions.

2.2. Topology Change

There is a growing evidence that skyrmions as nucleons could describe finite nuclei as well as infinite nuclear matter [19], but at present it is far from feasible to address dense compact star matter quantitatively in terms of the pure skyrmion structure. It is however found feasible to extract topological properties of dense baryonic matter by simulating skyrmions on crystal lattice. In doing this, it is assumed that the topological characteristics extracted from the skyrmion crystal can be taken as *robust* and be exploited for *making* the mapping of topology to density-functional (DF) theory. Of course, considering skyrmions on the crystal cannot be a good approximation for low-density matter. Clearly it would make little sense to think of nuclear matter, which is best described as a Fermi liquid, as a crystalline. However, at high density and in the large N_c limit, baryonic matter could very well be in the form of a crystal [11,20]. It appears quite reasonable—and it is assumed in this note—that the baryonic matter at a density greater than the putative baryon-quark transition density denoted as n_{BQC} could encapsulate certain characteristic features of topology that are *not* captured in *s*ChEFT-type approaches.

Consider skyrmions constructed with $\mathcal{L}_{\chi\mathrm{HLS}}$ put on FCC crystal lattice. The skyrmions in the system undergo interactions mediated by the DoFs in the way described, e.g., in [21,22]. What transpires from the calculation focusing on essentials without going into details are (see [7]):

- **Skyrmion-half-skyrmion "transition"**

There takes place a topology change from the state of matter with skyrmions to that of half-skyrmions at a density above the nuclear matter density. The transition density $n_{1/2}$ we equate to n_{BQC} is not predicted by the theory. It turns out from the detailed analysis of the data available by astrophysical observation [7] that

$$2 \lesssim n_{1/2}/n_0 < 4. \tag{6}$$

In the discussion that follows, we will take this range. In our approach, the transition density $n_{1/2}/n_0 = 4$ seems to be ruled out [23]. It is plausible that further development in astrophysical observations, e.g., the maximum mass of compact stars, might increase the upper limit of $n_{1/2}$ in (6).

A characteristic feature of this transition is the resemblance to the pseudo-gap phenomenon in superconductivity. (Possible pseudo-gap phase was also discussed at high temperature [24].). The quark condensate $\Sigma \equiv \langle\bar{q}q\rangle$, identified as the order parameter in the absence of baryonic matter background, goes to zero when space averaged (denoted $\bar{\Sigma}$), whereas the pion decay constant f_π remains nonzero. Thus, the changeover is not a phase transition in the Landau–Ginzburg–Wilson-type sense. For lack of a better term we will continue to refer to it as "transition" unless otherwise noted. This feature will play the crucial role in formulating GnEFT in the class of field theoretic density-functional approach.

- **Soft-to-hard transition in the equation of state**

One of the most important observations in the skyrmion-to-half-skyrmion transition is the cusp at $n_{1/2}$ in the symmetry energy E_{sym} [23]. The symmetry energy, the coefficient of the term proportional to $\zeta^2 = \left((N-P)/(N+P)\right)^2$ in the energy per particle of the system $E(n,\zeta)$, plays the key role in neutron stars with large excess of neutrons. The E_{sym} decreases as it approaches $n_{1/2}$ from below in density, providing attraction, and then after the cusp at $n_{1/2}$, increases rapidly, thus giving repulsion. Thus, the cusp provides the main—as it turns out—mechanism for the EoS going from soft-to-hard at $\sim n_{1/2}$. This feature will be found crucial for the maximum mass of neutron stars, as well as certain gravity-wave signals coming from merging neutron stars. It turns out also intricately connected to the onset of the pseudo-conformal sound speed at $n \gtrsim 3n_0$ [23].

An interesting observation to make here is that this cusp structure that appears at the leading N_c order in the skyrmion lattice treatment of the symmetry energy is present as an "inflection" at about the same density as $n_{1/2}$ in phenomenological energy density functional approaches [25].

What is given in the skyrmion-crystal simulation is, roughly speaking, a mean-field effect and correlation-fluctuations above the mean field would largely smoothen the cusp, but the soft-to-hard effect remains unaffected in the EoS. It has been shown that this topological feature can be translated into the nuclear tensor force in GnEFT, reproducing precisely the cusp structure [8,22,23]. Another important consequence of this cusp structure is that in going up in density from $n_{1/2}$, the HLS gauge coupling constant g_ρ is forced to move toward the "vector manifestation (VM)" fixed point [13] at the density $n_{\rm VM}$ at which the vector meson mass vanishes [14]

$$m_\rho \sim g_\rho \to 0 \text{ as } n \to n_{\rm VM} \gtrsim 25n_0. \tag{7}$$

- **Parity-doubling**

The skyrmion-1/2-skyrmion topology change exposes the emergence of another hidden symmetry in strong interactions, namely, the parity doubling. At high density $\gtrsim n_{1/2}$,

the effective nucleon mass deduced from the skyrmion mass tends to converge to the chiral-invariant mass $m_0 \sim (0.6 - 0.9)m_N$

$$m_N^* \to f_\chi^* \sim m_0 \not\to 0 \text{ as } \bar{\Sigma} \to 0. \tag{8}$$

It will be seen below that this implies that the trace of the energy-momentum tensor (TEMT) becomes—in the chiral limit—a function *solely* of $f_\chi \sim m_0$ *independent of density* at some density $> n_{1/2}$. This symmetry is not *explicit* in QCD, so one could say it is also emergent. This turns out to have a striking impact on the sound velocity of the massive stars.

2.3. Quasi-Free Composite Fermions

Though unproven yet, it seems very likely that the parity-doubling structure described above is closely related to that the "quasi-fermion" in the half-skyrmion phase is a nearly non-interacting quasiparticle of baryon number 1. This object can be classed neither as a pure baryon nor as a pure quark. In the absence of a better name, let us just, for simplicity, call it "quasi-fermion." The first indication came in the Atiyah–Manton approach to skyrmions on crystal lattice [26]. This turns out to be a highly pertinent observation in the current development in the context described below. This observation has been discussed in the reviews cited above, but it is worth recounting it here in terms of insights which were not recognized before. We revisit the result obtained in Figure 11 of [27] reproduced here in Figure 1.

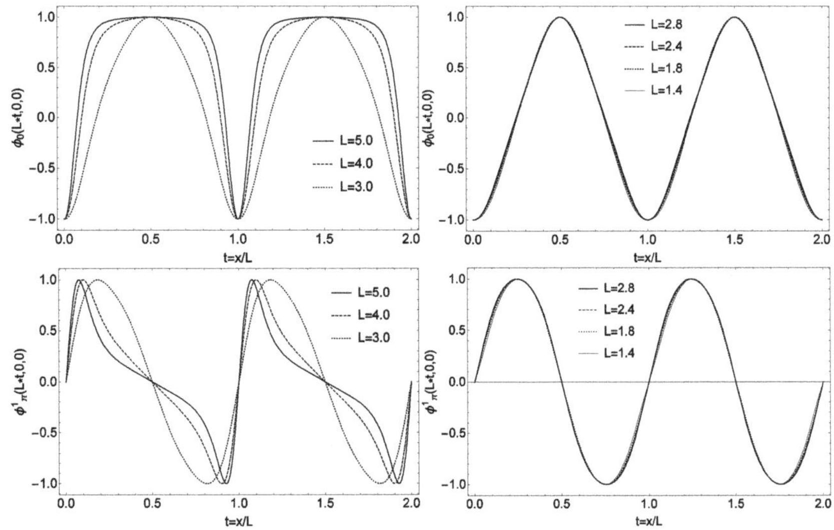

Figure 1. The field configurations ϕ_0 and ϕ_π^1 as a function of $t = x/L$ along the y = z = 0 line. The left panels correspond to $n < n_{1/2}$ and the right panels to $n \gtrsim n_{1/2}$. The half-skyrmion phase sets in when $L = L_{1/2} \lesssim 2.9$ fm where L is the crystal lattice size inversely related to the baryon density of the matter. What is to be noticed is the (near) density independence of the configurations ϕ_0 and ϕ_π^1 which engenders what signals "scale invariance" in the half-skyrmion phase.

What is given here are the field configurations as a function of $t = x/L$ along the y = z = 0 line. They figure in the energy density of the skyrmion matter given by the scale-symmetric HLS Lagrangian $\mathcal{L}_{\chi\text{HLS}}$. We focus on the chiral field as studied in [26]

$$U(\vec{x}) = \phi_0(x,y,z) + i\phi_\pi^j(x,y,z)\tau^j. \tag{9}$$

The χHLS fields lead to similar results. The field configurations $\phi_{0,\pi}$ correspond effectively to the mean fields in GnEFT which transcribed into Fermi-liquid theory [27], can be taken equivalent to the Fermi-liquid fixed point (FLFP) quantity in the baryonic χ-HLS Lagrangian $\mathcal{L}_{\psi\chi\mathrm{HLS}}$ with ψ standing for baryonic field. Those configurations, varying strongly due to nuclear correlations with increasing densities in the skyrmion phase at $n < n_{1/2}$ become *nearly density-independent in the half-skyrmion phase at* $n \gtrsim n_{1/2}$. This means that those configurations representing non-interacting quasiparticles behave scale-invariantly. This was reflected in the linear density dependence in the cusp for the symmetry energy in the half-skyrmion phase both in the crystal lattice and in V_{lowk}-RG calculations which go beyond the FLFP approximation described in [27] and below.

Particularly interesting is the density-independent configuration ϕ_0. Since this quantity is proportional to the pion decay constant f_π—and the dilaton condensate f_χ gets locked to f_π going toward the IR fixed point in Crewther's "genuine dilaton" scenario [15], this behavior of ϕ_0 impacts two important quantities at high density, first the sound velocity v_s of the compact-star matter, elaborated below, and second, what is significant at this point is its link to the possible Kohn–Sham energy density functional approach to topology.

One way to understand what we have is to notice that the half-skyrmion is attached with a monopole associated with a hidden $U(1)$ gauge field (say, the ω field in HLS) whose energy diverges when separated from the other half-skyrmion, but the divergence gets cancelled when the two half-skyrmions are bound, or more precisely, confined [28]. The resulting "composite" skyrmion (baryon) made up of two half-skyrmions in the crystal—with the kinetic energy suppressed—resembles the composite fermion (electron) in the FQHE with the kinetic energy absent in the limit of large magnetic field [10]. What the skyrmion crystal shows verifies that the quasi-fermions behave scale-invariantly when the lattice size is varied in the half-skyrmion regime $n \gtrsim n_{1/2}$.

It should be stressed at this point the quasi-fermion cannot be at or near the IR fixed point. The quasi-fermion that we have is identified as a quasiparticle in Fermi liquid away from the IR fixed point. It is generally considered likely that the Fermi-liquid structure breaks down in electronic systems in the "unitary limit" at which conformal invariance is present (see [29] for a recent discussion). This would be the case when the dilaton mass goes to zero in dense-matter systems. When we discuss pseudo-conformal sound speed in compact stars, $v_{pcs}^2 \approx 1/3$, we are not dealing with the, what is known as, "conformal sound speed" $v^2_{\mathrm{conformal}} = 1/3$ expected to set in at asymptotic density. The density involved in the stars is far from the IR fixed point density n_{sIR} with non-zero dilaton mass.

Pushing further the analogy to the FQHE, one may wonder whether the anyonic structure encountered in the FQHE as discussed in [10] has any relevance in the present problem. Indeed there are observations in skyrmion physics obtained with powerful mathematical techniques that there can be $1/q$ (baryon-)charged objets with q odd integer [30], and even other more exotic varieties. These may appear to be mathematical oddities, but in our approach they are physical. We will return to this issue in the second part of this note where fractional quantum Hall droplets (pancakes or pitas) could figure at high density.

3. Translating Topological Inputs into Effective Field Theory GnEFT

Given the topological inputs extracted from the skyrmion–half-skyrmion transition, the next step is to incorporate them into the EFT. To build the EFT concerned as an "analog" to the KS-DFT in the FQHE in [10] in the strategy to map topological properties to an EFT, we introduce baryon fields explicitly and couple them to χHLS fields chiral-scale symmetrically to $\mathcal{L}_{\chi\mathrm{HLS}}$. Let us denote it $\mathcal{L}_{\psi\chi\mathrm{HLS}}$ with ψ standing for baryons. It is of the form

$$\begin{aligned}
\mathcal{L}_{\mathcal{L}_{\psi\chi\mathrm{HLS}}} &= \bar{N} i\gamma^\mu D_\mu N - h f_\pi \frac{\chi}{f_\chi}\bar{N}N + g_{v\rho}\bar{N}\gamma^\mu\hat{\alpha}_{\parallel\mu}N \\
&+ g_{v0}\bar{N}\gamma^\mu \mathrm{Tr}\left[\hat{\alpha}_{\parallel\mu}\right]N + g_A\bar{N}\gamma^\mu\hat{\alpha}_{\perp\mu}\gamma_5 N. \qquad (10)
\end{aligned}$$

In the presence of explicit baryons, both the topological and "homogeneous" Wess–Zumino terms [13] (or "hidden WZ" terms of [31] with the coefficients vector-dominated and the FQH droplet "visible") in the mesonic Lagrangian $\mathcal{L}_{\chi\text{HLS}}$ from which topological baryons are built are absent. This is of course familiar in *s*ChEFT with the standard (nuclear) chiral Lagrangian with or without strangeness.

3.1. Density Functional via Fermi-Liquid Fixed-Point Theory

We confine to one unique Lagrangian defined *only* with the *relevant hadronic* variables $\mathcal{L}_{\psi\chi\text{HLS}}$ and eschew "hybridization" with non-hadronic degrees of freedom. The topology change will be encoded in the parameters of the Lagrangian $\mathcal{L}_{\psi\chi\text{HLS}}$ that change at the density $n_{1/2}$. Below the transition density, the Lagrangian endowed with the well-defined scaling of the pion and dilaton decay constants as dictated by the matching with QCD—i.e., via correlators—as proposed a long time ago [32] should reproduce *s*ChEFT. How well the predicted results fare with the established data of nuclear matter has been extensively reviewed (e.g., [7]). Of course with the extreme simplification, one cannot hope to match the precision enjoyed by *s*ChEFT treated at high chiral orders. However, the property of the equation of state (EoS), in particular the symmetry energy E_{sym} approaching $n_{1/2}$ from below, does have certain potentially nontrivial features associated with the properties, such as tidal polarizability (TP), measured in the recent gravity waves which differs from the prediction of *s*ChEFT.

As expected for possible hadron-quark continuity, the parameters of $\mathcal{L}_{\psi\chi\text{HLS}}$ are drastically affected by the topology change at $n_{1/2}$. They can differ qualitatively in the half-skyrmion phase from what was predicted in [32] and, hence, from naive extrapolation to $n \gtrsim n_{1/2}$ in *s*ChEFT. The most important impact is on the property of the bound half-skyrmions behaving as a scale-invariant quasi-fermion described in Section 2.3. It is represented in the Lagrangian $\mathcal{L}_{\psi\chi\text{HLS}}$ as an effective baryonic field ψ with its physical properties—the mass and coupling constants—dictated by the topological properties of the half-skyrmion phase, i.e., the near density independence of the effective mass and coupling constants and suppressed kinetic energy. Equally important is the (assumed) composite (HLS) gauge symmetry à la Suzuki theorem with the VM fixed point with the vanishing vector mass at $n_{VM} \gtrsim 25n_0$. It makes the vector field coupling to the quasi-fermion strongly weakened, leading to what we propose as the emergence of (pseudo-)conformal symmetry.

It should, in principle, be feasible to develop chiral-scale symmetric EFT in a parallel to high-order *s*ChEFT successful in nuclear physics at density $\sim n_0$. The cut-off could be set at $\Lambda_{G n\text{EFT}} \gtrsim m_V$. This feasibility in HLS was already discussed in [13] and initiated in χHLS [18]. Unfortunately, it is not in a form to perform high enough order scale-chiral perturbation calculations. Here, we will resort to a strategy resembling what is done in the FQHE [10], a sort of an application to dense nuclear matter of density functional theory consistent with Hohenberg–Kohn theorem [33].

In a nut-shell, the chain of reasonings goes as follows.

As well recognized in nuclear theory circles, the relativistic mean field theory as first formulated in Walecka's linear model [34] belongs to the class of density functional approaches. It has been extensively exploited in terms of the Kohn–Sham density functional in nuclear structure studies. Furthermore, in conjunction with *s*ChEFT, ab initio calculations in Kohn–Sham density functional are being explored, with the possibility of doing precision nuclear structure calculations. All these efforts are however limited at present to the density regime $\lesssim n_{1/2}$.

It is also known, though perhaps not so widely, that the Walecka model captures Landau Fermi-liquid theory [35]. Next, a chiral Lagrangian implemented with the HLS mesons is established [36,37] to lead to the Wilsonian renormalization-group approach to Fermi-liquid fixed-point theory [38,39]. It follows then that the Lagrangian $\mathcal{L}_{\psi\chi\text{HLS}}$ with the parameters encoding the topology change and matched to QCD in medium is expected to give in the mean field a highly powerful Fermi-liquid fixed point theory that can access densities $\gtrsim n_{1/2}$. The reliability of the large $\bar{N}$ approximation (where $\bar{N} = k_F/(\Lambda_{FS} - k_F)$

with Λ_{FS} the cut-off on top of the Fermi surface) the Fermi-liquid fixed point approximation indicates the validity of the mean field for densities near and higher than n_0 and more specially for high density [34]. In fact it is feasible to go beyond the mean-field by including $1/\bar{N}$ corrections taken into account in what is known as V_{lowk}RG.

3.2. *"Quenched g_A" as Precursor to Emergent Scale Symmetry at $n \gtrsim n_{1/2}$*

In nuclear physics, the V_{lowk} renormalization-group method has been applied to go beyond the FLFP approximation [27] along the line of the Wilsonian renormalization-group strategy [38,39]. What happens as the IR fixed point density $n_{\rm sIR}$ is approached has not been addressed. This density is never likely arrived at in compact stars where the dilaton mass is not zero, but it will become relevant at higher densities considered in what follows. In a recent development in condensed matter physics, the approach to soft modes—gapless phase—on top of the Fermi surface going beyond what amounts to the FLFP theory is being formulated by nonlinear bosonization of the quasiparticles on the Fermi surface leading to an order-by-order counting of the beyond-the-FLFP higher-order terms [40]. This would provide a systematic calculation of higher-order terms resembling that of chiral perturbation theory leading to sChEFT, illuminating the role of heavy DoFs in G*n*EFT. An approach of this type may give possible corrections to the solution to the quenched g_A problem, as well as the pseudo-conformal property of massive stars, both of which are discussed below.

The ultimate goal of G*n*EFT is to access the high density regime relevant to massive compact stars. Most intriguingly there is a hint however already at low density $n \lesssim n_0$ to what might be happening with the emergent scale symmetry at high density. Without going into details here (readily found in the reviews, such as [7]), we illustrate a case at low density $\sim n_0$ which shows a predictive power not shared (up-to-date) by sChEFT, with a close link to what takes place at $n \gtrsim n_{1/2}$. It is found to provide a simple and elegant resolution to a long-standing mystery lasting several decades of the "quenched" g_A observed in nuclear Gamow–Teller transitions in light nuclei [41]. The key element of the solution is that at the mean field level, i.e., at the Fermi-liquid fixed point, the superallowed Gamow–Teller transition described in G*n*EFT is precisely given by the soft-pion and soft-dilaton theorems, namely, the Goldberger–Treiman relation involving the density-dependent Landau Fermi-liquid fixed-point parameters F_1 that enter in the Landau effective mass.

The quantity identified as the Fermi-liquid fixed point axial constant $g_A^{\rm Landau}$ in EFT approach is found to encode "hidden scale invariance" emerging from strong nuclear correlations [42]. As discussed in [41] it can be equated to the effective g_A given by the "extreme single-particle shell-model (ESPSM)" defined for doubly closed-shell nuclei corresponding to the "scale-symmetric effective g_A" (denoted as $g_A^{\rm ss}$). To render a precise physical meaning to this quantity, imagine that one were able to obtain the up-to-date, non-existent, *exact* wave functions of the parent and daughter states involved in the superallowed (zero momentum transfer) Gamow–Teller transition. Now the exact transition matrix element in the doubly closed-shell nuclei would then be given by the Gamow–Teller matrix element given by the ESPSM multiplied by the constant $g_A^{\rm ss} = 1$ modulo scale-anomaly correction factor q_{ssb} [42]. This means that the total nuclear correlations are encapsulated in the constant $g_A^{\rm ss}$ multiplying the ESPSM matrix element $M^{\rm ESPSM}/g_A$ where g_A is the axial constant in neutron decay. In terms of Landau Fermi-liquid fixed point effective field theory, it corresponds to the quasiparticle sitting on top of the Fermi surface making the superallowed Gamow–Teller transition. Therefore, $g_A^{\rm Landau} = g_A^{\rm ss} = 1$. Now the expression for the Fermi-liquid fixed point relation for $g_A^{\rm Landau}$ which corresponds to the Ward identity is essentially the nuclear matter version of the Goldberger–Treiman relation established to hold within a few percentage accuracy in matter-free space [37]. What this means is the more correlations are taken into account, the closer the axial-vector constant that multiplies the single-particle Gamow–Teller operator will come to the *unquenched* value $g_A = 1.27$. In very light nuclei where powerful numerical techniques are available, this is in fact what is found with possible many-body (exchange-current) contributions amounting to $\lesssim 2\%$ [43].

In short, that $g_A^{ss} \approx 1$ in nature would indicate an emergence of quantum scale invariance as suggested in [42]. That the putative dilaton mass in nuclei is $\sim$600 MeV, so scale symmetry is certainly broken, at least spontaneously, so the quantum scale invariance must be hidden at low density. Deviation of the factor $q_{\rm ssb}$ from 1 would indicate that quantum scale invariance is broken in nature. A more precision re-measurement of the RIKEN experiment of the superallowed GT transition in the doubly magic nuclei discussed in [41] would be valuable on this issue.

At the dilaton-limit fixed point at high density that we discuss below, however, g_A does approach 1. What is particularly interesting is that the quantum scale invariance hidden at low density, presumably responsible for the quenching of g_A, seems to emerge in the "pseudo-conformal" sound velocity $v_s^2/c^2 \approx 1/3$ in massive compact stars as discussed below. Whether this result holds beyond the FLFP approximation could be addressed in this nonlinear bosonization technique [40]. One could also address neutrinoless double β decay where the momentum transfer could involve $\sim$100 MeV.

4. Going toward Massive Compact-Star Matter

As long as the density of the core of massive stars is not in the vicinity of the IR fixed point, we may assume that the notion of the Fermi-liquid fixed point applies to the density regime $n \gtrsim n_{1/2}$ as it does at $n \lesssim n_{1/2}$. One can then make simple calculations of the EoS to go toward the density relevant to massive compact stars. Given the paucity of the trustful knowledge about the structure of the state involved, the guidance available is the presumed constraints provided by the symmetries assumed to underlie the dynamics: The HLS (of the composite gauge symmetry [14]) with the vector manifestation (VM) density $n_{\rm VM} \gtrsim 25n_0$ and the HSS with the IR fixed point at $n_{\rm sIR}$ which we assume is near $n_{\rm VM}$. (It must be admitted that in the present framework there is no compelling reason to believe that $n_{\rm VM} \simeq n_{\rm sIR}$. The only thing one can say—and assumed here—is that both $n_{\rm (VM,sIR)}$ are higher than what is relevant to the maximum density supported by massive compact stars stable against gravitational collapse).

4.1. Dilaton-Limit Fixed Point

Consider $\mathcal{L}_{\psi\chi\rm HLS}$ for $n \geq n_{1/2}$ with the parameters of the Lagrangian taken density-dependent, but totally arbitrary, *unconstrained* by the topology change discussed above. Assume that the mean-field approximation holds at $n \gtrsim n_{1/2}$ in the sense defined in the large N_c and large $\bar{N} \sim k_F$ limit. Now what we would like to do is to take what corresponds to going toward the IR fixed point of [15]. This can be done by following Beane and van Kolck [44]: First, do the field re-parametrization $\mathcal{Z} = U\chi f_\pi/f_\chi = s + i\vec{\tau}\cdot\vec{\pi}$ in $\mathcal{L}_{\psi\chi\rm HLS}$, have the Lagrangian treated in the mean field and take the limit $\mathrm{Tr}(\mathcal{Z}\mathcal{Z}^\dagger) \to 0$. This limit is referred to as "dilaton-limit fixed point" (acronymed DLFP). Two qualitatively different terms appear from this manipulation: one is regular and the other singular in the limit. The singular part is of the form

$$\begin{aligned}\mathcal{L}_{\rm sing} &= (1-g_A)\mathcal{A}(1/\mathrm{Tr}(\mathcal{Z}\mathcal{Z}^\dagger)) \\ &+ (f_\pi^2/f_\chi^2 - 1)\mathcal{B}(1/\mathrm{Tr}(\mathcal{Z}\mathcal{Z}^\dagger)).\end{aligned} \tag{11}$$

The first (second) term is with (without) the nucleons involved. The requirement that there be no singularities leads to the "dilaton-limit fixed point (DLFP)" constraints

$$g_A \to g_A^{DL} = 1 \tag{12}$$

and

$$f_\pi \to f_\chi \neq 0. \tag{13}$$

We have denoted the g_A arrived at the DLFP as $g_A^{\rm DL}$ to be distinguished from the $g_A^{\rm Landau}$ in [41] arrived at the Landau Fermi-liquid fixed point at $n \sim n_0$. It turns out that the ρ meson decouples from fermions in dense matter going toward the DLFP even though the gauge coupling $g_\rho \not\to 0$ [27]. Therefore, the ρ meson drops out before reaching the VM fixed point.

These "constraints" are the same as what are in the genuine dilaton properties approaching the IR fixed point [15]. This suggests that the topological characteristics of the half-skyrmion phase are consistent with the genuine dilaton scenario. We should mention that there are other constraints in dense matter associated with the DLFP, among which highly relevant to the EoS at high density is the "emergence" of parity doubling in the nucleon structure mentioned above.

4.2. Emerging Pseudo-Conformal Symmetry

Next, consider $\mathcal{L}_{\psi\chi\rm HLS}$, in contrast to what is done above, with its parameters *constrained* by the skyrmion–half-skyrmion topology change at $n_{1/2}$. As shown, in the half-skyrmion phase, as the system flows toward the IR fixed point (or perhaps equivalently the VM fixed point)—although $n_{1/2} \ll n_{\rm sIR}$, the parameters set in *precociously* as

$$f_\pi \to f_\chi, \tag{14}$$

and

$$m_N \to f_\chi \propto \langle\chi\rangle \to m_0 \tag{15}$$

leading to the parity doubling (8).

In the mean field, that is, in the Landau Fermi-liquid fixed-point approximation in GnEFT, the energy-momentum tensor is easily calculable. It comes out to be [27]

$$\langle\theta_\mu^\mu\rangle = 4V(\langle\chi\rangle) - \langle\chi\rangle\frac{\partial V(\chi)}{\partial\chi}|_{\chi=\langle\chi\rangle} \neq 0 \tag{16}$$

where all the conformal anomaly effects (and also quark mass terms) are lumped into the dilaton potential $V(\chi)$. Thus, $\langle\theta_\mu^\mu\rangle$ is a function of *only* f_χ which does not depend on density (far below $n_{\rm VM}$) via (15). It follows that

$$\frac{\partial}{\partial n}\langle\theta_\mu^\mu\rangle = 0 \tag{17}$$

and, hence,

$$\frac{\partial\epsilon(n)}{\partial n}(1 - 3v_s^2/c^2) = 0 \tag{18}$$

where $v_s^2 = \frac{\partial P(n)}{\partial n}(\frac{\partial\epsilon}{\partial n})^{-1}$ is the sound velocity and ϵ and P are, respectively, the energy density and the pressure. It is fair to assume that there is no Lee–Wick-type anomalous nuclear state at the density involved, so $\frac{\partial\epsilon(n)}{\partial n} \neq 0$. Therefore, we have

$$v_{pc:s}^2/c^2 \approx 1/3. \tag{19}$$

Note that this is not to be identified with the "conformal sound velocity" $v_s^2/c^2 = 1/3$ expected at asymptotic density. The trace of the energy–momentum tensor is not zero at the compact-star density, so we call this pseudo-conformal (PC) sound velocity $v_{pc:s}$.

We should mention at this point a surprising observation in the V_{lowk}RG calculation that takes into account higher $1/\tilde{N}$ corrections in E/A of the A-nucleon ground state. The emergence of the PC symmetry is found to be intricately tied to whether the VM fixed point is lodged in the vicinity of the core of massive stars or at a much higher density. If it is taken far above the core density, say, at $n_{\rm VM} \gtrsim 25 n_0$, then E/A at $n \gtrsim n_{1/2}$ can be very accurately reproduced by the two-parameter formula [7,27]

$$E/A = -m_N + X^{\zeta}(n/n_0)^{1/3} + Y^{\zeta}(n/n_0)^{-1} \quad (20)$$

where $\zeta = (N-Z)/(N+Z)$ and X and Y are the constants to be fixed by equating (20) to the E/A given in V_{lowk}RG at $n = n_{1/2}$ by continuity in the chemical potential and pressure. X and Y depend on where $n_{1/2}$ is located. One can show that $\langle \theta_\mu^\mu \rangle$ is a constant independent of density for any values of X and Y. This then gives rise to the PC sound velocity (19). On the contrary, if $n_{\rm VM}$ were taken at $n \sim 6n_0$, say, in the center of massive stars, then the sound speed v_s^2/c^2 could not set in at the PC value in the range of star density but would exceed 1/3. This brings another surprise: That the PC symmetry is intimately tied to the VM property of hidden local symmetry. Why this is so is not understood.

Note that (19) is an approximate equality (with non-vanishing TEMT), not the equality which would hold at the asymptotic density $\gg n_{1/2}$. In the density regime concerned, $n \lesssim 7n_0$, there can, however, be deviations due to the quark mass term, and also higher-order terms of the anomaly-induced symmetry breaking involving the anomalous dimension β' that could make the sound speed deviate from (19). However, there are no reasons to suspect that the corrections would make v_s deviate appreciably from $v_{pc:s}$.

What is most glaringly different between the prediction of the GnEFT and that of *all* other models in the literature is the onset of the pseudo-conformal (PC) sound speed (19) at a relatively low density $\sim 3n_0$ which stays more or less constant up to the central density $\sim 6n_0$ of massive star $M_{max} \lesssim 2.3 M_\odot$.

As already stated, as far as we are aware, there are no observables so far measured with which the results of the GnEFT (including the recent GW observables) are at odds [5]. Because of the change of parameters of the Lagrangian $\mathcal{L}_{\psi\chi\rm HLS}$ controlling the EoS obtained from GnEFT, the most drastic of which is the cusp in E_{sym} at the leading order, there occur strong fluctuations in the density regime $\sim (2-4)n_0$ at which the topology change takes place. This gives rise to a spike in the sound velocity in that region after which the sound velocity v_{pcs}^2/c^2 stabilizes quickly to 1/3 above $\sim 3n_0$. The strength of the spike below the transition region can vary depending on the value of $n_{1/2}$. It can even overshoot the causality limit, for instance, for $n_{1/2} \gtrsim 4n_0$. This was the result that set the upper limit of $n_{1/2}$ to 4 n_0 in the bound (6. This strong enhancement in the sound speed going over the normal hadronic-to-non-hadronic crossing can also be seen with the transition mediated by hadronic-quarkyonic continuity [45]. Thus, this aspect of the sound speed could very well depend on how the changeover from hadronic to other forms of the state of matter takes place. This, of course, would be too difficult an issue to accurately sort out in the (over)simplified description. What is less unambiguous is the precocious onset of the PC sound velocity.

The robust takeaway from this result is that in the way the PC symmetry permeates from low density ($\lesssim n_0$) in the $g_A^{\rm Landau} \approx 1$ to high density ($> n_{1/2}$) in the $g_A^{\rm DL} = 1$, the PC sound velocity simply reflects the precocious emergence of the same PC symmetry. Among others it predicts that in the core of massive stars at a density $\sim 6n_0$, the objects found there are the composite quasi-fermions of bound half-skyrmions.

The question then is: Are these quasi-fermions unrelated to what might be described as "deconfined quarks"?

It has recently been argued in [46], based on detailed analyses combining astrophysical observations and theoretical calculations, that the matter in the core of maximally massive stars exhibits the characteristics of "deconfined phase" and suggests that the fermions residing in the core are most likely "deconfined quarks". The prediction of GnEFT differs from this interpretation: The objects found in the core are neither purely quarks nor purely baryons but quasi-fermions of the confined half-skyrmions [47]. The resemblance is, however, uncanny if one compares the predictions P/ϵ where P is the pressure and ϵ is the energy-density as function of density n and the polytropic index $\gamma = d(\ln P)/d(\ln \epsilon)$ made in the description given above with the analysis of [46]. This, we suggest, is the reflection of the topology change, a.k.a., baryon-quark continuity. We now turn to this issue in terms of what might be called "hadron-quark duality".

5. Hadron-Quark Duality and Cheshire Cat Phenomenon

The description given above involves composite fermions made of half-skyrmions in some sense "masquerading" as fractionally charged quarks. As noted above, however, half skyrmions are not the only objects that skyrmions can turn into. There could be other fractional objects, such as mentioned in [30], and others to be mentioned below.

Thus far we have ignored the possible role that the FQH droplet might play in dense matter. Although the skyrmion description applies to the octet baryons coming from the octet mesons for the flavor $SU(3)$, there is no skyrmion for the $U_A(1)$ meson η', i.e., the "dichotomy problem". A possible solution to this dichotomy is that at large N_c limit, the baryon coming from the η' is a FQH droplet or more appropriately "pancake" described in Chern–Simons field theory [48], not a skyrmion. In nature, the η' associated with the chiral anomaly is massive given that $N_f \ll \infty$, so could be ignored in low-energy/density dynamics of baryons. Indeed there seems to be no indication that it figures directly at least at low density. It does however figure, though indirectly, in the proton's suppressed flavor-singlet axial coupling-constant $g_A^{(0)} << 1$ [49] explained as due to the color anomaly [50,51] we will return to below. The question arises as to what happens at high density (and also at high temperature) where "deconfined" quarks are naively expected to intervene. There is a highly original and provocative—and, viewed from the point of view of our approach, compelling—argument [31,52,53] that indeed the FQH pancake is *essential* at some high density (and/or temperature) in the vicinity of chiral restoration with the possible restoration of $U_A(1)$ symmetry linked with the dropping η' mass [54]. It has, thus far unexplored, implications on how the vector mesons ρ and ω in χHLS behave near the chiral transition [53], for instance in heavy ion physics, with a possible paradigm change in the field. This could also be relevant to the inescapable question in confronting the theory with experiments in massive stars, say, as "deconfined quarks" [46] or pseudo-quarkonic phase suggested in the literature.

This brings us to the old question of what heavy nuclei are in QCD, an issue hotly discussed in the 1970s.

Consider $N_c = 3$ "confined quarks" in, say, the MIT bag for a nucleon in nuclear matter of mass number A. When squeezed in dense matter, as the bags overlap, one can visualize the quarks ultimately percolating from one bag to another bag and then coalesce into one big bag of $N_c \times A$ quarks. In the 1970s, this is the way some nuclear theorists thought of the ^{208}Pb nucleus as 624 quarks interacting via perturbative QCD confined within a giant bag. Such a picture was not—and still is not—a feasible one for the reason by now well-known at least for low density. Even so, incorporating the MIT bag structure with asymptotic free interactions at high density could make at least qualitative sense at asymptotic densities. Indeed many papers have been written where low-density hadronic description is hybridized with MIT bag description at increasing density. They typically involve phase transitions. We cite just a couple of the most recent of them [55,56] where other relevant references can be found.

One possible alternative was put forward for nuclear dynamics at low energy (and low density) by what was called the "Cheshire Cat Principle" [57–59] whereby N_c quarks in a bag transform into a topological soliton, skyrmion, so the quarks disappear into the "smiles" of the Cheshire Cat with the solitons interacting via fluctuating meson exchanges, in the way Weinberg admitted as what "nuclear physicists knew what they were doing" before the advent of the sChiEFT as prescribed in the Folk Theorem.

5.1. *"Infinite Hotel" for $N_f \geq 2$: Skyrmions*

What takes place can be imagined as a quark in a "jail" trying to escape from the jail, fully occupied, such as the filled Dirac sea This "jail-break" scenario is beautifully described in [60]. Actually N_c quarks are involved, but we focus on only one of them. A massless quark swimming on top of the sea, say, to the right in one spatial dimension, in an attempt to escape the jail, gets blocked at the "jail wall", so is unable to escape. It cannot swim back on top of the Dirac sea, because chiral symmetry forbids it. But it can plunge into the Dirac sea which is feasible, because the Dirac sea is infinite, and swim back to the left inside the sea. This infinite Dirac sea can be likened to an "infinite hotel (IH)" [60]. This exploitation of the infinity is a quantum effect known as "quantum anomaly".

There is one serious problem in this scenario, however. The fermion (baryon) charge carried by the quark disappears into the Dirac sea, so the baryon number is apparently "violated" in the process. In QCD, the baryon charge is absolutely conserved, so the fermion charge cannot disappear. Here takes place a miracle. The fermion charge is relayed to the "pion" that clouds the outside wall, with the pion (boson) turning into a baryon (fermion). This is by now the well-known story of skyrmions in (3 + 1)D mathematically characterized by the homotopy group $\pi_3(S^3) = \mathcal{Z}$ for the $N_f \geq 2$ systems.

This IH phenomenon can be considered to involve two domains, one the quark-gluon one and the other the hadronic one. There are two modes of a global symmetry, i.e., chiral symmetry, involved: Wigner–Weyl (WW) mode inside the bag and Nambu–Goldstone (NG) mode outside the bag. Therefore, the jail wall can be taken as a thin "domain wall" that delineates two vacua. This is the "jail-break" scenario for the $N_f = 2$ (i.e., proton and neutron) case.

The upshot is that the leaking baryon charge is taken up by the pion as a soliton. So in nuclear physics, we argue that for the given soliton chiral angle $\theta(R)$, the leaking baryon charge $1 - \theta(R)/\pi$ (in 1 spatial dimension) is lodged in the skyrmion cloud while the rest of the charge $\theta(R)/\pi$ remains in the bag, yielding the total baryon number 1 for a single baryon. When the bag is infinite the whole baryon charge is lodged inside the bag, while when the bag shrinks to zero size the whole baryon charge goes into the skyrmion cloud. So the size of the bag has no meaning for the property of the quark. The confinement size identified with the bag size R is, therefore, an unphysical quantity. One can think of this process as the pion fields giving rise to the baryons as solitons. This is what is referred to as the "Cheshire Cat Phenomenon" or "Cheshire Cat Principle (CCP)" [57]. This is akin to the disappearance of the Cheshire Cat in "Alice in the Wonderland" with the baryon number playing the role of the cat's smile. In fact it could be more appropriate to identify this phenomenon as a gauge artifact and formulate a gauge theory for the phenomenon [61].

This discussion of the CC "smile" applies straightforwardly to (3 + 1) dimensions. It has indeed been verified by Goldstone and Jaffe [62] in terms of the spectral asymmetry $\eta(s)$ (defined in (26) below) which renders the baryon charge lodged inside the bag for a given chiral angle $\theta(R)$. The fractionalization of the baryon charge is exact thanks to the topology involved. In (1 + 1)D, an exact bosonization allows an in-principle CCP also for non-topological processes. However, in the absence of bosonization, such exact CCP does not exist in nuclear processes in (3 + 1)D—except for the topological quantity, so much of what one can say of the processes in nature that are not topological is at best approximate.

5.2. No Infinite Hotel for $N_f = 1$ Baryons

The IH scenario discussed above famously turns out not to work when the number of flavors is one. This is because $\pi_3(U(1)) = 0$. One then wonders whether there is no soliton for baryon coming from the flavor singlet meson η'. As mentioned, this puzzle was recently resolved by ideas developed in condensed matter physics by Komargodski [48] who suggested that the η' can turn into a flavor singlet baryon—denoted from here on as $B^{(0)}$—as a fractional quantum Hall (FQH) droplet. At first sight this FQH droplet (pancake or pita) is unrelated to the usual skyrmion corresponding to the nucleon and, hence, the dichotomy.

There are two questions raised regarding this dichotomy [63] between the skyrmions and the FQH droplets. The first is: is there any relation between the two topological objects, the FQH droplet for $B^{(0)}$ and the skyrmion for nucleons? The second is: is the phenomenon of the FQH droplets relevant to the EoS at high density? Both questions are in some sense related.

5.3. *Baryon for $N_f = 1$*

Let us first discuss whether Komargodski's FQH pancake model can be given a formulation in terms of a Cheshire Cat phenomenon. This rephrases what was done in [64].

Suppose the quark in the bag is of $N_f = 1$ in the jailbreak scenario. Let the quark be coupled at the wall $x = R$ to the flavor-singlet meson η'. Again, the confinement leads to the breaking of the baryon charge and gives rise to an anomaly, but since $\pi_3(U(1)) = 0$, it *cannot* go into the infinite hotel because the topology does *not allow* it. We will come later to how and why the topology might "dictate" the flow. So where does it go? The answer [64] is that the quark moving in the x direction is allowed to escape by flowing in the y direction and go into a quantum Hall-type pancake, taking care of the anomaly generated by the boundary condition for what can be identified with a thin domain wall and keeping the baryon charge conserved. This is known as the "anomaly in-flow" mechanism leading to the Chern–Simons topological term [65], which in 3-form reads

$$\frac{N_c}{4\pi}\int_{2+1} ada \tag{21}$$

where a_μ is the Chern–Simons field which is to capture strong correlations in QCD—and will be identified later with the $U(1)$ field in HLS, namely the ω meson [31,52]. The Cheshire Cat Principle, if held, would imply that the baryon charge leaks *completely* into the FQH droplet, with the "smile" reducing to a $U(1)$ vortex line on the pancake. That the resultant FQH droplet correctly carries the baryon charge is assured by the gauge invariance of the Chern–Simons term (21). How this comes about can be explained in terms of a chiral bosonic edge mode [52] which will be found to play a key role in accessing the EoS for massive compact stars [7]). In accordance with the global symmetries of QCD, the $B = 1$ baryon with $N_c = 3$ quarks must then have spin $J = N_c/2 = 3/2$. This yields the high-spin baryon. Thus, when the bag is shrunk by fiat to zero size, the Cheshire Cat smile will go into the vortex line in the FQH droplet. For instance for $N_c = 3$, this picture yields the $\Delta(3/2, 3/2)$. The same $\Delta(3/2, 3/2)$ also appears in the rotational quantization of the skyrmion with $N_f = 2$ which comes from the ∞-hotel mechanism which does not work for the $N_f = 1$ baryons. These two descriptions present an aspect of the dichotomy problem: whether or how they are related?

Baryons for $N_f \geq 2$

Instead of a flavor-singlet quark, now consider the jail-breaking scenario of the doublet u and d quarks. There seems to be nothing to forbid the quark from flowing, instead of dropping into the infinite hotel giving rise to a skyrmion, into the y direction as the flavor-singlet quark did to compensate the anomaly generated by the bag wall. Or is there? This is the question raised.

Now, let us blindly apply the same anomaly-flow argument in CCP to the N_f-flavored quark. The spin-flavor symmetry for the flavor $N_f \neq 1$ will of course be different. Given $N_f = 2$, we expect to have a non-abelian Chern–Simons field $\mathbf{A}_\mu$ in place of the abelian a_μ [64],

$$\frac{N_c}{4\pi}\int_{2+1} \mathrm{Tr}\left(\mathbb{A}d\mathbb{A} + \frac{2}{3}\mathbb{A}^3\right). \tag{22}$$

This presents an alternative jail-break scenario to the infinite-hotel one.

However, there arises the question: What makes nuclear matter (at $n \sim n_0$) realized as a state of skyrmions as Nature seems to indicate, instead of stacks of fractional quantum Hall pancakes or pitas [52] or combinations of the two? Is the non-abelian Chern–Simons droplet a meta-stable state absent at low density but could figure at high density? This question is addressed below following the recent developments on the role of two hidden symmetries, flavor local and scale, intervening at high density involving fractional quantum Hall droplets [31,52,53].

5.4. Fermion Number and Hall Conductivity on Domain Wall

The Cheshire Cat Principle posits that physics should not depend on confinement size. It could be more appropriately phrased even as a gauge dependence in gauge theories [61]. This CCP on the confinement size was proven by showing that the baryon charge does not depend on R [57]. This follows from that the baryon charge is topological in any dimension. In (3+1)D, however, there is no exact bosonization and, hence, there is no exact CCP for other than the baryon charge, although approximate CCP holds for certain quantities like the flavor singlet axial charge of the proton [49]. In [64], the CCP was established also for the $N_f = 1$ baryon for the baryon charge with the fractional quantum Hall droplet replacing the skyrmion for $N_f \geq 2$. (A remark is in order for completeness. In [64], the bag boundary is taken as a domain wall. Whether the bag boundary can indeed be thought in terms of a domain wall is not clear and remains to be examined in detail. In modern developments in gauge theories, the concept of domain wall (together with "interface") plays a singularly important role. This is particularly so, in particular in QCD, and is a huge subject in the literature. The θ dependence in QCD with massless quarks makes the pertinent case in this note as will be elaborated below).

As is well known, the CP symmetry is spontaneously broken for the vacuum angle $\theta = \pi$. Suppose θ varies from 0 to 2π. There results a domain wall with Chern–Simons theory on it. Now, when quarks are massless, since the bulk property of the theory depends on $m^{N_f} e^{i\theta}$, the θ dependence is eliminated, replaced by a shift of η'. This is the anomaly cancelation restoring CP symmetry in 4D. Thus, the emergence of η' in the problem.

In order to understand what's going on, let us re-derive the CCP result for $N_f = 1$ baryon of [64] in (3 + 1)D by considering the bag boundary as an extremely thin "domain wall" located at $x_3 = 0$. Following [66], we will consider quantized Dirac fermions—say, "quarks"—in interaction with a background $U(1)$ gauge field a_μ, and scalar σ and pseudo-scalar π fields

$$\mathcal{L} = \bar{\psi}\mathcal{D}\psi \tag{23}$$

with

$$\mathcal{D} = i\gamma^\mu(\partial_\mu - iga_\mu) - (\sigma + i\gamma_5\pi),\ \sigma^2 + \pi^2 = 1. \tag{24}$$

a_μ, the $U(1)$ component of HLS, will be more precisely specified below.

Consider the background fields changing rapidly near $x_3 = 0$ and going to asymptotic values. One is interested in the *vacuum* baryon number B given by

$$B = -\frac{1}{2}\eta(0, H) \tag{25}$$

where $\eta(s, H)$ is the spectral asymmetry that was computed in [62] (for the infinite—hotel scenario)

$$\eta(s, H) = \sum_{\lambda>0} \lambda^{-s} - \sum_{\lambda<0} (-\lambda)^{-s} \tag{26}$$

where λ is the eigenvalues of the Dirac Hamiltonian H. With some reasonable approximations, it was obtained in [66] that

$$B = -\frac{g}{4\pi^2}\theta|_{x^3=-\infty}^{x^3=+\infty} \int d^2x f_{12} \tag{27}$$

where $\theta \equiv (\arctan(\pi/\sigma))$ and $f_{\mu\nu}$ is the gauge field tensor. Note that the vacuum fermion number B has two components, first the Goldstone–Wilczek fractionalized fermion number [67] and the other the magnetic flux through the (x,y) 2-d plane

Consider next a domain wall background defined by the fields σ and π that depend on x^3 only. The one-loop effective action in the non-static background is found to give the parity-odd action (why the parity-odd action becomes relevant is explained below)

$$S = \epsilon^{\mu\nu\rho 3} \int d^4x d^4y G(x,y) a_\mu(x) \partial^y_\nu a_\rho(y) \tag{28}$$

where G is a complicated non-local function of x^3, y^3 and $z^\alpha = x^\alpha - y^\alpha$, $\alpha = 0,1,2$. In the long-wavelength limit in the form factor G, the action can be written as a Chern–Simons term

$$S = g^2 \frac{k}{4\pi} \epsilon^{\mu\nu\rho 3} \int d^3y^\alpha a_\mu(y^\alpha, 0) \partial_\nu a_\rho(y^\alpha, 0) \tag{29}$$

with

$$g^2 \frac{k}{4\pi} = \int d^3x^\alpha dy^3 dx^3 G(z^\alpha, x^3, y^3). \tag{30}$$

Here, k can be identified as the "level" in the level-rank duality of the Chern–Simons term.

At this point, one can make contact with what was done in the CCP structure [64]. For this, consider the domain wall located at $x_3 = 0$ with the "quark" modes inside the bag $x_3 < 0$ corresponding to the Cheshire Cat smile. The $U(1)$ field in (23) could be considered, as suggested in [52,53], to be the ω field when the vector mesons ρ and ω in HLS are treated as the color–flavor locked $U(N_f)$ gauge fields dual to the gluon fields in QCD [68]. Then, the ω field can be taken as the Chern–Simons field that captures à la CCP the strongly-correlated excitations outside the bag. Now for $U(N_f)_{-N_c}$ dual to $SU(N_c)_{N_f}$ spontaneously broken, the vortex configurations in three dimensions made up of ρ and ω carry magnetic and electric charges of $U(1)^{N_f}$. The electric charge in the CS term can then be identified with the baryon charge [53,68]. This allows one to obtain the vector current from the action S (29), the time component of which is

$$J^0(x) = \frac{1}{g} \frac{\delta}{\delta a_0(x)} S. \tag{31}$$

The baryon number is [66]

$$B = \int d^3x J_0(x) = \frac{gk}{2\pi} \int f_{12} d^2x. \tag{32}$$

Setting the Dirac quantization for the magnetic flux threading the vortex [68]

$$\frac{g}{2\pi} \int f_{12} d^2x = 1 \tag{33}$$

one finds the baryon charge equal to the level

$$B = k. \tag{34}$$

This is the baryon charge lodged *in the vacuum.*

Now to make the connection à la [66] to the Cheshire Cat scenario discussed in [64], we identify the chiral angle for θ which is $= \eta'/f_{\eta'}$ in [64], and impose at $x^3 = 0$ the Cheshire Cat boundary condition

$$(1 - i\gamma^3 e^{i\gamma_5 \theta})\psi|_{x^3=0} = 0. \tag{35}$$

Then, the change in baryon charge is given by

$$\Delta B \approx \frac{\Delta\theta}{2\pi} \quad (36)$$

where $\Delta\theta$ is the jump of the η' field across the chiral bag boundary. This is the same result obtained in [64]. The Cheshire Cat dictates the baryon charge $B_{out} = 1 - B_{in}$ to be lodged in the Chern–Simons action

$$S' = g^2 \frac{k'}{4\pi} \epsilon^{\mu\nu\rho 3} \int d^3y^\alpha a_\mu(y^\alpha, 0)\partial_\nu a_\rho(y^\alpha, 0) \quad (37)$$

so it must be that

$$k' = 1 - k. \quad (38)$$

Here are two important points, among others, to note. First of all, as pointed out in [66], the Chern–Simons term (29) or (37) by itself is not topological. This is because the level k or k' separately as defined is not an integer so the action is not gauge invariant, hence unphysical, for $R \neq 0$ or ∞. The sum of the baryon charges of inside and outside is required by the anomaly cancellation. We believe this is related to the color anomaly found in the Cheshire Cat in (3 + 1)D [50] explaining the tiny flavor-singlet axial charge $g_A^{(0)} \lesssim 0.3$ [49].

Second, one could have *naively* done the same analysis for the $N_f = 2$ case with the pion fields included. That would have given rise to nonabelian CS theory with the same results as in the CCP strategy. So one is back to the dichotomy problem.

6. The Dichotomy Problem

6.1. Indispensable Role of Vector Mesons

We suggest that the key elements that provide the resolution of the dichotomy problem are the symmetries that led to the density-functional formalism GnEFT, namely, the hidden local symmetry and the scale/conformal symmetry. The degrees of freedom associated with these symmetries, the vector mesons and the dilaton, can be taken as emergent symmetries from strong nuclear correlations "dual" to QCD.

To see how one arrives at this aspect, let us incorporate the η' field in the two-flavor chiral field in HLS Lagrangian as

$$U = \xi^2 = e^{\eta'/f_\eta} e^{i\tau_a\pi_a}. \quad (39)$$

The notable observation made by Karasik [31,52] is that what is called "hidden" Wess–Zumino term in the HLS Lagrangian (replacing the "homogeneous" Wess–Zumino term of [13]) unifies the baryon currents for both the FQH droplet and the skyrmions [63]. It was noted that in the effective field theory that contains both η' and the HLS fields, the η' cusp that accounts for the jump from one vacuum to the other at $\eta' = \pi$ does not appear. Thus, the effective theory containing the HLS fields in the presence of η' captures the emergent theory on the η' domain wall. This suggests that it is more efficient and simpler to resort to the bosonic Lagrangian from which both the skyrmions and the FQH droplets emerge as solitons. It is not clear how to bring GnEFT to the density regime involved which could be higher than what is relevant in compact stars. It seems feasible to formulate this problem via nonlinear bosonization of the Fermi surface [40]) giving rise to soft modes of hidden symmetries at high density.

Now, let us consider tweaking the baryonic matter by increasing density in χHLS Lagrangian where the hWZ terms figure. The density for compact stars is $\lesssim 10n_0$. The relevant structure is more or less captured by the approach to the dilaton limit fixed point—Section 4.1—and the emergent pseudo-conformal symmetry (PCS)—Section 4.2—which say that $g_A^{DL} \to 1$ and $f_\pi \to f_\chi \sim m_0$ in the range of density involved in massive stars.

As argued in [63], the FQH droplet should become relevant at some high density at which $f_\chi \to 0$. This density must then be (much) higher than that reached at the DLFP close to the GD's IR fixed point. It is at this point the hWZ term exposes the Chern–Simons η' coupling term carrying the information on the FQH droplet with the correct baryons number. There the ω field in HLS Lagrangian can be identified as the Chern–Simons field.

In [53], a scenario different from that of [31,52] is suggested for the role of hidden local symmetry. There the coupling of the Chern–Simons fields in the bulk couple with the edge modes of vector mesons making the vector mesons gauge bosons. At the moment which scenario is preferred is not clear. However, what is absolutely clear is that hidden symmetries "dual" to QCD symmetries (e.g., HLS vector mesons as Seiberg-dual to the gluons [68]) must be essential for the correct description of the phase structure at high density. One cannot say whether the compact star density reaches the appropriate density beyond the PC regime. If future refined gravity wave observations were to indicate significant deviations from the PC sound speed predicted in our approach, this would give a hint to the possible role of the FQH droplets.

An interesting observation here is that the Chern–Simons field coupling to the FQH droplet in the hWZ term

$$\mathcal{L}_{\mathrm{CS}\eta'} = -\zeta \frac{N_c}{4\pi} J_{\mu\nu\alpha}\omega^\mu \partial^\nu \omega^\alpha \tag{40}$$

with the topological $U(1)$ 2-form symmetry current

$$J_{\mu\nu\alpha} = \frac{1}{2\pi}\epsilon_{\mu\nu\alpha}\partial^\beta \eta' \tag{41}$$

requires $\zeta = 1$ to have gauge invariance. (In the GD scheme [15] we are adopting, in the absence of the η' field, each of the four hWZ terms $a = 1,2,3,4$ can have a factor $\left(c_a + (1-c_a)(\frac{\chi}{f_\chi})^{\beta'}\right)$ with c_a an unknown constant and β' the anomalous dimension of the gluon stress tensor that multiplies the scale-invariant term.) This means that the "scale-symmetry breaking constant" c_{hWZ} in front of the hWZ term is $\zeta = c_{hWZ} = 1$ whereas in the absence of the FQH droplets it could be that $\zeta = c_{hWZ} << 1$ at scale-chiral symmetry restoration with $f_\chi = f_\pi = 0$ [69].

6.2. Dense Matter as "Sheets" of Pancakes/Pitas

As noted, at low density, the $N_f = 2$ quarks in the bag must be tending to fall into the infinite hotel, hence giving rise to skyrmions in (3 + 1)D. This may be "driven" by the parameters of the Lagrangian that unifies the $N_f \geq 2$ and $N_f = 1$ baryons to have the $B^{(0)}$ effect *suppressed* at low density. However, as density increases, the parameter change in the baryonic scale-symmetric Lagrangian $\mathcal{L}_{\psi\chi\mathrm{HLS}}$ in GnEFT that distorts the baryon current from the unified current to the $N_f = 1$ current could transform the EoS state toward the Chern–Simons QFT structure. One possible scenario for this is indicated in the recent skyrmion crystal analyses of dense matter where an inhomogeneous structure is found to be energetically favored over the homogeneous one at $n > n_{1/2}$ – but not asymptotically high density. It has been found that the baryonic dense matter consists of a layer of sheets of "lasagne" configuration with each sheet supporting half-skyrmions [70]. The constituents of this layer structure are quasi-fermions consisting of fractionalized quasiparticles of 1/2 baryon charge, possibly deconfined as conjectured below, appearing in baryon-quark continuity at a density $\sim n_{1/2}$, drastically different from those of the pasta structure discussed for the dilute outer layer of compact stars. In the Skyrme model (with pion field only) used in [70], the quartic (Skyrme) term effectively encodes massive degrees of freedom, including the hidden local fields, the monopole structure hidden in half-skyrmions, etc., described above. It appears feasible to formulate this "sheet dynamics" by a stack of FQH pancakes or pitas with tunneling half-skyrmions between the stacks, somewhat like arriving at the Chern–Simons field theory structure of FQHE in (2 + 1)D with a stack of (1 + 1)D quantum wires [71] (there are many papers on this matter in the literature. A good article with many relevant references is [71]). In the tunneling process it may be possible that the half-skyrmions transform to 1/3-charged quasiparticles resembling quasi-quarks as discussed in [72]. On the domain wall, the 1/3-charged objects could behave as "deconfined" quarks as discussed in [73]. This could then explain why the composite quasi-fermions given in GnEFT behave similarly to the "deconfined quarks" in the core of massive compact stars [46].

7. Conclusions

The question addressed in this note was: Is it feasible with a single unique effective Lagrangian to address the equation of state from normal nuclear matter to massive compact-star matter resorting to only one set of degrees of freedom with the vacuum sliding with density but without phase transitions? Put in another way, how far can one go with such a "unified" formalism without getting into fatal conflict with either empirical or theoretical constraints?

Influenced by strikingly successful developments in strongly correlated condensed matter physics together with impact on particle physics, an extremely simplified approach to the EoS for massive compact stars is formulated in terms of topology change to account for possible "continuity" from seemingly hadronic variables to QCD variables at high density $n_{1/2} \gtrsim 3n_0$. The single Lagrangian adopted in this study consists entirely of hadronic variables, the pion and the nucleon that figure in the standard nuclear EFT plus the massive degrees of freedom ρ, ω, and χ associated with hidden local and scale symmetries. The role of the topology change is to endow what could be identified as Kohn–Sham-type "density-functional" structure in the parameters of the effective Lagrangian that are supposed to capture the topological structure of QCD variables in dense medium. We further extended the scenario with the possible intervention of the FQH droplets $B^{(0)}$s going beyond the DLFP, brining in non-Fermi baryonic matter with scale-chiral restoration.

In this approach, there are neither explicit quark degrees of freedom nor strangeness flavor as in the standard approaches [74,75] and in other variations with bag models [55]. It is possible of course that there be corrections to the approximations made—given the admittedly drastic oversimplification—that could, quantitatively though not qualitatively, modify the results. There is, however, one serious potential obstruction to GnEFT. Should future measurements map out precisely the behavior of the sound velocity in the range of density $3 \lesssim n/n_0 \lesssim 7$ and falsify the precocious onset of, and the convergence to, the PC sound velocity, then that would bring a serious obstruction to the notion of the emergent symmetries, particularly hidden scale symmetry distinctive of the theory. That would then "torpedo" the GnEFT. If however it is not "torpedoed", then our approach with the encoded "duality" to QCD in approaching the chiral phase transition, as well as confinement, as argued recently by string-theory-inclined theorists [31,48,52,53,68] will bring a totally new perspective to nuclear physics, a paradigm almost totally foreign to nuclear theories.

A most interesting future direction would be to map the "generalized" sheet structure of Chern–Simons QFT in the topological sector conjectured above to an improved GnEFT phrased in Wilsonian-type Fermi-liquid theory more powerful and realistic than what has been achieved so far, perhaps along the line of the nonlinear bosonization approach accessing non-Fermi liquid state. It would offer a clear resolution of the dichotomy problem and escape the possible obstruction to the GnEFT.

Funding: This research received no external funding.

Institutional Review Board Statement: Not applicable.

Informed Consent Statement: Not applicable.

Data Availability Statement: Not applicable.

Acknowledgments: We are grateful for collaborations and/or discussions with Tom Kuo, Hyun Kyu Lee, Yong-Liang Ma, Maciej Nowak, Won-Gi Paeng, Sang-Jin Sin, and Ismail Zahed during and after the WCU/Hanyang Project.

Conflicts of Interest: The authors declare no conflict of interest.

References

1. Weinberg, S. What is quantum field theory, and what did we think it is? In *Conceptual Foundations of Quantum Field Theory*; Cambridge University Press: Boston, MA, USA, 1996; pp. 241–251.
2. Skyrme, T.H.R. A unified field theory of mesons and baryons. *Nucl. Phys.* **1962**, *31*, 556–569. [CrossRef]
3. Adam, C.; Martin-Caro, A.G.; Huidobro, M.; Vazquez, R.; Wereszczynski, A. Dense matter equation of state and phase transitions from a Generalized Skyrme model. *arXiv* **2022**, arXiv:2109.13946.
4. Tong, D. Lectures on the quantum Hall effect. *arXiv* **2016**, arXiv:1606.06687.
5. Ma, YL., Rho, M. Topology change, emergent symmetries and compact star matter. *AAPPS Bull.* **2021**, *31*, 16. doi: 10.1007/s43673-021-00016-1. [CrossRef]
6. Rho, M.; Ma, Y.L. Going from asymmetric nuclei to neutron stars to tidal polarizability in gravitational waves. *Int. J. Mod. Phys. E* **2018**, *27*, 1830006. [CrossRef]
7. Ma, Y.L.; Rho, M. Towards the hadron-quark continuity via a topology change in compact stars. *Prog. Part. Nucl. Phys.* **2020**, *113*, 103791. [CrossRef]
8. Rho, M.; Ma, Y.L. Manifestation of hidden symmetries in baryonic matter: From finite nuclei to neutron stars. *Mod. Phys. Lett. A* **2021**, *36*, 2130012. [CrossRef]
9. Zhao, J.; Thakurathi, M.; Jain, M.; Sen, D.; Jain, J.K. Density-functional theory of the fractional quantum Hall effect. *Phys. Rev. Lett.* **2017**, *118*, 196802. [CrossRef]
10. Hu, Y.; Jain, J.K. Kohn-Sham theory of the fractional quantum Hall effect. *Phys. Rev. Lett.* **2019**, *123*, 176802. [CrossRef]
11. Kaplunovsky, V.; Melnikov, D.; Sonnenschein, J. Holgraphic baryons and instanton crystal. In *Multifaceted Skyrmion*, 2nd ed.; Rho, M., Zahed, I., Eds.; World Scientific: Singapore, 2017.
12. Bando, M.; Kugo, T.; Uehara, S.; Yamawaki, K.; Yanagida, T. Is ρ meson a dynamical gauge boson of hidden local symmetry? *Phys. Rev. Lett.* **1985**, *54*, 1215–1218. [CrossRef]
13. Harada, M.; Yamawaki, K. Hidden local symmetry at loop: A New perspective of composite gauge boson and chiral phase transition. *Phys. Rep.* **2003**, *381*, 1–233. [CrossRef]
14. Suzuki, M. Inevitable emergence of composite gauge bosons. *Phys. Rev. D* **2017**, *96*, 065010. [CrossRef]
15. Crewther, R.J. Genuine dilatons in gauge theories. *Universe* **2020**, *6*, 96. arXiv:2003.11259.
16. Crewther, R.J.; Tunstall, L.C. $\Delta I = 1/2$ rule for kaon decays derived from QCD infrared fixed point. *Phys. Rev. D* **2015**, *91*, 034016. [CrossRef]
17. Debbio, L.D.; Zwicky, R. Dilaton and massive hadrons in a conformal phase. *arXiv* **2021**, arXiv:2112.11363.
18. Li, Y.L.; Ma, Y.L.; Rho, M. Chiral-scale effective theory including a dilatonic meson. *Phys. Rev. D* **2017**, *95*, 114011. [CrossRef]
19. Rho, M.; Zahed, I. (Eds.) *Multifaceted Skyrmion*, 2nd ed.; World Scientific: Singapore, 2017.
20. Rho, M.; Sin, S.J.; Zahed, I. Dense QCD: A holographic dyonic salt. *Phys. Lett. B* **2010**, *689*, 23–27. [CrossRef]
21. Park, B.Y.; Vento, V. Skyrmion approach to finite density and temperature. *arXiv* **2009**, arXiv:0906.3263.
22. Harada, M.; Ma, Y.-L.; Lee, H.K.; Rho, M. Fractionized skyrmions in dense compact-star matter. *arXiv* **2016**, arXiv:1601.00058.
23. Lee, H.K.; Ma, Y.L.; Paeng, W.G.; Rho, M. Cusp in the symmetry energy, speed of sound in neutron stars and emergent pseudo-conformal symmetry. *Mod. Phys. Lett.* **2022**, *37*, 2230003. [CrossRef]
24. Zarembo, K. Possible pseudogap phase in QCD. *JETP Lett.* **2002**, *75*, 59–62. [CrossRef]
25. Gil, H.; Papakonstantinou, P.; Hyun, C.H. Constraints on the curvature of nuclear symmetry energy from recent astronomical data within the KIDS framework. *Int. J. Mod. Phys. E* **2022**, *31*, 2250013. [CrossRef]
26. Park, B.Y.; Min, D.P.; Rho, M.; Vento, V. Atiyah-Manton approach to skyrmion matter. *Nucl. Phys. A* **2002**, *707*, 381–398. [CrossRef]
27. Paeng, W.G.; Kuo, T.T.S.; Lee, H.K.; Ma, Y.L.; Rho, M. Scale-invariant hidden local symmetry, topology change, and dense baryonic matter. II. *Phys. Rev. D* **2017**, *96*, 014031. [CrossRef]
28. Zhang, P.; Kimm, K.; Zou, L.; Cho, Y.M. Re-interpretation of Skyrme theory: New topological structures. *arXiv* **2017**, arXiv:1704.05975.
29. Rothstein, I.Z.; Shrivastava, P. Symmetry obstruction to Fermi liquid behavior in the unitary limit. *Phys. Rev. B* **2019**, **99**, 035101. [CrossRef]
30. Canfora, F. Ordered arrays of baryonic tubes in the Skyrme model in $(3+1)$ dimensions at finite density. *Eur. Phys. C* **2018**, *78*, 929. [CrossRef]
31. Karasik, A. Vector dominance, one flavored baryons, and QCD domain walls from the "hidden" Wess-Zumino term. *SciPost Phys.* **2021**, *10*, 138. [CrossRef]
32. Brown, G.E.; Rho, M. Scaling effective Lagrangians in a dense medium. *Phys. Rev. Lett.* **1991**, *66*, 2720–2723. [CrossRef]
33. Hohenberg, P.; Kohn, W. Inhomogeneous electron gas. *Phys. Rev.* **1964**, *136*, B864. [CrossRef]
34. Walecka, J.D. A theory of highly condensed matter. *Annals Phys.* **1974**, *83*, 491–529. [CrossRef]

35. Matsui, T. Fermi liquid properties of nuclear matter in a relativistic mean–field theory. *Nucl. Phys. A* **1981**, *370*, 365–388. [CrossRef]
36. Friman, B.; Rho, M. From chiral Lagrangians to Landau Fermi liquid theory of nuclear matter. *Nucl. Phys. A* **1996**, *606*, 303–319. [CrossRef]
37. Friman, B.; Rho, M.; Song, C. Scaling of chiral Lagrangians and Landau Fermi liquid theory for dense hadronic matter. *Phys. Rev. C* **1999**, *59*, 3357–3370. [CrossRef]
38. Shankar, R. Renormalization group approach to interacting fermions. *Rev. Mod. Phys.* **1994**, *66*, 129. doi: 10.1103/RevModPhys.66.129. [CrossRef]
39. Polchinski, J. Effective field theory and the Fermi surface. In Proceedings of the Recent Directions in Particle Theory, Boulder, CO, USA, 1–26 Junuary 1992.
40. Delacretaz, L.V.; Du, Y.H.; Mehta, U.; Son, D.T. Nonlinear bosonization of Fermi surfaces: The Method of coadjoint orbits. *arXiv* **2022**, arXiv:2203.05004.
41. Ma, Y.L.; Rho, M. Quenched g_A in nuclei and emergent scale symmetry in baryonic matter. *Phys. Rev. Lett.* **2020**, *125*, 142501. [CrossRef]
42. Rho, M. Multifarious roles of hidden chiral-scale symmetry: "Quenching" g_A in nuclei. *Symmetry* **2021**, *13*, 1388. [CrossRef]
43. Pastore, S.; Baroni, A.; Carlson, J.; Gandolfi, S.; Pieper, S.C.; Schiavilla, R.; Wiringa, R.B. Quantum Monte Carlo calculations of weak transitions in A = 6–10 nuclei. *Phys. Rev. C* **2018**, **97**, 022501. [CrossRef]
44. Beane, S.R.; van Kolck, U. The dilated chiral quark model. *Phys. Lett. B* **1994**, *328*, 137–142. [CrossRef]
45. McLerran, L.; Reddy, S. Quarkyonic matter and neutron stars. *Phys. Rev. Lett* **2019**, *122*, 122701. [CrossRef]
46. Annala, E.; Gorda, T.; Kurkela, A.; Nättilä, J.; Vuorinen, A. Evidence for quark-matter cores in massive neutron stars. *Nat. Phys.* **2020**, *16*, 907–910. [CrossRef]
47. Ma, Y.L.; Rho, M. What's in the core of massive neutron stars? *arXiv* **2020**, arXiv:2006.14173.
48. Komargodski, Z. Baryons as quantum Hall droplets. *arXiv* **2018**, arXiv:1812.09253.
49. Lee, H.J.; Min, D.P.; Park, B.Y.; Rho, M.; Vento, V. The proton spin in the chiral bag model: Casimir contribution and Cheshire Cat Principle. *Nucl. Phys. A* **1999**, *657*, 75–94. [CrossRef]
50. Nielsen, H.B.; Rho, M.; Wirzba, A.; Zahed, I. Color anomaly in a hybrid bag model. *Phys. Lett. B* **1991**, *269*, 389–393. [CrossRef]
51. Nielsen, H.B.; Rho, M.; Wirzba, A.; Zahed, I. The tale of the eta-prime from the cheshire cat principle. *Phys. Lett. B* **1992**, *281*, 345–350. [CrossRef]
52. Karasik, A. Skyrmions, quantum Hall droplets, and one current to rule them all. *SciPost Phys.* **2020**, *9*, 8. [CrossRef]
53. Kitano, R.; Matsudo, R. Vector mesons on the wall. *HEP* **2021**, *3*, 23. [CrossRef]
54. Nicola, A.G.; de Elvira, J.R.; Vioque-Rodr´
bibitemguez, A. Thermal hadron resonances in chiral and $U(1)_A$ restoration. *arXiv* **2022**, arXiv:2203.02612.
55. Lopes, L.L.; Biesdorf, C.; Marquez, K.D.; Menezes, D.P. Modified MIT bag models pt II: QCD phase diagram, hot quark stars and speed of sound. *arXiv* **2020**, arXiv:2009.13552.
56. Rather, I.A.; Rahaman, U.; Imran, M.; Das, H.C.; Usmani, A.A.; Patra, S.K. Rotating neutron stars with quark cores. *arXiv* **2021**, arXiv:2102.04067.
57. Nadkarni, S.; Nielsen, H.B.; Zahed, I. Bosonization relations as bag boundary conditions. *Nucl. Phys. B* **1985**, *253*, 308–322. [CrossRef]
58. Nadkarni, S.; Zahed, I. Nonabelian Cheshire Cat bag models in (1+1)-dimensions. *Nucl. Phys. B* **1986**, *263*, 23;
59. Rho, M. Cheshire cat hadrons. *Phys. Rept.* **1994**, *240*, 1–142. [CrossRef]
60. Nielsen, H.B.; Wirzba, A. The Cheshire Cat applied to hybrid bag models. In *Springer Proceedings in Physics*; Springer: Berlin, Germany, 1988; Volume 26. [CrossRef]
61. Damgaard, P.H.; Nielsen, H.B.; Sollacher, R. Smooth bosonization: The Cheshire cat revisited. *Nucl. Phys. B* **1992**, *385*, 227–250.
62. Goldstone, J.; Jaffe, R.L. The baryon number in chiral bag models. *Phys. Rev. Lett.* **1983**, *51*, 1518. [CrossRef]
63. Ma, Y.L.; Rho, M. Dichotomy of baryons as quantum Hall droplets and skyrmions in compact-star matter. *Symmetry* **2021**, *13*, 1888. [CrossRef]
64. Ma, Y.L.; Nowak, M.A.; Rho, M.; Zahed, I. Baryon as a quantum Hall droplet and the hadron-quark duality. *Phys. Rev. Lett.* **2019**, *123*, 172301. [CrossRef]
65. Callan, C.G., Jr.; Harvey, J.A. Anomalies and fermion zero modes on strings and domain walls. *Nucl. Phys. B* **1985**, *250*, 427–436. [CrossRef]
66. Guilarte, J.M.; Vassilevich, D. Fractional fermion number and Hall conductivity of domain walls. *Phys. Lett. B* **2019** *797*, 134935. [CrossRef]
67. Goldstone, J.; Wilczek, F. Fractional quantum numbers on solitons. *Phys. Rev. Lett.* **1981**, *47*, 986. [CrossRef]
68. Kan, N.; Kitano, R.; Yankielowicz, S.; Yokokura, R. From 3d dualities to hadron physics. *arXiv* **2019**, arXiv:1909.04082. [CrossRef]
69. Ma, Y.L.; Rho, M. Scale-chiral symmetry, ω meson and dense baryonic matter. *Phys. Rev. D* **2018**, *97*, 094017.
70. Park, B.Y.; Paeng, W.G.; Vento, V. The Inhomogeneous phase of dense skyrmion matter. *Nucl. Phys. A* **2019**, *989*, 231–245. [CrossRef]
71. Fontana, W.B.; Gomes, P.R.S.; Hernaski, C.A. From quantum wires to the Chern-Simons description of the fractional quantum Hall effect. *Phys. Rev. B* **2019**, *99*, 201113. [CrossRef]
72. Sarti, V.M.; Vento, V. The half-skyrmion phase in a chiral-quark model. *Phys. Lett. B* **2014**, *728*, 323–327. [CrossRef]

73. Sulejmanpasic, T.; Shao, H.; Sandvik, A.; Unsal, M. Confinement in the bulk, deconfinement on the wall: Infrared equivalence between compactified QCD and quantum magnets. *Phys. Rev. Lett.* **2017**, *119*, 091601. [CrossRef]
74. Baym, G.; Hatsuda, T.; Kojo, T.; Powell, P.D.; Song, Y.; Takatsuka, T. From hadrons to quarks in neutron stars: A review. *Rept. Prog. Phys.* **2018**, *81*, 056902. [CrossRef] [PubMed]
75. Alford, M.G.; Sedrakian, A. Compact stars with sequential QCD phase transitions. *Phys. Rev. Lett.* **2017**, *119*, 161104. [CrossRef]

symmetry

Article

Universal Nuclear Equation of State Introducing the Hypothetical X17 Boson

Martin Veselský [1,†], Vlasios Petousis [1,*,†], Jozef Leja [2,†] and Laura Navarro [3,†]

1 Institute of Experimental and Applied Physics, Czech Technical University, 110 00 Prague, Czech Republic
2 Faculty of Mechanical Engineering, Slovak University of Technology in Bratislava, 812 31 Bratislava, Slovakia
3 Department of Physics and Mathematics, University of Zaragoza, 50009 Zaragoza, Spain
* Correspondence: vlasios.petousis@cern.ch
† These authors contributed equally to this work.

Abstract: Within the scope of the Symmetry journal special issue on: "The Nuclear Physics of Neutron Stars", we complemented the nuclear equation of state (EoS) with a hypothetical 17 MeV boson and observed that only instances with an admixture of 30%–40% satisfy all of the constraints. The successful EoS resulted in a radius of around 13 km for a neutron star with mass $M_{NS} \approx 1.4M\odot$ and in a maximum mass of around $M_{NS} \approx 2.5M\odot$. The value of the radius is in agreement with the recent measurement by NICER. The maximum mass is also in agreement with the mass of the remnant of the gravitational wave event GW190814. Thus, it appears that these EoSs satisfy all of the existing experimental constraints and can be considered as universal nuclear equations of state.

Keywords: X17; EoS; neutron star

Citation: Veselský, M.; Petousis, V.; Leja, J.; Navarro, L. Universal Nuclear Equation of State Introducing the Hypothetical X17 Boson. *Symmetry* **2023**, *15*, 49. https://doi.org/10.3390/sym15010049

Academic Editor: Charalampos Moustakidis

Received: 3 November 2022
Revised: 6 December 2022
Accepted: 20 December 2022
Published: 24 December 2022

1. Introduction

In 2016, Krasznahorkay et al. [1] reported an anomaly in the angular correlation of the electron–positron decay of the 1^+ excited level of a ^{8}Be nucleus at 18.15 MeV. An enhancement at a folding angle close to 140 degrees was interpreted as a signature of decay via the emission of a neutral boson with a mass of around $m_X = 17$ MeV. Subsequently, a similar effect was reported by the same group in the decay of the lower 1^+ excited state of ^{8}Be at 17.6 MeV [2] and later in the 0^- excited state of ^{4}He at 21.01 MeV [3], at a folding angle close to 115 degrees. Also recently, the same group investigated the 17.2 MeV $1^- \to 0^+$ transition of the ^{12}C nucleus, resulting in an excess in the folding angle of around 155 degrees [4]. These reported observations placed the hypothetical X17 boson as a dark matter candidate, and, in that spirit, since then, several theoretical works pursued this claim [5,6].

However, an explanation relating this particle to the QCD vacuum was also proposed [7]. In this picture, the 17 MeV particle mediates nucleon–nucleon interactions at large distances between nucleons in the otherwise unbound cluster configuration. A corresponding equation of state was obtained, which was also applied to neutron stars [8].

Since the assumption that the 17 MeV boson is the only carrier of nuclear interactions is somewhat extreme, we explored the possibility of constructing a nuclear equation of state (EoS), introducing both an ω meson with mass 782.5 MeV and a 17 MeV boson in an admixture, which were then tested using experimental constraints on nuclear matter, finite nuclei and heavy ion collisions. The presented analysis falls within the scope of the Symmetry journal special issue on: "The Nuclear Physics of Neutron Stars".

The paper is organized as follows: in Section 2, we introduce the universal nuclear EoS, in Section 3 we present our findings and in Section 4, we discuss the results.

2. Tolman–Oppenheimer–Volkoff Equations and the Equation of State

The structure of a neutron star is usually described using the Tolman–Oppenheimer–Volkoff (TOV) equations (Equations (1) and (2)) based on general relativity:

$$\frac{dP}{dr} = \frac{-G}{c^2}\frac{(P+\epsilon)(m+\frac{4\pi r^3 P}{c^2})}{r(r-\frac{2Gm}{c^2})} \tag{1}$$

$$\frac{dm}{dr} = 4\pi r^2 \frac{\epsilon}{c^2} \tag{2}$$

where $m(r)$ is the total mass contained within radius r and pressure P. The only model-dependent input is the EoS of nuclear matter, which is what makes the neutron star an ideal laboratory for nuclear physics. The EoS of nuclear matter can be described by relativistic mean field theory [9]. The corresponding equations for infinite symmetric nuclear matter are:

$$\begin{aligned}\epsilon = {} & \frac{g_v^2}{2m_v^2}\rho_N^2 + \frac{m_s^2}{2g_s^2}(m_N - m_N^*)^2 + \frac{\kappa}{6g_s^3}(m_N - m_N^*)^3 \\ & + \frac{\lambda}{24g_s^4}(m_N - m_N^*)^4 + \frac{\gamma}{(2\pi)^3}\int_0^{k_F} d^3k\sqrt{k^2 + (m_N^*)^2}\end{aligned} \tag{3}$$

$$\begin{aligned}P = {} & \frac{g_v^2}{2m_v^2}\rho_N^2 - \frac{m_s^2}{2g_s^2}(m_N - m_N^*)^2 + \frac{\kappa}{6g_s^3}(m_N - m_N^*)^3 \\ & + \frac{\lambda}{24g_s^4}(m_N - m_N^*)^4 + \frac{1}{3}\frac{\gamma}{(2\pi)^3}\int_0^{k_F} d^3k\frac{k^2}{\sqrt{k^2 + (m_N^*)^2}}\end{aligned} \tag{4}$$

where ϵ is the energy density, P is the pressure for pure neutron matter, g_s and g_v are the couplings of the scalar and vector boson, respectively, m_s and m_v are the rest masses of scalar and vector bosons, κ and λ are the couplings of the cubic and quartic self-interaction of the scalar boson, m_N and m_N^* are the rest mass and the effective mass of the nucleon, ρ_N is the nucleonic density, k_F is the Fermi momentum of nucleons at zero temperature and γ is the degeneracy (with value γ = 4 for symmetric nuclear matter and γ = 2 for neutron matter).

The EoSs (Equations (3) and (4)), which are regularly used with the ω-meson in the role of the vector boson, were used in [8] for TOV calculations under the assumption that the nuclear force is being mediated by a 17 MeV boson, as reported in the study of the anomalous electron–positron pair production in the excited states of ^{8}Be [1,2], ^{4}He [3] and ^{12}C [4]. Here, we extended our previous work by using the assumption that both the ω-meson and the 17 MeV boson mediate the nuclear force as vector bosons.

After writing the corresponding relativistic mean field (RMF) Lagrangian:

$$\begin{aligned}L_{MFT} = {} & \overline{\psi}\{i\partial^\mu\gamma_\mu - g_v V_0\gamma_0 - (M - g_s\Phi_0)\}\psi - \frac{1}{2}m_s^2\Phi_0^2 - \frac{1}{3!}k\Phi_0^3 - \frac{1}{4!}\lambda\Phi_0^4 \\ & + \frac{1}{2}m_\omega^2(1-q)^2V_0^2 + \frac{1}{2}m_X^2 q^2 V_0^2\end{aligned} \tag{5}$$

with duplicate vector boson terms, we conclude that the resulting EoS will be identical to the above, with "effective" vector boson mass:

$$m_v^{*2} = q^2 m_X^2 + (1-q)^2 m_\omega^2 \tag{6}$$

where q is the admixture coefficient of the $m_X = 17$ MeV boson to the total vector potential. Depending on the value of q, the effective mass can range from $m_\omega = 782.5$ MeV to 17 MeV. We decided to test this theory using various available constraints, ranging from properties of finite nuclei, through heavy ion collisions all the way to the neutron stars.

3. Analysis Results

As a first step, we generated the EoS of infinite symmetric nuclear matter using values of the vector boson effective mass corresponding to an admixture of a 17 MeV boson ranging between 20% to 50% and choosing the values of couplings within corresponding ranges depicted in Table 1. Each set of parameters was tested for binding energy (16 MeV) and saturation density ρ_0 = 0.15–0.16 fm^{-3}. Successful sets of parameters were further tested for incompressibility within the range: $K_0 = 250 \pm 20$ MeV.

Table 1. Constrained parameter sets for three EoSs with three admixtures q and incompressibility $K_0 = 250 \pm 20$ MeV.

K_0	q	κ	λ	g_v	g_s	m^*_v [MeV]	m_σ [MeV]
235.95	0.3	21.50	−163.33	8.38	9.20	547.77	482.16
269.14	0.4(A)	11.00	−50.00	6.85	7.23	469.55	391.44
257.50	0.4(B)	11.50	−60.00	6.85	7.23	469.55	391.44

The parameter sets that passed the first step were used to calculate properties of the finite nucleus ^{208}Pb; in particular, its binding energy (1636 MeV) and neutron skin $\Delta R_{PREX2} = 0.283 \pm 0.071$ fm. The latter value is of special interest since recent measurements [10] reported a value larger than the predictions of theory. The RMF code of Ring, Gambhir and Lalazissis from CPC [11] was used for calculation. The code also uses the ρ-meson as a mediator of the isovector interaction and thus a measure of the symmetry energy. We kept the ρ-meson coupling identical to the NL3 EoS [12]. The NL3 EoS can reproduce the values of the binding energy and neutron skin of ^{208}Pb; however, the incompressibility is unrealistically high and constraints from nuclear reactions are not satisfied.

A typical picture is shown in Figure 1, where the values of the binding energy and the neutron skin $\Delta R = R_n - R_p$ are plotted. The main sequence does not seem to fulfill both constraints; nevertheless, several combinations of parameters appeared to satisfy both constraints. These were parameter sets with the 17 MeV boson admixture ranging between 20% and 40%. However, the parameter sets with a 20% admixture fail to satisfy constraints from heavy ion collisions [13], and thus only parameter sets with an admixture of 30% to 40% remain, signalling that there is some range of admixtures that satisfies all of the constraints. Such an observation can have physical meaning.

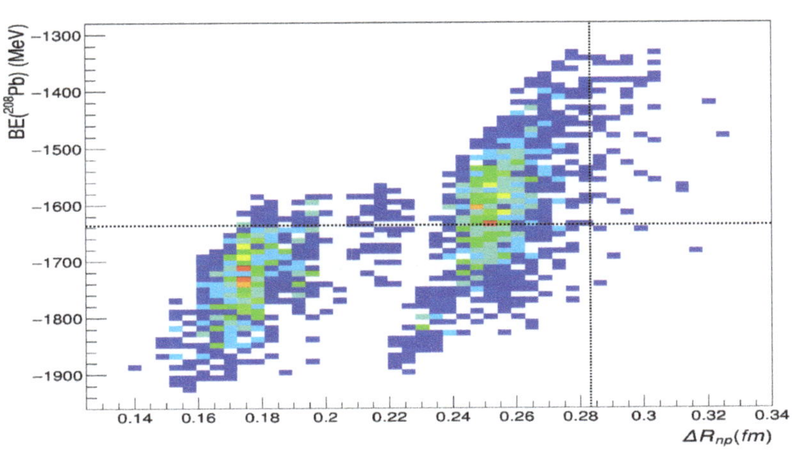

Figure 1. (Color online). Binding energy (BE) of the ^{208}Pb versus its neutron skin using 30% admixture of the 17 MeV boson in an EoS.

For the TOV calculations, the equation of state $P(\rho)$ needed to be expressed in the form of polytropes. For that reason, three transition densities were defined—ρ_1 =

2.8×10^{14} g/cm^3, $\rho_2 = 10^{14.7}$ g/cm^3 and $\rho_3 = 10^{15}$ g/cm^3—and four parameters were calculated: three exponents of the power law polytropes $\Gamma_1, \Gamma_2, \Gamma_3$, respectively, and the value a_0 (where $a_0 = log(p(\rho_1)) + \Gamma_1(log(\rho_2) - log(\rho_1))$. In the last step, the remaining equations of state, specifically their versions for pure neutron matter, were used as an input to the TOV equation, and the resulting mass–radius plot is shown in Figures 2 and 3. The three EoSs listed in Table 2 result in a radius of the neutron star of 1.4 solar masses around 13 km and a maximum mass of the neutron star of around 2.5 solar masses. The value of the radius is in agreement with the recent measurement by NICER [14,15], and the value of the maximum mass is in agreement with the recently reported mass of pulsar 2.35 solar masses [16] and potentially also with the mass of the remnant of the gravitational wave event GW190814 [17]. Thus, it appears that these three EoSs satisfy all of the existing experimental constraints and can be considered as universal equations of state of nuclear matter.

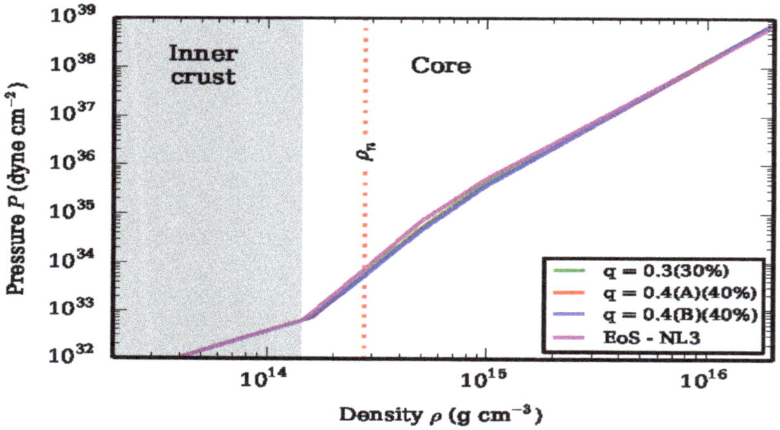

Figure 2. (Color online). The pressure as function of nuclear density for three EoSs with admixtures of 30% and 40% of the 17 MeV boson plus the NL3 EoS. The parameters are defined in Table 2.

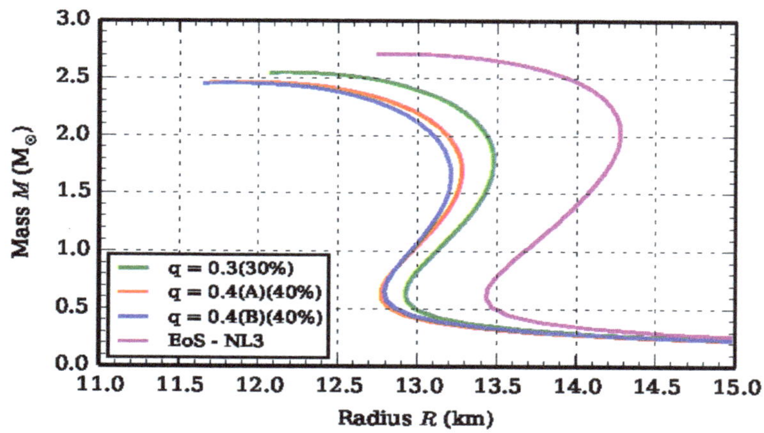

Figure 3. (Color online). The mass–radius relation for three EoSs plus the NL3 with admixtures of 30% and 40% of the 17 MeV boson plus the NL3 EoS.

Table 2. Polytropes for three EoSs plus the NL3 EoS used for the Tolman–Oppenheimer–Volkoff calculations. The 0.3 EoS represents a 30% admixture of the $X17$ boson and the 0.4A and 0.4B EoSs represent a 40% admixture with different values of parameters κ and λ.

EoS q-Admixture (%)	a_0	Γ_1	Γ_2	Γ_3	K_0 (MeV)
0.3 (30%)	34.703	3.741	3.118	2.497	235.95
0.4(A) (40%)	34.673	3.744	3.036	2.517	269.14
0.4(B) (40%)	34.653	3.643	3.095	2.540	257.50
NL3 (0%)	34.846	3.872	2.925	2.394	332

4. Conclusions

In summary, within the scope of the Symmetry journal special issue on: "The Nuclear Physics of Neutron Stars", we implemented a hypothetical 17 MeV boson to a nuclear EoS complementing the ω meson and observed that only instances with an admixture of 30–40% satisfy all of the experimental constraints. When applied to TOV equations, the successful EoSs result in a radius of around 13 km for a neutron star with a mass of $M_{NS} \approx 1.4M\odot$ and in a maximum mass of around $M_{NS} \approx 2.5M\odot$. The values of our results are in good agreement with the recent measurement reported by NICER [14,15]. The obtained value of the maximum mass is also in agreement with the recently reported mass of a pulsar [16] and potentially also with the mass remnant of the gravitational wave event GW190814 [17]. Thus, it appears that these EoSs satisfy all of the existing experimental constraints and can be considered as universal EoSs of nuclear matter.

Author Contributions: M.V., Conceptualization; V.P., Writing-review-editing and scientific research; J.L., Software; L.N., investigation. All authors have read and agreed to the published version of the manuscript.

Funding: This research was funded by Czech Science Foundation-GACR Contract No. 21-24281S.

Institutional Review Board Statement: Not applicable.

Informed Consent Statement: Informed consent was obtained from all subjects involved in the study.

Data Availability Statement: Not applicable.

Acknowledgments: This work is supported by the Czech Science Foundation (GACR Contract No. 21-24281S). The simulations were performed at the Supercomputing facility of Czech Technical University in Prague.

Conflicts of Interest: The authors declare no conflict of interest.

References

1. Krasznahorkay, A.J.; Csatlós, M.; Csige, L.; Gácsi, Z.; Gulyás, J.; Hunyadi, M.; Kuti, I.; Nyakó, B.M.; Stuhl, L.; Timár, J.; et al. Observation of Anomalous Internal Pair Creation in ^{8}Be: A Possible Indication of a Light, Neutral Boson. *Phys. Rev. Lett.* **2016**, *116*, 042501. [CrossRef] [PubMed]
2. Krasznahorkay, A.J.; Csatlós, M.; Csige, L.; Gulyás, J.; Hunyadi, M.; Ketel, T.J.; Krasznahorkay, A.; Kuti, I.; Nagy, A.; Nyakó, B.M.; et al. New experimental results for the 17 MeV particle created in ^{8}Be. *EPJ Web Conf.* **2017**, *137*, 08010. [CrossRef]
3. Krasznahorkay, A.J.; Csatlós, M.; Csige, L.; Gulyás, J.; Koszta, M.; Szihalmi, B.; Timár, J.; Firak, D.S.; Nagy, Á.; Sas, N.J.; et al. A new anomaly observed in ^{4}He supporting the existence of the hypothetical X17 particle. *J. Phys. Conf. Ser.* **2020**, *1643*, 012001. [CrossRef]
4. Krasznahorkay, A.J.; Krasznahorkay, A.; Begala, M.; Csatló, M.; Csige, L.; Gulyás, J.; Krakó, A.; Timár, J.; Rajta, I.; Vajda, I.; et al. New anomaly observed in ^{12}C supports the existence and the vector character of the hypothetical X17 boson. *arXiv* **2022**, arXiv: 2209.10795
5. Feng, J.F.; Fornal, B.; Galon, I.; Gardner, S.; Smolinsky, J.; Tait, T.M.P.; Tanedo, P. Protophobic Fifth-Force Interpretation of the Observed Anomaly in ^{8}Be Nuclear Transitions. *Phys. Rev. Lett.* **2016**, *117*, 071803. [CrossRef]
6. Feng, J.F.; Fornal, B.; Galon, I.; Gardner, S.; Smolinsky, J.; Tait, T.M.P.; Tanedo, P. Particle physics models for the 17 MeV anomaly in beryllium nuclear decays. *Phys. Rev. D* **2017**, *95*, 035017. [CrossRef]
7. Veselský, M.; Petousis, V.; Leja, J. Anomaly in the decay of ^{8}Be and ^{4}H—Can an observed light boson mediate low-energy nucleon-nucleon interactions? *J. Phys. G Nucl. Part. Phys.* **2021**, *48*, 105103. [CrossRef]

8. Petousis, V.; Veselský, M.; Leja, J. Neutron star structure with nuclear force mediated by hypothetical X17 boson. *EPJ Web Conf.* **2021**, *252*, 04008. [CrossRef]
9. Serot, B.D.; Walecka, J.D. Recent Progress in Quantum Hadrodynamics. *Int. J. Mod. Phys. E* **1997**, *6*, 515. [CrossRef]
10. Adhikari, D.; Albataineh, H.; Androic, D.; Aniol, K.; Armstrong, D.S.; Averett, T.; Barcus, S.; Bellini, V.; Beminiwattha, R.S.; Benesch, J.F.; et al. Accurate Determination of the Neutron Skin Thickness of ^{208}Pb through Parity-Violation in Electron Scattering. *Phys. Rev. Lett.* **2021**, *126*, 172502. [CrossRef]
11. Ring, P.; Gambhir, Y.K.; Lalazissis, G.A. Computer program for the relativistic mean field description of the ground state properties of even-even axially deformed nuclei. *Comput. Phys. Commun.* **1997**, *105*, 77–97. [CrossRef]
12. Lalazissis, G.A.; König, J.; Ring, P. New parametrization for the Lagrangian density of relativistic mean field theory. *Phys. Rev. C* **1997**, *55*, 540. [CrossRef]
13. Danielewicz, P.; Lacey, R.; Lynch, W.G. Determination of the Equation of State of Dense Matter. *Science* **2002**, *298*, 1592. [CrossRef] [PubMed]
14. Riley, T.E.; Watts, A.L.; Bogdanov, S.; Ray, P.S.; Ludlam, R.M.; Guillot, S.; Arzoumanian, Z.; Baker, C.L.; Bilous, A.V.; Chakrabarty, D.; et al. A NICER View of PSR J0030+0451: Millisecond Pulsar Parameter Estimation. *Astrophys. J. Lett.* **2019**, *887*, L21. [CrossRef]
15. Miller, M.C.; Lamb, F.K.; Dittmann, A.J.; Bogdanov, S.; Arzoumanian, Z.; Gendreau, K.C.; Guillot, S.; Harding, A.K.; Ho, W.C.G.; Lattimer, J.M.; et al. PSR J0030+0451 Mass and Radius from NICER Data and Implications for the Properties of Neutron Star Matter. *Astrophys. J. Lett.* **2019**, *887*, L24. [CrossRef]
16. Romani, R.W.; Kel, D.; Filippenko, A.V.; Brink, T.G.; Zheng, W. PSR J0952-0607: The Fastest and Heaviest Known Galactic Neutron Star. *Astrophys. J. Lett.* **2022**, *934*, L18. [CrossRef]
17. Abbott, R.; Abbott, T.D.; Abraham, S.; Acernese, F.; Ackley, K.; Adams, C.; Adhikari, R.X.; Adya, V.B.; Affeldt, C.; Agathos, M.; et al. GW190814: Gravitational Waves from the Coalescence of a 23 Solar Mass Black Hole with a 2.6 Solar Mass Compact Object. *Astrophys. J.* **2020**, *896*, L44. [CrossRef]

symmetry

MDPI

Article

Dense Baryonic Matter Predicted in "Pseudo-Conformal Model"

Mannque Rho

Institut de Physique Théorique, Université Paris-Saclay, CNRS, CEA, 91191 Gif-sur-Yvette, France; mannque.rho@ipht.fr

Abstract: The World-Class University/Hanyang Project launched in Korea in 2008 led to what is now called the "pseudo-conformal model" that addresses dense compact star matter and is confronted in this short note with the presently available astrophysical observables, with focus on those from gravity waves. The predictions made nearly free of parameters by the model involving "topology change" remain more or less intact "un-torpedoed" by the data.

Keywords: massive compact stars; EoS; topology change; hadron-quark continuity; Cheshire Cat Principle; hidden symmetries dual to gluons; G*n*EFT; psuedo-conformality

Citation: Rho, M. Dense Baryonic Matter Predicted in "Pseudo-Conformal Model". *Symmetry* **2023**, *15*, 1271. https://doi.org/10.3390/sym15061271

Academic Editor: Charalampos Moustakidis

Received: 25 May 2023
Revised: 12 June 2023
Accepted: 14 June 2023
Published: 16 June 2023

1. Introduction

In 2008, the Korean Government launched a five-year "World-Class University (WCU)" Project, and the Hanyang University in Seoul was chosen as one of the projects to be under the directorship of Hyun Kyu Lee in the Physics Department. The objective of the WCU/Hanyang was to elevate the university in basic science to the world-class level, in anticipation of the forthcoming establishment of an ambitious research institute called the Institute of Basic Science (IBS). The subject matter picked was "Baryonic Matter under Extreme Conditions in the Universe", which focused on the superdense matter expected to be found in massive compact stars on the verge of gravitational collapse. This subject matter was already one of the major themes at the Korea Institute of Advanced Studies (KIAS) in late 1990s and early 2000s while I was an invited professor in its School of Physics, working in collaboration with Hyun Kyu Lee, Dong-Pil Min, and Byung-Yoon Park of Korea and Vicente Vento of Spain, all at the KIAS as visiting scholars.

The property of dense baryonic matter in compact stars is in the realm of QCD involving both low and high densities. However, QCD cannot access the density regimes, famously non-perturbative, of nuclear and compact star matter. Therefore, there was no reliable theoretical tool to access the regimes concerned. Neither could it be accessed experimentally, since no accelerators probing dense matter at low temperature involved were available then. What started at the WCU/Hanyang Project was the construction of a *single* unified theoretical framework to explore these uncharted density regimes starting with what was explored in KIAS. The objective was to formulate an effective field theory approach with a minimal number of unknown parameters, *post-dict correctly* the known nuclear matter properties at $n \sim n_0 \simeq 0.16\ \text{fm}^{-3}$, and *predict* the terrestrial nuclear and compact star properties that were yet to be measured. It was, in our mind, in anticipation of what is to be studied at the costly RIB machine "RAON" approved to be constructed at the IBS.

The status of the model in nuclear physics and astrophysics up to early 2017 before the advent of the recent gravity wave measurements was sketched in [1]. The gravity wave data made feasible the direct confrontation of some of the predictions made then with the oncoming observables.

To the utter surprise—and perhaps more to the incredulity—of the workers in the field, what appears to be an over-simplified "coarse-grained framework" with no parameter fiddling—in stark contrast to the currently favored approaches of hybridizing with "artificially revamped" quark descriptions—has met, so far, with no serious tension in explaining satisfactorily *all* up-to-date available data. In this paper, I will list the most relevant observables—there are too many to be fair to all—both nuclear and astrophysical, just to show how the predictions that followed from the formulation initiated at the KIAS and pursued at the WCU/Hanyang fare, and how the possible discrepancies, if any, between what is predicted and what is measured can be reconciled within the model. It should be stressed that the spirit of this presentation is basically different from the current activities in the field where various sophisticated statistical analyses in the theoretical inputs and experimental results are focused on. All the results I will give are found essentially in the two papers [2,3], which constituted an important part of the PhD thesis of Won-Gi Paeng and are extensively reviewed in [4,5]. Some trivial numerical errors committed in [3] that remained in [4,5] will be corrected in the predicted results cited in what follows. Only if necessary will I refer to the specific articles for more precision or explanation. Otherwise I will avoid entering into details as much as possible.

2. G*n*EFT

In going from nuclear matter to dense compact star matter, as is commonly believed, there must be present a transition, either a phase change or just a continuous crossover, from the (low-)density regime, say, $\sim 2n_0$, of hadrons to the (high-)density regime, say, $\gtrsim 6n_0$, of compact stars. This transition is commonly referred to as "hadron-quark continuity (HQC)", presumed—but not proven—to be encoded in QCD. The strategy that was adopted in the WCU/Hanyang was that this HQC could be effectuated by a change in topology from baryons in the baryonic matter to fractionally charged objects in the compact star matter, an idea anchored on what is referred to as the "Cheshire Cat Principle (CCP)". This idea followed from the notion that in QCD, a nucleon can be described as a topological object, say, a skyrmion and half-skyrmions at large N_c and at high density when put on a crystal lattice. An early review on this matter can be found in [6].

The key idea of how to implement the skyrmion–half-skyrmion transition—referred to in what follows as "topology change" **(**The topology change involved here could be different in character from what is taking place in condensed matter systems.)—as a mechanism for the HQC was worked out first in early 2000 but appeared in the literature a decade later [7]. The publication of this work was delayed so long due to the referees' objections to the novel ideas developed in the paper, dismissing them as mere "conjectures" without any counter arguments.

The topology change involved here is best described in terms of skyrmions put on crystal lattice, although it is well known that the skyrmion-half-skyrmion changeover actually makes no sense. This is because whereas the 1/2-skyrmion phase can be justified on crystal lattice at high density (and large N_c limit), low-density matter cannot be in crystal, so the transition, whether bona-fide phase transition or smooth cross-over, cannot be established with skyrmions on crystal lattice [8]. This of course does not mean that the crossover in the skyrmion description in a more general setting does not exist. In fact it is this point that was resolved in [7]; it involves hidden local symmetry (HLS) and hidden scale symmetry (HSS) entering into the baryonic structure. The details given in [4,5] on how the hidden symmetries must figure appear to be somewhat complicated at first sight, but the basic structure is rather simple as I will try to explain. See [9] for more details. I believe this accounts for the lack of attention paid to this development in nuclear and astrophysical communities.

The most crucial ingredient for the topology change is the cusp structure in the symmetry energy (denoted E_{sym}). It reflects the isospin asymmetry in the energy functional $E(n)$. The cusp is seen when the nucleons are put on a crystal lattice. It appears at the density, denoted $n_{1/2}$, lying above the normal matter density n_0. Identified as the putative HQC density, it is found to be in the range

$$n_{\rm HQC} \sim n_{1/2} \approx (2-4)n_0. \quad (1)$$

This cusp is displayed by the dotted red curve in the schematic figure, Figure 1.

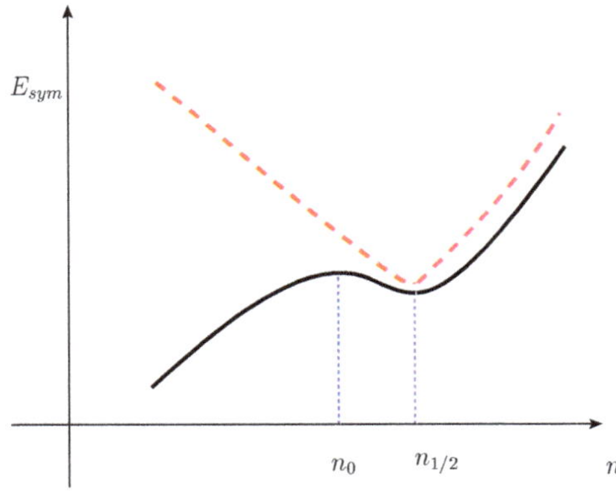

Figure 1. Schematic illustration of the symmetry energy $E_{sym}(n)$ by the skyrmion crystal (red dashed line)) and by nucleon correlations dominated by the nuclear tensor forces (solid line).

While one can reasonably assume that $E_{sym}(n)$ for $n \gtrsim n_{1/2}$ makes sense on the crystal lattice, the behavior for $n < n_{1/2}$, however, cannot be taken seriously, as mentioned above.

First, how does the topology change take place?

It is triggered on the crystal lattice by the bilinear quark condensate $\langle\bar{q}q\rangle$, when averaged, going to zero, whereas the pion decay constant remains non-zero. So it does not involve chiral symmetry restoration. It implies that the non-vanishing order parameter may be the quartic quark condensate $\langle\bar{q}\bar{q}qq\rangle$. There are arguments in the literature that such a symmetry structure is at odds with 't Hooft anomaly constraints. However, it remains controversial whether such a no-go theorem holds in the present case. In fact there are some cases in condensed matter where this no-go theorem does not seem to hold [10].

To exploit this cusp structure requires knowing how the topology change can be modified in reality. Now how can the topology change be incorporated in a realistic theory?

As shown in [7], it is the hidden symmetries that bring in heavy degrees of freedom to an effective Lagrangian. It is now recognized that Weinberg's chiral effective field theory (chiEFT) with the nucleons and pions as the *only* relevant degrees of freedom, which will be called SchiEFT in what follows, with "S" standing for "standard", works well with a suitable cutoff $\Lambda_{sEFT} \lesssim m_\rho$ up to the density n_0 and slightly higher. It is bound to break down at higher densities, say, $\gtrsim 2n_0$. This success can be considered as a case where Weinberg's "Folk Theorem" on EFT is "proven". What was shown in [7] is that the vector mesons $V = (\rho, \omega)$ and the scalar meson (σ_d), which is related to what is known as "conformal compensator", with the former endowed with "vector manifestation fixed point" (VMFP) (at which the mass $m_\rho \to \epsilon \approx 0$ [11]) and the latter with the "genuine dilaton" with an IR fixed point (at which the dilaton mass tends to $m_{\sigma_d} \to 0$), enable one to go across, in the E_{sym} (more generally the EoS), "smoothly" from below to above $n_{1/2}$. What is in action is the interplay between the VMFP and the nuclear tensor force that leads to $m_V \to 0$ at high density ($\gtrsim 25n_0$) and the "genuine dilaton" with an IR fixed point at $n_{IR} \gtrsim 25n_0$ at which scale symmetry is restored. The net effect of the interplays at $n_{1/2}$ is displayed by the black solid curve in Figure 1. It will be shown later that the cusp, smoothed to an inflection, will play an important role in confronting some of the important gravity wave data, such as the tidal deformability, sound velocity, etc.

The effective theory GnEFT detailed in the reviews [4,5] is formulated with the Lagrangian $\mathcal{L}_{\psi\pi\text{HLS}\chi}$ with the HLS mesons and the genuine dilaton (GD) scalar included as the relevant degrees of freedom in addition to the ψ, π that figure in SchiEFT. The heavy degrees of freedom (HDFs for short) are to mediate the crossover from hadrons to quark/gluons. In our approach, it is here that the topology change enters as a mechanism for hadron–quark continuity.

Given the Lagrangian $\mathcal{L}_{\psi\pi\text{HLS}\chi}$, there can be several ways of setting up a GnEFT. What is required is the implementation of the HQC at a density $n_{\text{HQC}} > n_0$.

It turns out to be feasible to set up a scale-HLS-invariant Lagrangian (with HLS gauge-equivalent to non-linear sigma model, so chiral symmetry is encoded therein [11].) with a power expansion going beyond the chiral expansion employed in the standard chiral EFT by taking into account the hidden symmetries including HLS. The expansion has been worked out to NLO in scale-chiral expansion following [11–13]. Unfortunately there are much too many parameters even at NLO that it has remained unexplored.

An alternative approach exploited in the WCU/Hanyang program was to use the "double-decimation" strategy developed in [14], which is to apply (Wilsonian) renormalization group approach to the strongly correlated fermions on the Fermi sphere. The first decimation is made to the Landau(-Migdal) Fermi liquid fixed point (FLFP) with the cutoff Λ_{FL} on top of the Fermi sea along the line developed in [15] for electrons. Then we carry out the second decimation going beyond the FLFP. It was shown a long time ago that a chiral Lagrangian of the $\mathcal{L}_{\psi\pi\text{HLS}\chi}$-type, somewhat simplified, can be mapped to Landau(-Migdal) Fermi liquid structure, which worked remarkably well at the FLFP level [16,17]. This structure is incorporated into the GnEFT with the possibility of going beyond the FLFP in the V_{lowK}-RG approach, as developed by Tom Kuo with Gerry Brown and collaborators at Stony Brook. Tom Kuo's role in the initial development resulted in the crucial publication of [2,3]. In the predictions discussed below, it will be primarily at the level of the FLFP approximation. The corrections in the V_{lowK}-RG will be quoted to justify the FLFP approximation.

For those who are not familiar with the GnEFT strategy sketched above, let me just mention that this approach can be considered as a "refined" version of covariant density functional approaches anchored on the Hohenberg–Kohn theorem on DFT. The refinement, among others, has to do with the replacement of the high-dimension field operators, injected (arbitrarily) to improve the Walecka-type linear model (e.g., the excessively high nuclear matter compression modulus K_0), by the parameters of the Lagrangian with the dilaton condensate $\langle\chi\rangle$ encoded by the scale-chiral symmetry. The approach is free of arbitrariness and thermodynamically consistent [17].

3. Predictions

Here I will give the predictions obtained in [3] and listed in [4,5]. What is given involves no fiddling in the parameters in the Lagrangian $\mathcal{L}_{\psi\pi\text{HLS}\sigma_d}$. Only some numerical errors committed in [3] will be corrected.

3.1. Density Regime $n \lesssim n_0$

First up to $n_{1/2}$, at which the HQC intervenes, what is more or less equivalent to what is given in SchiEFT is reproduced by the mean-field of $\mathcal{L}_{\psi\pi\text{HLS}\chi}$, the parameters of which are controlled by BR scaling Φ sliding in density in the dilaton condensate $\langle\chi\rangle^*$ (where $*$ stands for the density dependence) known up to n_0. At the equilibrium density n_0, one post-dicts

$$n_0 = 0.16\ \text{fm}^{-3},\ E/A = -16.7\ \text{MeV},\ K_0 = 250\ \text{MeV}.$$

Just to give an idea what the significance of this result is, let me quote what the present state-of-the-art high-order ($\text{N}^{\gtrsim 2}$LO) SchiEFT calculation obtains: $n_0 = 0.164 \pm 0.07$ and $E/A = -15.86 \pm 0.37 \pm 0.2$ MeV. All other nuclear matter properties at the equilibrium density n_0 (including the symmetry energy $J = E_{sym}(n_0)$) do come out essentially the same as what are calculated in SchiEFT at $N^{\geq 2}LO$. The only parameter needed is the mass of the "genuine dilaton" identified with $f_0(500)$. The BR scaling relates the scaling of the dilaton condensate to that of the pion condensate

$$\Phi(n) = f^*_{\sigma_d}/f_{\sigma_d} \simeq f^*_\pi/f_\pi \tag{2}$$

which is measured in deeply bound atomic nuclei $\Phi(n_0) \approx 0.8$. This is given by chiral symmetry, so is not a parameter. Roughly speaking, the linear HLS with the BR scaling does what covariant density functional models with higher-dimension operators do. The power of this approach over covariant density functional models is that the parameters of higher derivative terms are fixed by hidden local symmetry with thermodynamic consistency [17]. It also captures higher chiral power terms, say, N^3LO in SchiEFT.

On the other hand, the symmetry energy slope L could be different from what one obtains in SchiEFT. This is because of the onset of the cusp, as shown in Figure 1 at $n > n_{1/2}$. The cusp as discussed in [7] involves the tensor force structure controlled by the behavior of the HLS gauge coupling g_ρ running in the RG flow toward the vector manifestation $g_\rho \to 0$. If $n_{1/2}$ were not too far above n_0, then the slope of $E_{sym}(n)$ at n_0 would be inevitably affected by the hidden cusp structure. I will not go for higher derivatives of E_{sym}—such as $K_{\rm sym}$ with two derivatives—since they will depend more sensitively on where n_{12} lies.

The GnEFT predicts for $n_{1/2} \sim (2-3)n_0$

$$J \equiv E_{sym}(n_0) = 30.2 \text{ Mev}, \ L = 67.8 \text{ MeV} \tag{3}$$

to be compared with the SchiEFT results

$$J = 32.0 \pm 1.1 \text{ Mev}, \ L = 51.9 \pm 7.9 \text{ MeV}. \tag{4}$$

It should be noted that while J is more or less the same as what SchiEFT gives, "soft" in the EoS, L is significantly greater than that of SchiEFT, showing the (smooth) onset of hardness, tending toward what is observed in the PREX/Jefferson experiment $L = 106 \pm 37$ [18,19]. What is noteworthy is that the behavior of $E_{sym}(n)$ near n_0 in GnEFT manifesting the "pseudo-gap" behavior of the chiral condensate in the topology change *predicts naturally* the soft-to-hard crossover tendency of the EoS at $\sim n_{1/2}$, which is attributed to the putative HQC in QCD.

3.2. Density Regime $n > n_{1/2}$

Although the slope L given in (3) can be considered as a prediction, not as a post-diction, of the PCM, one cannot, however, have a great confidence in its precision. The reason is that it is the most difficult density regime in the EoS to theoretically control. At $n_{1/2}$, EFT valid at low density and perturbative QCD valid at high density "meet". Therefore, the slope L will be sensitive to the location with interplay of different degrees of freedom that can be treated with the least confidence. This aspect will appear significantly in the tidal deformability Λ measured at 1.4 $M_\odot$ and also in the sound velocity of the star.

While the $n \lesssim n_{1/2}$ region is controlled essentially by the scaling factor Φ, accessible both by theory and experiment, the topology change brings in major modifications in the properties of the Lagrangian $\mathcal{L}_{\psi\pi\text{HLS}\chi}$. This is explained in terms of a series of "Propositions" in [4]. I admit that some of them are superfluous or redundant and could be largely weeded out.

Basically what happens is rather simple.

Phenomenology in nuclear processes suggests the crossover density regime overlaps with the point n_{DD} at which the double decimation is to be made [14]. It has been taken to be [2,3]

$$n_{DD} \simeq n_{1/2}. \tag{5}$$

The primary mechanism that produces the cusp in the symmetry energy E_{sym}, namely, the skyrmion-1/2-skyrmion transition density, is driven in GnEFT by the nuclear tensor forces sliding with density, going to $\sim$0 at the range most effective, say, $\sim$1 fm in nuclear interactions. What was required was that the VM fixed point density n_{VM} be $n_{VM} \gtrsim 25n_0$ [3], much greater than the $\sim(6–7)n_0$ thought to be present in the core of massive stars. This feature required that while the pion decay constant f_π does not go to zero at n_{VM}, it is the gauge coupling g_ρ that should tend to zero [11]. (I note as a footnote that this feature, which presumably takes place also in temperature, was not taken into account in heavy-ion experiments looking for the dropping ρ mass near the chiral restoration temperature T_c. It led to the erroneous "ruling out of BR scaling" following the NA60 data.) The scenario with $n_{VM} \gtrsim 25n_0$ differs from $n_{VM} \sim 6n_0$ [2] in the prediction for the sound speed v_s in compact stars. How the VM density n_{VM} intervenes in the pseudo-conformal behavior of the sound velocity however remains mysterious.

Another important property in $n \gtrsim n_{1/2}$ is that the dilaton decay constant** (**From here on, I will use the linear conformal compensator field χ instead of the nonlinear field σ_d for the dilaton field, $\chi = f_\chi e^{\sigma_d/f_\chi}$.) f^*_χ gets locked to the pion decay constant f^*_π in the GD scheme [12] and remains more or less constant:

$$\mathbf{I}: f^*_\chi \simeq f^*_\pi \propto m_0 \text{ for } n > n_{1/2} \tag{6}$$

where m_0 is a chiral symmetric mass of the quasiparticle in the 1/2-skyrmiom phase. This follows from the *emergent* parity doubling in the baryon structure. This feature differs from other parity-doubling scenarios where the symmetry is present intrinsically, not emergent, in the effective Lagrangian [20]. It is not clear at the moment how this difference impacts on the properties of compact stars. One of the crucial consequences of this parity doubling is that the $U(2)$ symmetry for the ρ and ω, fairly good in $n < n_{1/2}$, gets broken by the dynamics involved in the quasiparticle interactions with ω and χ exchanges in the 1/2-skyrmion phase

$$m^*_\rho/m_\rho \neq m^*_\omega/m_\omega \tag{7}$$

and leads to weakly interacting quasiparticles of two 1/2-skyrmions bound or confined by hidden monopoles [21] with the quasiparticle mass

$$\mathbf{II}: m^*_Q \to f^*_\chi \to m_0. \tag{8}$$

I will speculate below how the suppression of the monopoles could liberate the half-skyrmions and transform them into fractionized quasiparticles that mimic fractionally charged quarks.

What is given in (6) can also be obtained in what is referred to as "dilaton-limit fixed point" [22] when $\mathrm{Tr}(\Sigma\bar{\Sigma}) \to 0$ where $\Sigma = \frac{f_\pi}{f_\chi} e^{i\pi/f_\pi}\chi$ in the mean field of GnEFT. In that limit one finds

$$\mathbf{III}: g^*_A \to 1,\ f^*_\chi \to f^*_\pi. \tag{9}$$

Since QCD cannot be solved nonperturbatively for the various limiting conditions, the locations of the DLFP, the vector manifestation (VM) fixed point, the IR fixed point, etc., though not too far apart, are not precisely known. For the issue concerned, i.e., the physics of compact stars, whether or not and where they overlap cannot be addressed. They may, however, be irrelevant for the qualitative properties we are interested in near the density regime of HQC.

To be more quantitative, one needs to go beyond the mean-field-level approximation of GnEFT. To do this, the $1/\bar{N}$ corrections to the Landau Fermi-liquid fixed point approximation—in V_{lowK} RG in the double-decimation strategy [14]—could be made as described in [3]. In this reference, a rather involved scaling behavior of the ρ gauge coupling constant g_ρ^* in the vicinity of the crossover density $n_{1/2}$ was used. Although it has not been checked in detail, it seems most likely that such a complicated scaling behavior is unnecessary because it simply reflects how the gauge coupling moves toward the vector manifestation density n_{VM} that lies way above the density involved in the star. This is indeed supported in the "pseudo-conformal model (PCM)" (defined below) used for making predictions.

In listing the predictions made in GnEFT, there are two additional remarks to make. First, the predictions have been made for the range of the crossover density

$$\mathbf{IV}: 2 < n_{1/2}/n_0 < 4. \tag{10}$$

The predictions are roughly the same within that range so I will not favor any specific values in between. The extremes $n_{1/2}/n_0 = 2$ and 4 are somewhat disfavored, although they cannot be dismissed, as we will see. Second, the prediction made in the PCM is checked with the double decimation V_{lowK} RG only for $n_{1/2}/n_0 \sim 2$. It was concluded that the same should hold for the range (10).

Second, the PCM** ** I must admit that the term "pseudo-conformal" could be a misnomer. It simply indicates that conformal symmetry, both explicitly and spontaneously broken, *emerges* in dense matter driven by nuclear interactions.) was constructed by replacing the V_{lowK} RG for $n \geq n_{1/2}$ in the energy density of the nucleon by two-parameter analytic form matched at $n = n_{1/2}$ to the V_{lowK} for $n \leq n_{1/2}$. The matched energy density (PCM) is found to precisely reproduce V_{lowK} RG data for the whole range of density. For example, in Figure 2, the symmetry energy E_{sym} in the PCM (solid line) is shown to match exactly the full V_{lowK}RG. It also shows that the higher-order terms beyond the mean-field approximation do indeed smoothen the cusp singularity—schematically indicated in Figure 1—as well as correctly treat the density regime $> n_{1/2}$.

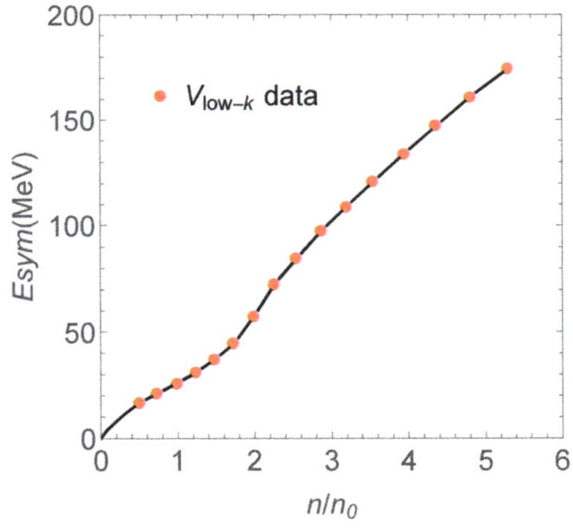

Figure 2. E_{sym} (solid circle) obtained in the *full* V_{lowk} RG approach for $n_{1/2} = 2n_0$. It is reproduced exactly by the pseudo-conformal model (solid line). Idem for $n_{1/2} \sim (2\text{–}4)n_0$.

This result strongly suggests that the complicated scaling for the HLS gauge coupling used in the V_{lowK} RG calculation could well be made much simpler, as mentioned above.

As mentioned, the smooth matching of the skyrmion-1/2-skyrmion property at $n_{1/2}$ could be deceptive given the oversimplified joining of hadron–quark degrees of freedom. The PCM may therefore hide complex structure lying just above the crossover density, say, in the density regime $\sim(2\text{–}4)n_0$. I will point this out in connection with some of the astrophysical observables to be discussed below.

What transpires from the properties (**I**)–(**IV**) incorporated into GnEFT for the PCM is that the trace of the energy-momentum-tensor θ_μ^μ for $n \gtrsim n_{1/2}$ goes as

$$\theta_\mu^\mu \propto \chi^4 \to \text{constant}. \tag{11}$$

This feature, the crucial element in the theory, is reproduced in the V_{lowK}RG double-decimation approach, as shown in Figure 2.

3.3. Predictions vs. Observables

Avoiding extensive references, both theory and experiment, I will list only those considered to be well-determined to the extent that it is feasible with the source from [18,19].

- **Smoothed cusp of $E_{sym}(n)$ at $n \gtrsim n_{1/2}$:**
  The bending-over of E_{sym} influences the slope L and induces the "soft-to-stiff" changeover. It also plays a crucial role in giving rise to the pseudo-conformal sound velocity (to be addressed below). Although as stressed the detailed structure and magnitude cannot be precisely pinned down, its simplicity with intricate topology change in the jungle of theories (as depicted in [9]) is a distinctive prediction of the PCM. It is at odds with the PREX-II/Jefferson data, which give generally stiff EoS, although there are some caveats [18,19]. To date, there are no trustworthy experimental data to quantitatively compare with.
- **Maximum mass star:** $M^{\max}$:

$$\begin{aligned} \text{PCM prediiction}: \quad & M^{max} \approx 2.05 M_\odot, \\ & R_{2.0} \approx 12.8 \text{ km}, \\ & (n_{central} \approx 5.1 n_0), \end{aligned} \tag{12}$$

$$\begin{aligned} \text{PSRJ0740}+6620: \quad & M^{max} = 2.08 \pm 0.07 M_\odot, \\ & R_{2.0} = 12.35 \pm 0.75 \text{ km}, \\ & (n_{central} = \,??), \end{aligned} \tag{13}$$

  No empirical data are known to be available at present for the central density $n_{central}$. The only information on this quantity *inferred*—not extracted—from PSR J0740 + 6620 is violently at odds with the PCM prediction. I will address this issue below.
- **1.44 $M_\odot$ star:**

$$\begin{aligned} \text{PCM prediction} \quad &: R_{1.44} \approx 12.8 \text{ km} \\ \text{PSRJ0030}+0451 \quad &: R_{1.44} = 12.45 \pm 0.65\, km. \end{aligned} \tag{14}$$

  The stunning agreements between the PCM predictions and the NICER and XMM-Newton measurements—with the exception of the sound velocity to be addressed below—could not be accidental. Not only does the maximum star mass come out the same, but also the radii agree. Furthermore, the difference $\Delta R = R_{2.0} - R_{1.4} \approx 0$, in agreement with the data. We will note later that this support of the PCM by the NICER/XMM-Newton has an even more surprising implication on scale-chiral symmetry in nuclear medium so far unsuspected.

  Let me make some further comments here on the PCM results.

What is given falls in the range of $n_{1/2} \sim (2.5\text{–}3.0)n_0$. The maximum mass comes out to be $\sim$2.4 $M_\odot$ for $n_{1/2} = 4n_0$. However, at this crossover density, although other global properties are not drastically different from the lower values of $n_{1/2}$, the sound speed overshoots the causality bound with a more pronounced bump and the pressure greatly exceeds what is indicated in heavy-ion data. It seems to be ruled out in the PCM of GnEFT.

One observes that the radius comes out to be $\sim$12.8 km in the wide range of the star mass and central density involved. Thus, the stars of masses $\sim$1.4 solar mass and $\sim$2.4

solar mass have almost the same radius. This is in agreement with what is being observed in the gravity wave data.

- **Tidal deformability** $\Lambda_{1.4}$

  The $\Lambda_{1.4}$ predicted in the PCM comes out to be $\sim$550, to be compared with $\Lambda_{1.4} = 190^{+390}_{-120}$ (GW1700817). This may seem to signal a tension. However, there is a basic difficulty in theoretically pinning down $\Lambda_{1.4}$. In the PCM, the density at which $\Lambda_{1.44}$ is measured is $\sim$2.4n_0. This density sits very close to where the topology change takes place. It is here the SchiEFT is most likely to start breaking down as the cusp in E_{sym} indicates and the pQCD cannot access. This is an "uncharted wilderness" for theory. As can be seen in [4], a small increase in the central density, say, from 2.3n_0 to 2.5n_0 (or increase in corresponding star mass), makes Λ drop to 420 while involving no change at all in radius. This means that the location of the HQC will strongly influence the Λ. *One can associate this behavior with the increase in attraction in going from n_0 toward $n_{1/2}$ in the cusp structure as one can see in the schematic plot Figure 1.* This clearly suggests that it would be extremely difficult to theoretically pin down Λ in the vicinity of the crossover regime.

  As noted below, the sound velocity has a complex "bump" structure in the vicinity of the topology change density. This is due to the interplay, encoding the putative HQC, between the hadronic degrees of freedom and the "dual quark-gluon" degrees of freedom. This would complicate significantly the linking of $\Lambda_{1.4}$ to the structure of the sound velocity below or near $n_{1/2}$. To give an example, let me quote [23] where the bump structure—"the slope, the hill, the drop, the swoosh, etc."—associated with the possible phase structure of QCD is proposed to pin down $\Lambda_{1.4}$ by up-coming measurements. The hope here is to determine the possible phase transition near the HQC density. Given the theoretical wilderness inevitably involved, this seems a far-fetched endeavor.

  In short, contrary to what is claimed by some workers in the field, ruling out an EoS based on the precise value of $\Lambda_{1.4}$ would be premature.

- **Sound speed** v_s

  The most striking prediction of the PCM, so far not shared by other models, is the sound speed for $n \gtrsim n_{1/2}$. It *predicts* the pseudo-conformal sound speed

$$v_s^{pcss}/c^2 \approx 1/3 \text{ for } n \gtrsim n_{1/2}. \tag{15}$$

  It is not to be identified with the conformal sound speed $v_s^{conform}/c^2 = 1/3$ because the energy-momentum tensor is *not* traceless, i.e., scale symmetry is spontaneously broken. This prediction can be understood as follows.

  As noted above, the quasiparticle mass m_Q^* goes $\propto \langle\chi\rangle^*$ as the density goes above $n_{1/2}$ and the dilaton condensate becomes independent of density, reaching m_0. This has to do with a delicate interplay between the attraction associated with the dilaton exchange and the ω repulsion, which leads to the parity doubling. Where this interplay starts taking place cannot be pinned down precisely but it must be in the density regime where the symmetry energy is involved, going from $n_{1/2}$ to the core of massive stars, say, $\gtrsim$6n_0. In this density regime, the Landau fixed-point approximation with $\bar{N}^{-1} = (\Lambda_F - k_f)/k_F \sim 1/k_F \to 0$ can be taken to be reliable. One can then calculate the trace of the energy-momentum tensor in the mean-field approximation of GnEFT, i.e., LFL fixed-point approximation, which will become density-independent as given by (11). In this density range we will have

$$\frac{\partial}{\partial n}\langle\theta_\mu^\mu\rangle = \frac{\partial\epsilon(n)}{\partial n}\left(1 - 3\frac{v_s^2}{c^2}\right) \approx 0 \tag{16}$$

  where $\epsilon(n)$ is the energy density and $v_s^2/c^2 = \frac{\partial P(n)}{\partial n}/\frac{\partial\epsilon(n)}{\partial n}$. It is approximate since there can easily be terms that are compounded with EFT and pQCD at the point where the symmetry energy has the cusp structure. Since there is no Lee-Wick-type state, one must have

$$\left(1 - 3\frac{v_s^2}{c^2}\right) \approx 0 \tag{17}$$

which gives the pseudo-conformal sound speed

$$(v_s^{pcs}/c)^2 \approx 1/3. \tag{18}$$

The "approximate zero" here stands for the fact that it is pseudo-conformal with scale symmetry broken both explicitly and spontaneously, the dilaton mass and the ω mass balancing so as to lead to parity doubling in the dense system. The true conformal velocity, within the model, should be reached only at a density much higher than that of the core density of the massive stars. Where precisely the conformality sets in is not relevant to the compact star physics.

In Figure 3 is shown the sound speed v_s/c for $\alpha = 0$ (nuclear matter) and 1 (neutron matter) calculated in V_{lowK} RG for $n_{1/2}$. They are of the same form for $2 < n_{1/2}/n_0 < 4$ except for the slight shift in the density and the height of v_s. This result serves as an illustration of the arguments to follow.

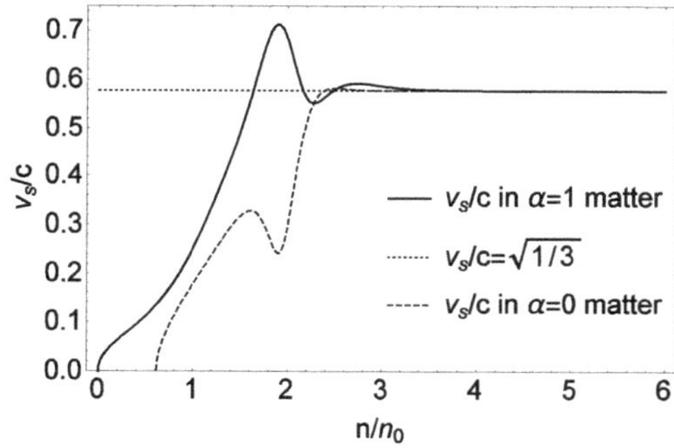

Figure 3. v_s vs. density for $\alpha = 0$ (nuclear matter) and $\alpha = 1$ (neutron matter) in V_{lowk} RG for $n_{1/2} = 2n_0$ and $v_{\text{vn}} = 25n_0$.

What is noticeable is the large bump in v_s in the vicinity of $n_{1/2}$ and the rapid convergence to the speed 1/3. The approximation involved on top of the pseudo-conformality would of course give fluctuations on top of $v_{pcs}^2/c^2 \approx 1/3$ but the point here is it is the pseudo-conformality that "controls" the general structure. The large bump signals a complex interplay between hadronic and non-hadronic degrees of freedom manifested through the pseudo-gap structure of the chiral condensates. I will discuss below how the degrees of freedom in the core of the massive stars could masquerade as "deconfined quarks".

• **Quenched g_A in nuclei**

Though it is not directly connected with the star properties, a relevant and intriguing observation is what I would call "quasibaryon" g_A in nuclear matter. It follows from the possible existence of the IR fixed point associated with the "genuine dilaton (GD)." The effective g_A in the Gamow–Teller transitions in nuclei, g_A^{eff}, is observed to be $g_A^{eff} \approx 1$ from light nuclei to heavy nuclei and even to the dilaton-limit fixed point at $n \gtrsim 25n_0$. It has been argued that an approximate scale invariance "emerges" in nuclear interactions [24], in a way most likely related to the way $(v_s^{pcs}/c)^2 \approx 1/3$ sets in precociously.

Returning to v_s, is there any indication in recent astrophysical observations for such a precocious onset of the pseudo-conformal sound velocity?
To date, there is no known "smoking-gun" signal for the sound velocity from observations. In the literature, however, there are a gigantic number of articles on the structure of sound velocity deduced from the gravity wave data as well as theoretically. Some argue for phase transitions or continuous ones or simply no crossovers, etc. Some extreme cases are discussed in [9]. I will no go into this *wilderness* here. Let me just describe one case which illustrates most transparently what can very well be involved.

Let us take the case of NICER and XMM-Newton observables (NXN for short) discussed, namely (13) and (14). This case brings out how puzzling the problem can be.

In [25], the properties of high-density matter were inferred in most detailed analyses of the NXN data. Ruling out essentially all other scenarios, with or without phase changes, the authors arrive at the sound velocity ("H-bump") plotted in Figure 4.

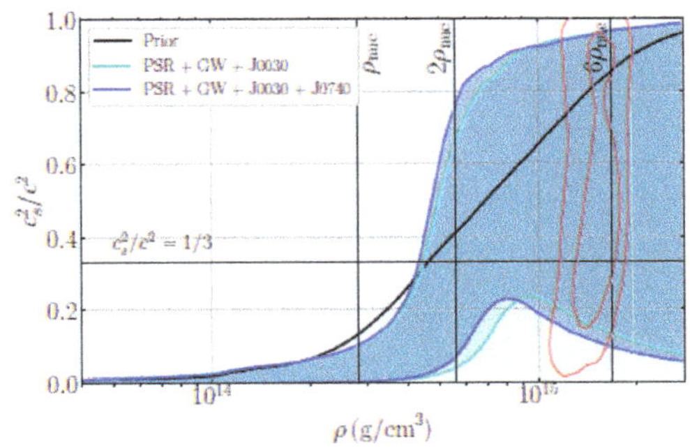

Figure 4. ρ in unit of $\mathrm{g/cm^3}$ (the "H-bump" scenario) taken from [25]. The red contour stands for 50% and 90% inferred sound speed and central density.

The central density and the maximum sound velocity inferred were

$$\begin{aligned} n_{\rm cent}/n_0 &= 3.0^{+1.6}_{-1.6}\,, \\ v_s^2/c^2 &= 0.79^{+021}_{-0.20}\,. \end{aligned} \quad (19)$$

While the star properties they took into account are exactly those reproduced by the PCM, i.e., (13) and (14), the central density and the sound velocity are totally different from the PCM predictions. One can understand the low central density accounting for the sound speed overshooting the conformal bound, characteristic of strongly interacting hadronic phase. In fact there are in the literature numerous scenarios anchored on a variety of density-functional approaches giving rise to the wilderness of one form or other in the sound velocity—including bumps similar to the H-bump—but I am not aware of any that can survive the battery of bona fide constraints coming from the current observations both in theory and experiment as claimed by [25].

A puzzle immediately raised is this: How can the PCM with an emergent (pseudo-)conformality and the strong H-bump with no hint of conformal symmetry give the almost *identical* global star properties (13) and (14)? *The only statement one can make at this point is (A) either the sound velocity and the global star properties are totally unrelated or (B) there is something wrong either in the strong H-bump scenario or in the simple PCM structure.* Option (A) is hard to accept, so perhaps option (B) is a plausible possibility. My bet is option (B) and the H-bump scenario are at odds with nature.

• **Conformailty**

In this connection, let me make a remark on the role of conformal symmetry in the sound velocity currently being discussed in the literature. This issue is a focused topic in MDPI's Special Issue on "Symmetries and Ultra Dense Matter of Compact Stars" being edited with contributions devoted to the issue. Without going into detail, let me just mention that there are a variety of models hybridizing hadronic degree of freedom and "revamped" quark/gluon degrees of freedom at a density $n_{HQC} \gtrsim 2n_0$. Some of the models such as quarkyonic and holographic QCD do tend to see the conformal symmetry (perhaps involving percolation, etc.) emerge at certain density $\gtrsim n_{HQC}$ in going up in density [26–28]. Going down the density ladder from asymptotic density where $v^2_{conf}/c^2 = 1/3$, one seems to observe the approximate conformality which persists down to the crossover regime where the big bump develops as it does in the PCM [29]. This may represent a microscopic rendition of HQC in contrast to the PCM, which presents a coarse-grained picture permeating in dense medium. This point is evidenced in Figure 6 in [30] where the results of quarkyonic models are compared with the PCM prediction Figure 3.

4. Conclusion: The Duck Story

Briefly summarized, I have shown how to go from low density to high density capturing the *putative* hadron-quark continuity (HQC) by formulating baryonic matter as Landau-Migdal Fermi-liquid matter resulting via renormalization group [15]. It is a sort of generalized density-functional approach (à la the Hohenberg–Kohn theorem), implementing heavy degrees of freedom in terms of hidden symmetries involving a mass scale above that given by standard chiral EFT, which is shown to be valid at nuclear matter density. The resulting effective field theory, GnEFT, exploits the possibility of simulating via duality the HQC in terms of a topology change from skyrmions at low density $\sim n_0$ to 1/2-skyrmions at high density $\sim 6n_0$. The resulting EoS has so far successfully accounted for nuclear matter as well as dense compact star matter. The structure that is arrived at in compact star matter, coined as pseudo-conformality, can be considered as a coarse-grained description of the hadron-to-quarks changeover, e.g., quarkyonic *"IdylliQ"* [31], captured in terms of "emergent" scale symmetry permeating from low to high density.

The formulation made so far is valid at zero temperature. Upcoming terrestrial laboratory observations complimentary to astrophysical data, e.g, at FAIR of GSI, however, will necessarily involve relatively high temperature. It remains to be formulated in the GnEFT framework to meet the conditions of the terrestrial laboratories. How topology enters in the hot *and* dense matter is a totally open issue as indicated in recent puzzling manifestations of scale invariance at high temperature [32].

Finally, I touch on fractionalzed "quasibaryon" structure inside the core of the massive star.

When a paper appeared in 2020 [33] with the suggestion that the cores of the most massive neutron stars are characterized by approximate conformal symmetry, with the speed of sound $v_s^2/c^2 \to 1/3$, the polytropic index $\gamma = \mathrm{d}\ln p/\mathrm{d}\ln\epsilon \to 1$, and the normalized trace anomaly $\Delta = (\epsilon - 3p)/(3\epsilon) \to \delta \approx 0$, indicating that the cores are most likely populated by deconfined fractionally charged objects, identified as quarks, those quantities were quickly calculated in the PCM formulated in 2017 [3]. I considered this as a prediction of the PCM. The predicted results [34] were quite consistent with the conclusion of [33].

Now the question was this: Given that the degrees of freedom in the PCM are quasibaryons, albeit fractionalized, how do they carry the characteristics of fractionally charged quarks?

I do not have an immediate answer to this question. However, there are certain ideas that could lead to an understanding of this puzzle [35]. One of them is this: In the skyrmion-half-skyrmion crystal simulation, the half-skyrmions "confined" into a skyrmion by monopoles [21] could be liberated at high density and propagate freely with little interactions as seen in skyrmion crystals [3]. Two half skyrmions can then be rearranged into three 1/3-charged objects as in a schematic model [36]. In fact, in condensed matter physics, with domain walls, there can be stacks of sheets containing deconfined fractionally charged objects behaving like "deconfined quarks" coming from the bulk in which the objects are confined [37].

This reminds one of the "duck test": *"If it looks like a duck, swims like a duck, and quacks like a duck, it probably is a duck."*

Funding: This research received no external funding.

Data Availability Statement: Not applicable.

Conflicts of Interest: Not applicable.

References

1. Rho, M. 'Mass without mass' and nuclear matter. *arXiv* **2017**, arXiv:1703.02277.
2. Paeng, W.G.; Kuo, T.T.S.; Lee, H.K.; Rho, M. Scale-invariant hidden local symmetry, topology change and dense baryonic matter. *Phys. Rev. C* **2016**, *93*, 055203.
3. Paeng, W.G.; Kuo, T.T.S.; Lee, H.K.; Ma, Y.L.; Rho, M. Scale-invariant hidden local symmetry, topology change, and dense baryonic matter. II. *Phys. Rev. D* **2017**, *96*, 014031.
4. Ma, Y.L.; Rho, M. Towards the hadron–quark continuity via a topology change in compact stars. *Prog. Part. Nucl. Phys.* **2020**, *113*, 103791.
5. Rho, M.; Ma, Y.L. Manifestation of hidden symmetries in baryonic matter: From finite nuclei to neutron stars. *Mod. Phys. Lett. A* **2021**, *36*, 2130012.
6. Park, B.Y.; Vento, V. Skyrmion approach to finite density and temperature. In *The Multifacetted Skyrmion*; Brown, G.E., Rho, M., Eds.; World Scientific: Singapore , 2010.
7. Lee, H.K.; Park, B.Y.; Rho, M. Half-skyrmions, tensor forces and symmetry energy in cold dense matter. *Phys. Rev. C* **2011**, *83*, 025206; Erratum in *Phys. Rev. C* **2011**, *84*, 059902.
8. Adam, C.; García Martín-Caro, A.; Huidobro, M.; Vázquez, R.; Wereszczynski, A. Quantum skyrmion crystals and the symmetry energy of dense matter. *Phys. Rev. D* **2022**, *106*, 114031.
9. Lee, H.K.; Ma, Y.L.; Paeng, W.G.; Rho, M. Cusp in the symmetry energy, speed of sound in neutron stars and emergent pseudo-conformal symmetry. *Mod. Phys. Lett. A* **2022**, *37*, 2230003.
10. Garaud, J.; Babaev, E. Effective model and magnetic properties of the resistive electron quadrupling state. *Phys. Rev. Lett.* **2022**, *129*, 087602.
11. Harada, M.; Yamawaki, K. Hidden local symmetry at loop: A New perspective of composite gauge boson and chiral phase transition. *Phys. Rep.* **2003**, *381*, 1.
12. Crewther, R.J. Genuine dilatons in gauge theories. *Universe* **2020**, *6*, 96
13. Crewther, R.J.; Tunstall, L.C. $\Delta I = 1/2$ rule for kaon decays derived from QCD infrared fixed point. *Phys. Rev. D* **2015**, 91, 034016.
14. Brown, G.E.; Rho, M. Double decimation and sliding vacua in the nuclear many body system. *Phys. Rep.* **2004**, *396*, 1–39.
15. Shankar, R. Renormalization group approach to interacting fermions. *Rev. Mod. Phys.* **1994**, *66*, 129–192.
16. Friman, B.; Rho, M. From chiral Lagrangians to Landau Fermi liquid theory of nuclear matter. *Nucl. Phys. A* **1996**, *606*, 303.
17. Song, C. Dense nuclear matter: Landau Fermi liquid theory and chiral Lagrangian with scaling. *Phys. Rep.* **2001**, *347*, 289–371.
18. Piekarewicz, J. The nuclear physics of neutron stars. *arXiv* **2022**, arXiv:2209.14877;
19. Salinas, M.; Piekarewicz, J. Building an equation of state density ladder. *Symmetry* **2023**, *15*, 994;
20. He, B.R.; Harada, M. Parity doubling among baryons in a holographic QCD model at finite density. *PoS* **2013**, *192*, arXiv: 10.22323/1.205.0192.
21. Zhang, P.; Kimm, K.; Zou, L.; Cho, Y.M. Re-interpretation of Skyrme theory: New topological structures. *arXiv* **2017** , arXiv:1704.05975.
22. Beane, S.R.; van Kolck, U. The dilated chiral quark model. *Phys. Lett. B* **1994**, *328*, 137–142.
23. Tan, H.; Dexheimer, V.; Noronha-Hostler, J.; Yunes, N. Finding structure in the speed of sound of supranuclear matter from binary love relations. *Phys. Rev. Lett.* **2022**, *128*, 161101.
24. Rho, M. Multifarious roles of hidden chiral-scale symmetry: "Quenching" g_A in Nuclei. *Symmetry* **2021**, *13*, 1388.
25. Legred, I.; Chatziioannou, K.; Essick, R.; Han, S.; Landry, P. Impact of the PSR J0740+6620 radius constraint on the properties of high-density matter. *Phys. Rev. D* **2021**, *104*, 063003.
26. Fujimoto, Y.; Fukushima, K.; McLerran, L.D.; Praszalowicz, M. Trace anomaly as signature of conformality in neutron stars. *Phys. Rev. Lett.* **2022**, *129*, 252702.
27. Marczenko, M.; McLerran, L.; Redlich, K.; Sasaki, C. Reaching percolation and conformal limits in neutron stars. *Phys. Rev. C* **2023**, *107*, 025802.

28. Rojas, J.C.; Demircik, T.; Järvinen, M. Popcorn transitions and approach to conformality in homogeneous holographic nuclear matter. *Symmetry* **2023**, *15*, 331.
29. Annala, E.; Gorda, T.; Hirvonen, J.; Komoltsev, O.; Kurkela, A.; Nättilä, J.; Vuorinen, A. Strongly interacting matter exhibits deconfined behavior in massive neutron stars. *arXiv* **2023**, arXiv:2303.11356.
30. Ma, Y.L.; Rho, M. The sound speed and core of massive compact stars: A manifestation of hadron-quark duality. arXiv **2021**, arXiv:2104.13822.
31. Fujimoto, Y.; Kojo, T.; McLerran, L.D. Momentum shell in quarkyonic matter from explicit duality: A solvable model nalysis. *arXiv* **2023**, arXiv:2306.04304.
32. Meng, X.L.; Sun, P.; Alexandru, A.; Horváth, I.; Liu, K.-F.; Wang, G.; Yang, Y.-B. Separation of infrared and bulk in thermal QCD. *arXiv* **2023**, arXiv:2305.09459.
33. Annala, E.; Gorda, T.; Kurkela, A.; Nättilä, J.; Vuorinen, A. Evidence for quark-matter cores in massive neutron stars. *Nat. Phys.* **2020**, *16*, 907–910.
34. Ma, Y.L.; Rho, M. What's in the core of massive neutron stars? *arXiv* **2020**, arXiv:2006.14173.
35. Rho, M. Fractionalized quasiparticles in dense baryonic matter. *arXiv* **2020**, arXiv:2004.09082.
36. Sarti, V.M.; Vento, V. The half-skyrmion phase in a chiral-quark model. *Phys. Lett. B* **2014**, *728*, 323–327.
37. Sulejmanpasic, T.; Shao, H.; Sandvik, A.; Unsal, M. Confinement in the bulk, deconfinement on the wall: Infrared equivalence between compactified QCD and quantum magnets. *Phys. Rev. Lett.* **2017**, *119*, 091601.

symmetry

MDPI

Article

Nuclear Matter Properties and Neutron Star Phenomenology Using the Finite Range Simple Effective Interaction

Xavier Viñas [1,2,*,†], Parveen Bano [3,†], Zashmir Naik [3,†] and Tusar Ranjan Routray [3,*]

1 Departament de Física Quàntica i Astrofísica (FQA), Institut de Ciències del Cosmos (ICCUB), Universitat de Barcelona (UB), Martí i Franquès 1, E-08028 Barcelona, Spain
2 Institut Menorquí d'Estudis, Camí des Castell 28, E-07702 Maó, Spain
3 School of Physics, Sambalpur University, Jyotivihar, Burla, Sambalpur 768 019, India; mailme7parveen@suniv.ac.in (P.B.); z.naik@suniv.ac.in (Z.N.)
* Correspondence: xavier@fqa.ub.edu (X.V.); trr1@rediffmail.com (T.R.R.)
† These authors contributed equally to this work.

Abstract: The saturation properties of symmetric and asymmetric nuclear matter have been computed using the finite range simple effective interaction with Yukawa form factor. The results of higher-order derivatives of the energy per particle and the symmetry energy computed at saturation, namely, Q_0, K_{sym}, K_τ, Q_{sym}, are compared with the corresponding values extracted from studies involving theory, experiment and astrophysical observations. The overall uncertainty in the values of these quantities, which results from a wide spectrum of studies described in earlier literature, lies in the ranges $-1200 \lesssim Q_0 \lesssim 400$ MeV, $-400 \lesssim K_{sym} \lesssim 100$ MeV, $-840 \lesssim K_\tau \lesssim -126$ MeV and $-200 \lesssim Q_{sym} \lesssim 800$ MeV, respectively. The ability of the equations of state computed with this simple effective interaction in predicting the threshold mass for prompt collapse in binary neutron star merger and gravitational redshift has been examined in terms of the compactness of the neutron star and the incompressibility at the central density of the maximum mass star. The correlations existing between neutron star properties and the nuclear matter saturation properties have been analyzed and compared with the predictions of other model calculations.

Keywords: nuclear matter saturation properties; landau parameters; incompressibility of nuclear matter; neutron star properties; binary neutron star merger; gravitational redshift

Citation: Viñas, X.; Bano, P.; Naik, Z.; Routray, T.R. Nuclear Matter Properties and Neutron Star Phenomenology Using the Finite Range Simple Effective Interaction. *Symmetry* **2024**, *16*, 215. https://doi.org/10.3390/sym16020215

Academic Editor: Charalampos Moustakidis

Received: 13 January 2024
Revised: 28 January 2024
Accepted: 30 January 2024
Published: 10 February 2024

1. Introduction

The empirical properties of infinite nuclear matter (NM) at saturation are key features in the study of any phenomenon resulting from the nucleon–nucleon (NN) interactions in a many-body system. Ideally, the solution to the many-body problem with the NN interaction would be the way to determine the NM properties. But in the absence of comprehensive knowledge of NN interaction, microscopic many-body models, such as Dirac-Brueckner-Hartree-Fock (DBHF) and its non relativistic counterpart BHF [1–9], variational methods [10,11], chiral effective field theory [12,13], etc., use realistic potentials whose parameters are fitted to phase shift data in different partial wave channels and properties of few-body systems (deuteron and triton). The inadequacy of our understanding of the in-medium NN interaction is reflected by the fact that in the aforementioned many-body calculations, the saturation density ρ_0 of symmetric nuclear matter (SNM) is over-predicted. This could be brought within the empirical range by incorporating the three-body and higher-order many-body effects in an ad hoc phenomenological manner. The consensus range for the value of saturation density $\rho_0 = 0.17 \pm 0.03$ fm^{-3} has been estimated from studies of various kinds, which include different variants of the liquid drop model, optical model of NN scattering, muonic atoms, and Hartree–Fock (HF) calculations of nuclear density distributions [14]. The values of energy per nucleon at saturation density $e(\rho_0)$ have been extracted to be ~ -16 MeV from the mass analysis over the periodic table.

An alternate method, adopted in contrast to the microscopic realistic calculations to handle the many-body problem, is to use phenomenological effective interactions. Though less fundamental compared to the microscopic calculations, its ability in the analytical calculations of nuclear properties makes it highly popular. Skyrme [15], Gogny [16], and M3Y effective forces [17] are some of those successful forces. The basic idea to build up these interactions is to parameterize the effects of the microscopic NN interaction upon averaging over the spin, parity and isospin of the interacting nucleons and constrain the parameters from the ground state properties of finite nuclei and nuclear matter. A drawback of this approach is that the parameter set is not unique, and higher-order nuclear matter properties at saturation cannot be unambiguously predicted. Moreover, many parameters of the effective force become strongly correlated in the course of their fixation. Of late, another effective force, the so-called finite range simple effective interaction (SEI) [18], which has a similar predicting ability as that of Skyrme and Gogny in the NM and finite nuclei domain, has been established. The parameter fitting protocol adopted in the case of SEI makes it different from other effective forces, minimizing the correlation effects. The SEI parameters are systematically fitted in SNM and pure neutron matter (PNM), which allows the study of both SNM and asymmetric nuclear matter (ANM). Moreover, the parameters responsible for the momentum dependence of the mean-field are fixed from the experimental/empirical constraints exclusively so that each of the two aspects of the mean-field, the density dependence and the momentum dependence, could be studied independently without altering the predictions of the other one [19]. Further, in the determination of nine parameter combinations of the eleven SEI parameters required for the study of ANM, one is required to assume only three standard values of saturation properties, namely, ρ_0, $e(\rho_0)$ and symmetry energy $E_{sym}(\rho_0)$. Within this parameter fixation protocol, we shall use the SEI to compute higher-order derivatives of the energy per particle and the symmetry energy at saturation density and compare these values with the results extracted from various different studies. The stability conditions, in terms of Landau parameters of the interacting Fermi liquid model, serve as an acid test for the reliability of an effective force for its applicability in the different channels of spin and isospin. We shall check to which extent the observance of the Landau stability conditions is fulfilled by our SEI.

In Section 2, we have given a brief account of SEI and its parameter-fitting protocol. In Section 3, we have obtained different properties of SNM at saturation, which are re-evaluated in the framework of the interacting Fermi liquid model by computing the Landau parameters using SEI. In the same section, the high-order derivatives of the energy per particle and the symmetry energy are calculated and compared with the empirical range of values extracted from theory, experiment and astrophysical observations. In Section 4, we explore the predictive power of the SEI EoSs in the domain of high-density neutron-rich matter pertaining to the recent NS phenomenology associated with the binary neutron star merger (BNSM) and gravitational redshift. Finally, Section 5 contains the summary and outlook.

2. Formalism

The SEI in this work was initially proposed by Behera et al. [18] and has the following explicit expression if a form-factor of Yukawa type is used.

$$\begin{aligned} V_{eff} &= t_0(1+x_0P_\sigma)\delta(\vec{r}) + \frac{t_3}{6}(1+x_3P_\sigma)\left(\frac{\rho(\mathbf{R})}{1+b\rho(\mathbf{R})}\right)^\gamma \delta(\vec{r}) \\ &+ (W + BP_\sigma - HP_\tau - MP_\sigma P_\tau)\frac{e^{-r/\alpha}}{r/\alpha} + \text{Spin-orbit part} \end{aligned} \tag{1}$$

We denote this force as SEI-Y thereafter.

The SEI-Y in Equation (1) has 12 parameters in total, namely, $\alpha, \gamma, b, x_0, x_3, t_0, t_3, W, B, H,$ and M plus the spin-orbit strength parameter W_0, which enters in the description of finite nuclei. The energy density in isospin asymmetric nuclear matter for the SEI-Y interaction in Equation (1) is given by,

$$
\begin{aligned}
H_Y(\rho_n,\rho_p) &= \frac{3\hbar^2}{10m}(k_n^2\rho_n + k_p^2\rho_p) + \frac{\varepsilon_0^l}{2\rho_0}(\rho_n^2+\rho_p^2) + \frac{\varepsilon_0^{ul}}{\rho_0}\rho_n\rho_p \\
&+ \left[\frac{\varepsilon_\gamma^l}{2\rho_0^{\gamma+1}}(\rho_n^2+\rho_p^2) + \frac{\varepsilon_\gamma^{ul}}{\rho_0^{\gamma+1}}\rho_n\rho_p\right]\left(\frac{\rho(\mathbf{R})}{1+b\rho(\mathbf{R})}\right)^\gamma \\
&+ \frac{\varepsilon_{ex}^l}{2\rho_0}\left\{\rho_n^2\left[\left(\frac{3\Lambda^6}{32k_n^6}+\frac{9\Lambda^4}{8k_n^4}\right)\ln\left(1+\frac{4k_n^2}{\Lambda^2}\right) - \frac{3\Lambda^4}{8k_n^4} + \frac{9\Lambda^2}{4k_n^2} - \frac{3\Lambda^3}{k_n^3}\tan^{-1}\left(\frac{2k_n}{\Lambda}\right)\right]\right. \\
&+ \left.\rho_p^2\left[\left(\frac{3\Lambda^6}{32k_p^6}+\frac{9\Lambda^4}{8k_p^4}\right)\ln\left(1+\frac{4k_p^2}{\Lambda^2}\right) - \frac{3\Lambda^4}{8k_p^4} + \frac{9\Lambda^2}{4k_p^2} - \frac{3\Lambda^3}{k_p^3}\tan^{-1}\left(\frac{2k_p}{\Lambda}\right)\right]\right\} \\
&+ \frac{\varepsilon_{ex}^{ul}}{\rho_0}\rho_n\rho_p\left\{\frac{3}{32}\left[\left(\frac{\Lambda^6}{k_n^3k_p^3}+\frac{6\Lambda^4}{k_nk_p^3}+\frac{6\Lambda^4}{k_n^3k_p}\right) - \frac{3\Lambda^2(k_p^2-k_n^2)^2}{k_n^3k_p^3}\right]\ln\left[\frac{\Lambda^2+(k_n+k_p)^2}{\Lambda^2+(k_p-k_n)^2}\right]\right. \\
&+ \left[\frac{3}{2}\left(\frac{\Lambda^3}{k_n^3}-\frac{\Lambda^3}{k_p^3}\right)\tan^{-1}\left(\frac{k_p-k_n}{\Lambda}\right) - \frac{3}{2}\left(\frac{\Lambda^3}{k_p^3}+\frac{\Lambda^3}{k_n^3}\right)\tan^{-1}\left(\frac{k_n+k_p}{\Lambda}\right)\right] \\
&+ \left.\left[\frac{9}{8}\left(\frac{\Lambda^2}{k_p^2}+\frac{\Lambda^2}{k_n^2}\right) - \frac{3}{8}\frac{\Lambda^4}{k_n^2k_p^2}\right]\right\}
\end{aligned} \tag{2}
$$

where ρ_n, ρ_p are neutron (n) and proton (p) densities, $\rho = \rho_n + \rho_p$ is the total NM density, $\Lambda(=1/\alpha)$ is the inverse of the range of the Yukawa form-factor, and $k_i = (3\pi^2\rho_i)^{1/3}$ (i = n, p) is the respective Fermi momentum. The study of ANM involves altogether nine parameters, γ, b, α, ε_0^l, ε_0^{ul}, ε_γ^l, ε_γ^{ul}, ε_{ex}^l, ε_{ex}^{ul}. The connection of these new parameters to the interaction parameters of Equation (1), which was derived in Ref. [20], is also reported in Appendix A. Here, the indexes l and ul denote the interaction between like and unlike pairs of nucleons, respectively.

The Fitting Procedure of SEI

The formulation of NM and PNM using SEI and the parameter fixation protocol has been discussed at length in Refs. [20,21]. We briefly outline it in the following. The SNM only requires the following three combinations of the strength parameters,

$$
\left(\frac{\varepsilon_0^l+\varepsilon_0^{ul}}{2}\right) = \varepsilon_0,\ \left(\frac{\varepsilon_\gamma^l+\varepsilon_\gamma^{ul}}{2}\right) = \varepsilon_\gamma,\ \left(\frac{\varepsilon_{ex}^l+\varepsilon_{ex}^{ul}}{2}\right) = \varepsilon_{ex}, \tag{3}
$$

which, together with γ, b and α, are the six parameters needed to determine the SNM completely. For a given value of the exponent γ, which characterizes the stiffness parameter and determines the incompressibility K in SNM, the remaining five parameters ε_0, ε_γ, ε_{ex}, b and α of SNM are determined as follows assuming the standard values of the nucleon mass (mc^2 = 939 MeV), the saturation density ρ_0 and the energy per particle at saturation $e(\rho_0)$. The range α and the exchange strength ε_{ex} are determined simultaneously by adopting an optimization procedure [18], using the condition that the nuclear mean-field in SNM at saturation density vanishes for the kinetic energy of the nucleon of 300 MeV, a result extracted from the optical model analysis of nucleon–nucleus scattering data [22–25]. The parameter b is determined to avoid supra-luminous behavior [26]. The two remaining parameters, namely ε_γ and ε_0, are obtained from the saturation conditions, $T_{f_0} = \frac{\hbar^2k_{f_0}^2}{2m}$ = 37 MeV, which allow us to obtain k_{f_0} and, therefore, the saturation density ρ_0, and $e(\rho_0) = -16$ MeV. The stiffness parameter γ, kept as a free parameter, is chosen from the range of values for which the pressure–density relation in SNM lies within the region extracted from the analysis of flow data in heavy-ion collision experiments at intermediate energies [27]. It is verified that γ = 1 is the upper limit for which the pressure-density relation is obeyed, which corresponds to the nuclear matter incompressibility $K(\rho_0)$ = 269 MeV for the SEI-Y model. Therefore, we can study the nuclear matter properties by assuming different values of γ up to a limiting value γ = 1. In this work, we will use three EoS corresponding to $\gamma = 1/3, 1/2$, and $2/3$. In order to study ANM, we need to know how the strength parameters ε_{ex}, ε_γ and ε_0 of Equation (3) split into the like and unlike components. The splitting of ε_{ex} into ε_{ex}^l and ε_{ex}^{ul} is decided to be $\varepsilon_{ex}^l = 2\varepsilon_{ex}/3$ [28] using the condition that the entropy in PNM does not exceed that of the SNM [28]. The splittings of the remaining two strength parameters, ε_γ and ε_0, are decided from the values of the symmetry energy parameter $E_{sym}(\rho_0)$ and its derivative $E_{sym}^{'}(\rho_0) = \rho_0 \frac{dE_{sym}(\rho_0)}{d\rho_0}$ at saturation density ρ_0. For a given $E_{sym}(\rho_0)$ within its empirical range [29], we can produce different density dependence of symmetry energy $E_{sym}(\rho)$ by assigning arbitrary values to $E_{sym}^{'}(\rho_0)$. The slope parameter in each case will be $L(\rho_0) = 3E_{sym}^{'}(\rho_0)$. In the study where the variation in $L(\rho_0)$ is not an explicit requirement, the value of $E_{sym}^{'}(\rho_0)$ is fixed from the condition that the asymmetric contribution of the nucleonic part of the energy density in charge-neutral β-equilibrated neutron star $n + p + e + \mu$ matter (NSM), i.e., $S^{NSM}(\rho) = [H(\rho, Y_p) - H(\rho, Y_p = 1/2)]$ is maximum, where Y_p is the equilibrium proton fraction. The characteristic $E_{sym}^{'}(\rho_0)$ value thus obtained predicts a density dependence of the symmetry energy, which is neither stiff nor very soft [30]. With the parameters determined in this way, the SEI is able to reproduce the trends of the EoS and the properties of the momentum dependence of the mean-field with similar quality as predicted by microscopic calculations [10,28,31–33]. As a consequence of this fitting procedure, one can also vary the n and p effective mass splitting, which only depend on the ε_{ex}^l and ε_{ex}^{ul} parameters, while the density dependence of $E_{sym}(\rho)$, i.e., the slope parameter L, which depends on the splitting of ε_γ and ε_0, remains invariant and the vice-versa [28,33].

We now have three open parameters that we have chosen as t_0, x_0 and W_0. However, to describe ANM, the explicit value of the t_0 and x_0 parameters is not necessary because they enter as specific combinations that can be determined from the ε_0^l and ε_0^{ul} [21]. In our work [20], we further constrained x_0 by using the DBHF predictions on the effective mass splitting between spin-up and spin-down neutrons in spin-polarized neutron matter. It was found that the SEI predictions agree well with the DBHF ones [34] for $\varepsilon_{ex}^{l,l} = \varepsilon_{ex}^l/3$, where the superscript ("l,l") (and its counterpart ("l,ul")) symbolizes the exchange strength for parallel (and anti-parallel) spin orientations in polarized neutron matter. This consideration allows us to determine x_0 in a unique way as [20], $x_0 = 1 - \frac{2\varepsilon_0^l - \varepsilon_{ex}^l}{\rho_0 t_0}$, if t_0 is known. The two remaining free parameters, t_0 and W_0, have to be fitted to finite nuclei data, as explained in detail in Refs. [21,35], in the case of a SEI with a Gaussian form factor. For the sake of completeness, the twelve numbers of parameters of the SEI-Y EoSs corresponding to γ = 1/3, 1/2 and 2/3 are given in Table 1. The inputs corresponding to the saturation density, energy per particle in SNM, and symmetry energy needed to obtain these parameters are given in Table 2.

Table 1. The twelve parameters for the SEI-Y EoSs corresponding to γ = 1/3, 1/2 and 2/3.

γ	b [fm^3]	α [fm]	ε_{ex} [MeV]	ε_{ex}^l [MeV]	ε_0 [MeV]
1/3	0.4161	0.4232	−129.344	−86.229	−82.245
1/2	0.5880	0.4242	−127.707	−85.138	−50.600
2/3	0.7796	0.4250	−126.390	−84.260	−34.904
ε_0^l [MeV]	ε_γ [MeV]	ε_γ^l [MeV]	t_0 [$MeV fm^3$]	x_0	W_0 [$MeV fm^5$]
−47.189	104.428	74.006	333.5	1.151	119.3
−27.509	73.124	54.250	566.7	0.664	118.4
−17.859	58.095	44.690	647.4	0.520	118.2

Table 2. Nuclear matter properties at saturation density ρ_0 for three EoSs—SEI-Y ($\gamma = 1/3$), SEI-Y ($\gamma = 1/2$) and SEI-Y ($\gamma = 2/3$).

	SEI-Y (γ = 1/3)	SEI-Y (γ = 1/2)	SEI-Y (γ = 2/3)
ρ_0 [fm^{-3}]	0.161	0.158	0.156
e_0 [MeV]	−16.04	−16	−16
(m_s^*/m)	0.664	0.686	0.666
(m_v^*/m)	0.685	0.621	0.622
K_0 [MeV]	220.346	237.643	253.219
Q_0 [MeV]	−478.763	−461.807	−437.529
$\rho_{sat,2}$ [fm^{-3}]	−0.163	−0.149	−0.138
E_{sym} [MeV]	35.5	35	34.5
L [MeV]	74.4	74.7	74.7
K_{sym} [MeV]	−103.487	−101.471	−99.252
Q_{sym} [MeV]	273.008	252.462	234.0525
K_τ [MeV]	−388.232	−404.509	−418.381
e_{PNM} [MeV]	19.46	19	18.5
K_{PNM} [MeV]	116.858	136.172	153.966
Q_{PNM} [MeV]	−205.754	−209.345	−203.476

3. Symmetric and Asymmetric Nuclear Matter Properties

The equation of state of ANM can be expressed as a power series in the isospin asymmetry $\delta = (\rho_n - \rho_p)/\rho$, as given by

$$e(\rho, \delta) = e_0(\rho) + E_{sym}(\rho).\delta^2 + O(\delta^4) \tag{4}$$

where $e_0(\rho)$ is the energy per nucleon in SNM, while $E_{sym}(\rho)$ is the symmetry energy. The energy per nucleon in SNM can also be Taylor expanded around the saturation density as:

$$e_0(\rho) = e_0(\rho_0) + \frac{K_0}{2!}\chi^2 + \frac{Q_0}{3!}\chi^3 + O(\chi^4), \tag{5}$$

where $\chi = \left(\frac{\rho-\rho_0}{3\rho_0}\right)$ and $K_0 = 9\rho_0^2 \frac{\partial^2 e_0(\rho)}{\partial \rho^2}|_{\rho=\rho_0}$ and $Q_0 = 27\rho_0^3 \frac{\partial^3 e_0(\rho)}{\partial \rho^3}|_{\rho=\rho_0}$ are the incompressibility and skewness parameter, respectively, in SNM. Notice that the first derivative does not appear in this expansion due to the saturation condition.

The symmetry energy is the energy cost to convert SNM in PNM [6,36,37]. It is defined as

$$E_{sym}(\rho) = \frac{1}{2!}\frac{\partial^2 e(\rho, \delta)}{\partial \delta^2}|_{\delta=0} \tag{6}$$

The odd-order terms in δ will not appear in Equation (4) due to the isospin invariance of nuclear force in nuclear matter when one neglects the Coulomb interaction. The nuclear symmetry energy $E_{sym}(\rho)$ corresponds to the lowest-order coefficient in the expansion of the energy per particle in ANM in terms of the isospin asymmetry. The contribution from higher-order terms δ^n, n ≥ 4 is very small and has been estimated to be less than 1 MeV in microscopic many-body, as well as effective model calculations [37]. Keeping up to the δ^2-term in Equation (4) refers to the parabolic approximation (PA) of the EoS of ANM. The density dependence of $e_0(\rho)$ is much better understood than that of $E_{sym}(\rho)$, which is still elusive, even more so in the supra saturation regime that makes the study of ANM an important area of contemporary nuclear research. The analysis of the density dependence of $E_{sym}(\rho)$ is performed in terms of the various coefficients in its Taylor expansion about normal NM density ρ_0, given by,

$$E_{sym}(\rho) = E_{sym}(\rho_0) + L\chi + \frac{K_{sym}}{2!}\chi^2 + \frac{Q_{sym}}{3!}\chi^3 + O(\chi^4) \tag{7}$$

where χ has been defined before. The coefficients

$$L = 3\rho_0 \frac{\partial E_{sym}(\rho)}{\partial \rho}|_{\rho=\rho_0}, \; K_{sym} = 9\rho_0^2 \frac{\partial^2 E_{sym}(\rho)}{\partial \rho^2}|_{\rho=\rho_0}, \text{ and } Q_{sym} = 27\rho_0^3 \frac{\partial^3 E_{sym}(\rho)}{\partial \rho^3}|_{\rho=\rho_0} \quad (8)$$

are the slope parameter, curvature parameter and skew symmetry parameter, respectively, and they characterize the density dependence of the nuclear symmetry energy around the normal nuclear density ρ_0, and thus carry important information about the properties of nuclear symmetry energy at both high and low density regions.

The incompressibility of ANM, which depends on both the density and isospin asymmetry, is given by

$$\begin{aligned} K(\rho,\delta) &= 9\frac{\partial P(\rho,\delta)}{\partial \rho} = 18\rho\frac{\partial e(\rho,\delta)}{\partial \rho} + 9\rho^2\frac{\partial^2 e(\rho,\delta)}{\partial \rho^2} \\ &= 18\frac{P(\rho,\delta)}{\rho} + 9\rho^2\frac{\partial^2 e(\rho,\delta)}{\partial \rho^2}, \end{aligned} \quad (9)$$

where $P(\rho,\delta) = \rho^2\frac{\partial e(\rho,\delta)}{\partial \rho}$ is the pressure in ANM, and $e(\rho,\delta)$ is given in Equation (4). At saturation density $\rho_{sat}(\delta)$ of ANM, the pressure $P(\rho_{sat},\delta) = 0$ and the incompressibility in Equation (9) becomes a function of δ only, $K_{sat}(\delta) = 9\rho_{sat}^2\frac{\partial^2 e(\rho,\delta)}{\partial \rho^2}|_{\rho=\rho_{sat}}$, and is referred to as the isobaric incompressibility coefficient. The saturation density in ANM, $\rho_{sat}(\delta)$, is a function of isospin asymmetry and differs from normal NM density ρ_0. The corrections to ρ_0 on account of the isospin asymmetry δ have been worked out in Ref. [37] in terms of expansion in even powers of δ. For the lowest-order correction $\rho_{sat}(\delta) = \rho_0 + \rho_{sat,2}\delta^2 = \rho_0 - \left(\frac{3L}{K_0}\rho_0\right)\delta^2$. Under the PA of EoS of ANM, the isobaric incompressibility coefficient reads $K_{sat}(\delta) = K_0 + K_\tau(\rho_0)\delta^2$, where K_0 is the incompressibility of SNM at saturation density and $K_\tau = \left(K_{sym} - 6L - \frac{Q_0}{K_0}L\right)$ is the isospin part of $K_{sat}(\delta)$ [37,38]. The value of the nuclear matter saturation properties, such as isoscalar effective mass (m_s^*/m), isovector effective mass (m_v^*/m), energy per particle in PNM (e_{PNM}), neutron matter incompressibility ($K_{PNM} = K_0 + K_{sym}$), neutron matter skewness ($Q_{PNM} = Q_0 + Q_{sym}$), etc., for the SEI-Y (γ = 1/3, 1/2, and 2/3) EoSs at saturation density are given in Table 2.

The incompressibility of SNM at saturation density $K_0(\rho_0)$ is mainly obtained from the analysis of experimental data of the giant monopole resonance (GMR) in open- and closed-shell nuclei. Theoretical studies using non-relativistic and relativistic mean-field models and the analysis of PREX-II and CREX data also provide useful information about $K_0(\rho_0)$. Relevant works using these techniques, which predict different values of $K_0(\rho_0)$, are collected in Table 3. The $K_0(\rho_0)$ values predicted by the SEI-Y models, which are given in Table 2, lie in the range 220–253 MeV within the limiting values extracted from the analysis of the experimental GMR [39,40] and PREX-II and CREX [41] data as well from the compilation of the predictions of a large set of relativistic and non-relativistic mean-field models [29,42].

Table 3. Symmetric nuclear matter incompressibility.

Expt./Theory	**K_0 [MeV]**
Analysis of experimental data of the GMR	
S. Shlomo et al. (2006) [39]	240 ± 20
U. Garg et al. (2018) [40]	240 ± 20
J.R. Stone et al., 2014 [43]	250–315
P. Avogadro and C.A. Bertulani, 2013 [44]	200

Table 3. *Cont.*

Expt./Theory	K_0 [MeV]
Relativistic and non-relativistic mean-field calculations of the GMR	
E. Khan et al., 2012 [45]	230 ± 40
D. Vretenar et al., 2003 [46]	250–270
M. Dutra et al., 2012 [29]	246 ± 41
M. Dutra et al., 2014 [42]	271 ± 86
Analysis of PREX-II and CREX data	
S. Tagami et al., 2022 [41]	210–275

3.1. Landau Parameters

In the framework of the Landau theory of normal Fermi liquids [47–50], the bulk properties of nuclear matter can be written in terms of a two-body interaction expressed as a functional of the second derivative of the energy per particle with respect to the occupation numbers at the Fermi surface. The interaction energy has the following form [51,52],

$$\begin{aligned} \langle k_1k_2|V|k_1k_2\rangle &= N_0^{-1}\{F(\theta) + F'(\theta)\tau_1.\tau_2 + G(\theta)\sigma_1.\sigma_2 \\ &+ G'(\theta)\sigma_1.\sigma_2\tau_1.\tau_2\}, \end{aligned} \tag{10}$$

where $N_0^{-1} = \frac{\hbar^2\pi^2}{2k_F m_s^*}$ is the inverse of the level density at the Fermi surface. The quantity m^* is the effective mass associated with the interaction, and σ and τ are the Pauli matrices in spin and isospin space, respectively. Since both particles are on the Fermi surface, F, F', G, and G' are functions of θ, the angle between k_1 and k_2, which are expanded in terms of Legendre polynomials [53] as,

$$F = \sum_l F_l P_l(cos\theta) \tag{11}$$

and likewise for F', G, and G'. For a spherical Fermi surface to be stable against any deformation, the parameters must satisfy the relations

$$F_l > -(2l+1) \tag{12}$$
$$F'_l > -(2l+1) \tag{13}$$
$$G_l > -(2l+1) \tag{14}$$
$$G'_l > -(2l+1). \tag{15}$$

We have calculated the Landau parameters for the SEI-Y interaction. The corresponding analytical expressions for F_l, F'_l, G_l, and G'_l with l = 0, 1 and 2 are given in Appendix B. Landau parameters at ρ_0 predicted by the SEI-Y ($\gamma = 1/3, 1/2$, and $2/3$) EoSs together with the results of different Skyrme, Gogny, and chiral effective interactions are shown in Figure 1. The density dependence of different Landau parameters is shown for the SEI-Y ($\gamma = 1/2$) parameter set in Figure 2. All the Landau parameters satisfy the condition given in Equations (13)–(15) at all densities, except F_0, which violates the condition Equation (13) for densities less than 0.09 fm^{-3}. The violation of the stability condition of F_0 in the density range $\rho \leq 0.09$ fm^{-3}, as shown in Figure 2, has no significance, as the physical quantity associated with it, the incompressibility K_0, has no physical meaning in this range of density.

The values of the Landau parameters at saturation for the three SEI-Y sets for ($\gamma = 1/3, 1/2$, and $2/3$) are given in Table 4.

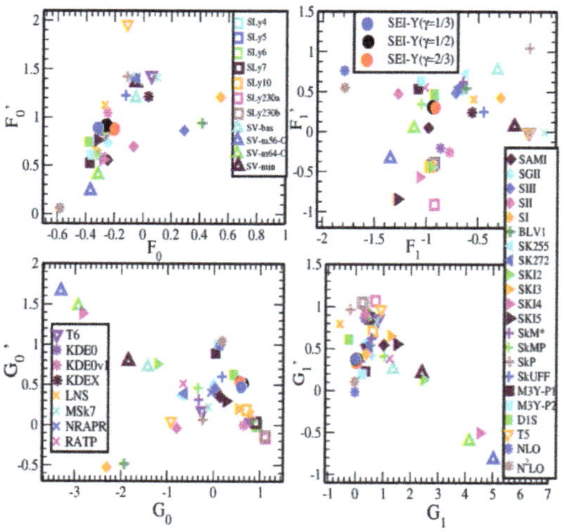

Figure 1. Landau parameters for SEI-Y ($\gamma = 1/2$) parameter sets compared with different Skyrme sets (SLy4, SLy5, SLy6, SLy7, SLy10, SLy230a, SLy230b, SI, SII, SIII, BLV1, SGII, SkM*, RATP, SkP, T6, KDE0, KDE0v1, SK255, SKI2, SKI3, SKI4, SKI5, SkMP, LNS, SV-bas, SV-m56-O, SV-m64-O, SV-min, MSk7, T5, KDEX, NRAPR, SAMi, SK272, SkUFF [54–63]), Gogny interaction (M3Y-P1, M3Y-P2, D1S [64]), and chiral effective interactions [65].

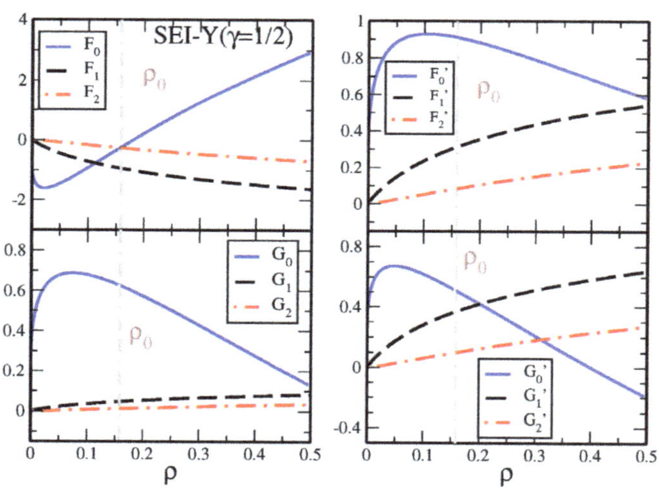

Figure 2. Landau parameters as a function of density using the SEI-Y ($\gamma = 1/2$) set. Saturation density $\rho_0 = 0.158\ \text{fm}^{-3}$ is denoted by a grey dashed vertical line.

These dimensionless parameters F_l, F_l', G_l, and G_l' are directly related to quantities describing SNM and ANM properties, such as effective mass, incompressibility, symmetry energy, the speed of sound, etc., through relationships [66,67],

$$\text{Incompressibility, } K = 3\frac{\hbar^2 k_F^2}{m_s^*}(1 + F_0) \tag{16}$$

$$\text{Isoscalar effective mass, } \frac{m_s^*}{m} = 1 + \frac{F_1}{3}, \tag{17}$$

$$\text{Isoscalar/Isovector effective mass ratio, } \frac{m_s^*}{m_v^*} = 1 + \frac{F_1'}{3}, \tag{18}$$

$$\text{Symmetry energy, } E_{sym} = \frac{\hbar^2 k_F^2}{6m_s^*}(1 + F_0') \tag{19}$$

$$\text{Spin asymmetry coefficient, } E_\sigma = \frac{\hbar^2 k_F^2}{6m_s^*}(1 + G_0) \tag{20}$$

$$\text{Spin-Isospin asymmetry coefficient, } E_{\sigma\tau} = \frac{\hbar^2 k_F^2}{6m_s^*}(1 + G_0') \tag{21}$$

The sound velocity is directly related to the compression modulus $K(\rho)$, which can be expressed in terms of the Landau parameters F_0 and F_1 :

$$mv_s^2 = \frac{\hbar^2 k_F^2}{3m_s^*}(1 + F_0) = \frac{1}{9}K = \frac{\hbar^2 k_F^2}{3m}\frac{1 + F_0}{1 + \frac{F_1}{3}}. \tag{22}$$

The values of these NM properties at saturation are listed in Table 5 for the three EoSs of SEI-Y, corresponding to $\gamma = 1/3, 1/2,$ and $2/3$. These results are in agreement with the saturation properties predicted under the parameter fitting protocol given in Table 2.

Table 4. Landau parameters at the saturation density for SEI-Y ($\gamma = 1/3$), SEI-Y ($\gamma = 1/2$), and SEI-Y ($\gamma = 2/3$) EoS.

EoS	F_0	F_1	F_2	F_0'	F_1'	F_2'
SEI-Y ($\gamma = 1/3$)	−0.31	−0.913	−0.247	0.89	0.304	0.0826
SEI-Y ($\gamma = 1/2$)	−0.257	−0.939	−0.253	0.914	0.313	0.0845
SEI-Y ($\gamma = 2/3$)	−0.195	−0.9091	−0.2447	0.8771	0.303	0.0816
EoS	G_0	G_1	G_2	G_0'	G_1'	G_2'
SEI-Y ($\gamma = 1/3$)	0.617	0.0321	0.0087	0.474	0.3739	0.1013
SEI-Y ($\gamma = 1/2$)	0.627	0.0479	0.0129	0.514	0.3690	0.0998
SEI-Y ($\gamma = 2/3$)	0.584	0.0744	0.02001	0.5401	0.3297	0.0887

Table 5. Nuclear matter properties predicted using Landau parameters at saturation density for SEI-Y ($\gamma = 1/3$), SEI-Y ($\gamma = 1/2$), and SEI-Y ($\gamma = 2/3$) sets.

SEI-Y (γ) (γ)	ρ_0 [fm^{-3}]	K_0 [MeV]	$\frac{m_s^*}{m}$	$\frac{m_s^*}{m_v^*}$	E_{sym} [MeV]	E_σ [MeV]	$E_{\sigma\tau}$ [MeV]	mv_s^2 [MeV]
(1/3)	0.161	230.59	0.695	1.101	35.10	30.02	27.38	24.47
(1/2)	0.158	237.74	0.686	1.104	34.048	28.95	26.94	26.38
(2/3)	0.156	263.14	0.696	1.101	34.10	28.79	27.97	27.94

3.2. High-Order Derivatives of the Energy per Particle in Asymmetric Nuclear Matter

The symmetry energy is an important quantity in nuclear physics, which rules many properties in the isovector sector of the energy density and has a relevant impact in nuclear astrophysics. The symmetry energy parameter, $E_{sym}(\rho_0)$, is constrained somewhat less rigorously as compared to the energy per particle, $e(\rho_0)$, of SNM. It is determined from the analysis of the predictions of a large set of mean-field models [29] and from data of astrophysical observations [68–70] but also using experimental nuclear data, such as charged pion spectra at high transverse momenta [71] or charge exchange and elastic nuclear reactions [72]. The values of $E_{sym}(\rho_0)$ extracted from these works are reported in the upper panel of Table 6. The symmetry energy values used in the SEI-Y models lie between 34.5 and 35.5 MeV (see Table 2), which are within the ranges predicted by almost all the analyses displayed in Table 6. Heavy-ion collision (HIC) studies have provided relevant constraints on the EoS of SNM at supra-saturation densities, which allow us to predict $e_0(\rho)$ up to about 4.6 times the normal nuclear matter density [27]. In the recent work of Ref. [73], the value for $e_0(4\rho_0)$ is constrained in the range $63.3^{+19.7}_{-6.6}$ at a 68% confidence level. However, the EoS of the high-density neutron-rich matter is still highly uncertain due to the limited progress in the analysis of isospin-sensitive observables of HIC experiments. It also remains an open question whether the symmetry energy is stiffer or super-soft at supra-saturation densities.

In Table 6, the available data found in the literature on nuclear symmetry energy at two times the saturation density, $E_{sym}(2\rho_0)$, together with the corresponding predictions provided by the SEI-Y ($\gamma = 1/3, 1/2,$ and $2/3$) EoSs, are given. These results are based on theoretical analysis of data from laboratory experiments, such as the ASY-EoS experiment at GSI [74] and HIC [75,76], astrophysical data of different types, as outlined in Refs. [73,77–83], theoretical calculations within chiral EFT [84,85] and effective mean-field models [86]. From this table, we see that the SEI-Y predictions are in good agreement with the available data reported in Table 6, in particular with the predictions of Refs. [75,77–82] obtained using different techniques. The symmetry pressure at twice saturation density predicted by the SEI-Y ($\gamma = 1/3, 1/2,$ and $2/3$) EoSs are 10.659, 10.572, and, 10.488 MeV fm^{-3}, respectively, which lies within the range $P_{sym}(2\rho_0) = (35 \pm 32)$ MeV fm^{-3} extracted from the experimentally derived density functional [87]. The nuclear symmetry energy at three times the saturation density, $E_{sym}(3\rho_0)$, for SEI-Y ($\gamma = 1/3, 1/2,$ and $2/3$) EoS are 69.64, 69.38, and 69 MeV, respectively. These SEI-Y values are consistent with the results extracted from the GW170817 data ($76.91^{+25.96}_{-25.96}$ MeV) [79] but are slightly higher than the predictions of Dutra et al., 2012, which range from 33.65 to 60.92 MeV [29].

Table 6. Symmetry energy at several densities.

Expt./Observation/Theory	$E_{sym}(\rho_0)$ **[MeV]**
Mean-field calculations and Astrophysical Observations	
Dutra et al., 2012 [29]	27–36
B A Li and Han, 2013 [68]	31.6 ± 0.92
Oertel et al., 2017 [69]	31.7 ± 3.2
PREX II Experiment Reed et al., 2021 [70]	38.1 ± 4.7
Charged Pion Spectra at high momenta Estee et al., 2021 [71]	32.5–38.1
Charge exchange and elastic scattering data Danielewicz et al., 2017 [72]	33.5–36.4
Expt./Observation/Theory	$E_{sym}(2\rho_0)$ [MeV]
SEI-Y ($\gamma = 1/3$)	55.74
SEI-Y ($\gamma = 1/2$)	55.38
SEI-Y ($\gamma = 2/3$)	54.93
HIC and Transport Calculations	
ASY-EoS experiment at GSI [74]	46–54
UrQMD transport calculation [75]	55 ± 5
Zhang et al., 2020 [76]	35–55
Gravitational Waves Zhang and Li, 2019 [77]	46.9 ± 10.1
Xie and Li, 2019 [73]	$39.2^{+12.1}_{-8.2}$
Tong et al., 2020 [78]	60.7 ± 10.9
Chiral Effective Field Theory	
Drischler et al., 2020 [84]	45 ± 3
Lonardoni et al., 2020 [85]	45 ± 5
Neutron Star Observables B A Li et al., 2021 [79]	51 ± 13
Nakazato and Suzuki, 2019 [80]	40–60

Table 6. *Cont.*

Expt./Observation/Theory	$E_{sym}(\rho_0)$ [MeV]
Yue et al., 2022 [81]	62.8 ± 15.9
Xie and Li, 2020 [82]	47^{+23}_{-22}
Zhou et al., 2019 [83]	$[39.4^{-6.4}_{+7.5}, 54.5^{-3.2}_{+3.1}]$
Mean-Field Calculations Chen et al., 2015 [86]	40.2 ± 12.8

The parameters associated with higher-order derivatives of the energy per particle and symmetry energy at saturation, specifically Q_0, L, K_{sym}, and Q_{sym}, remain poorly constrained and present challenges for experimental measurements. Among these parameters, the slope of the symmetry energy L is of particular relevance, as this quantity is nicely correlated with some finite nuclei properties; for example, the neutron skin thickness in heavy neutron-rich nuclei is ^{208}Pb. This parameter has been estimated using information extracted from the analysis of terrestrial nuclear experiments and astrophysical observations [68,69,73,82,88], the analysis of the PREX-II data [70], from results of charge exchange and elastic scattering involving isobaric analog states [72], charged pion spectra [71] and isospin diffusion [89], and, very recently, from the charge radii difference in mirror nuclei [90]. The range of the values of the slope parameter L is relatively large and covers from about 40 MeV to 120 MeV depending on the inputs used in the different analyses carried out, as can be seen in the upper panel of Table 7. The L value predicted by the SEI-Y models is about 75 MeV, which lies approximately in the middle of the range of the various estimates considered.

Table 7. L, K_{sym} and K_τ at saturation density from different experimental and theoretical analyses along with the results of SEI-Y ($\gamma = 1/3, 1/2,$ and $2/3$) EoS.

Expt./Observation/Theory	L [MeV]
Terrestrial Experiments and Astrophysical Observations	
Li and Han, 2013 [68]	58.9 ± 16.5
Oertel et al., 2017 [69]	58.7 ± 28.1
Lattimer and Lim, 2013 [88]	40.5–61.9
Xie et al., 2019, 2020 [73,82]	66^{+12}_{-20}
PREX-II Experiment	
Reed et al., 2021 [70]	106 ± 37
Charge exchange and elastic scattering data Danielewicz et al., 2017 [72]	70–101
Charged Pion Spectra at high momenta Estee et al., 2021 [71]	42–117
Isospin Diffusion Data Chen et al., 2005 [89]	63–113
Charge radii difference in mirror pairs	
Bano et al., 2023 [90]	70–100
Expt./Observation/Theory	K_{sym} [MeV]
Analysis of Different Neutron Star Observables Li et al., 2020 [91]	-120^{+80}_{-100}
d'Etivaux et al., 2019 [92]	-85^{+82}_{-70}
Carson et al., 2019 [93]	-259 to $+32$
Choi et al., 2021 [94]	-128 to -33
Chiral Effective Field Theory	
Drischler et al., 2016 [95]	-240 to -70
Newton and Crocombe, 2021 [96]	-209^{+270}_{-182}
Grams et al., 2022 [97]	-200 to $+50$
Terrestrial Nuclear Experiments and Mean-Field Predictions	
Sagawa et al., 2019 [98]	-120 ± 40
Tews et al., 2017 [99], Zhang et al., 2017 [100]	-400 to $+100$
Mondal et al., 2017 [101]	-118.8 ± 71.3

Table 7. *Cont.*

Expt./Observation/Theory	L [MeV]
Expt./Theory	K_τ [MeV]
Experimental data of Isoscalar Giant Monopole Resonances	
Sagawa et al., 2008 [102]	-500 ± 50
Li et al., 2010 [103,104]	-550 ± 100
Stone et al., 2014 [43]	-840 to -350
Theoretical calculations of GMR with MDI interactions Chen et al., 2009 [37]	-370 ± 120
Cozma, 2018 [105]	-354 ± 228
Neutron skin sizes across the mass table	
Centelles et al., 2009 [106]	-500^{+125}_{-100}

The incompressibility parameter in ANM, K_{sym}, has been estimated from astrophysical inputs provided by astrophysical observations [91–94], from nuclear and neutron matter calculations using chiral effective field theory [95–97], from terrestrial experiments [98] and from the analysis of mean-field predictions [99–101]. This parameter is, in general, negative and of the order of a few hundred MeV. The SEI-Y predictions, given in Table 2, are in line with the estimates obtained from terrestrial experiments and astrophysical observations reported in the middle panel of Table 7.

In Ref. [101], a correlation between the K_{sym} and $3E_{sym} - L$ parameters is obtained from the analysis of 500 Skyrme and RMF models. However, in the case of SEI-Y, K_{sym} is also strongly correlated with L in addition to the $3E_{sym} - L$ correlation. In particular, in the case of SEI-Y ($\gamma = 1/2$), we find the linear relation $K_{sym} = 4.1165L - 408.98$ MeV. The isovector incompressibility parameter K_τ is mainly extracted from experimental data of the isoscalar giant monopole resonance [43,102–104] from theoretical mean-field model calculations for different MDI interactions [37,105] and from information extracted from measurements of neutron skins across the mass table [106]. These estimates have an average value of approximately -500 MeV but with large error bars, as can be seen in the lower panel of Table 7. The values predicted by the SEI-Y models, given in Table 2, are in agreement with the values extracted using different techniques given in the lower panel of Table 7.

Experimental constraints on the skewness parameter in both symmetric and asymmetric nuclear matter, Q_0 and Q_{sym}, respectively, are currently lacking in precision. Based on the analysis of different experimental and observational data [99,107–110], it is found that the skewness in SNM is negative, and its range is approximately between $\simeq -1200$ and 400 MeV. The situation is similar for the skewness parameter of the symmetry energy, where different estimates constrain its value within the range between -200 and 800 MeV [77,99–101]. The values of Q_0 in SNM predicted by SEI-Y ($\gamma = 1/3, 1/2$, and $2/3$) EoSs are listed in Table 2 and lie consistent with the values extracted from other different analyses. In the same table, we display the skewness parameter of the symmetry energy computed with the SEI-Y models, which are in the range 234–273 MeV, which is in good agreement with the value of $Q_{sym} = 296.8 \pm 73.6$ MeV suggested in Ref. [101]. With the SEY-Y model, we also find a strong anti-correlation between the Q_{sym} and L parameters, which in the case of SEI-Y ($\gamma = 1/2$) EoS reads $Q_{sym} = -8.805L + 910.26$ MeV.

4. Neutron Star Phenomenology

4.1. The Radius of Neutron Stars and the Slope of the Isoscalar Incompressibility

The density derivative of the isoscalar incompressibility of symmetric nuclear matter, which is defined as $K'(\rho) = 3\rho\frac{dK(\rho)}{d\rho}$, can be written at saturation density as a combination of the skewness and the incompressibility of SNM as $K_0'(\rho_0) = Q_0 + 12K_0$ [111]. The value of this parameter, estimated from a large set of non-relativistic and relativistic mean-field models, lies in the range $K_0' = 1800 - 2400$ MeV [112]. A relatively wider range, $1556 \leq K_0' \leq 4971$ MeV, is extracted from the analysis of the tidal deformability measurement in the BNSM event GW170817 [93]. In Ref. [112], the correlation of the radius of the NS with the linear combinations of the slopes of the nuclear matter incompressibility and the symmetry energy coefficients at saturation, which is almost independent of the NS mass in the range 0.6 $M_\odot$–1.8 $M_\odot$, is shown. Here, we explore, using the SEI-Y EoSs with ($\gamma = 1/3, 1/2,$ and $2/3$), the possible correlations between the NS radius and the slope of the incompressibility at saturation for NS of 1.4 $M_\odot$ and 1.6 $M_\odot$. To obtain the radii predicted by these EoSs, we solve the Tolman–Oppenheimer–Volkoff (TOV) equation, where the BPS-BBP EoS [113,114] is used up to 0.07468 fm^{-3} (the crust-core transition density for SEI) and our EoS thereafter.

The values of the radii of these 1.4 $M_\odot$ and 1.6 $M_\odot$ NSs, $R_{1.4}$ and $R_{1.6}$ are shown as a function of the slope of the incompressibility parameter $K_0^{'}$ in Figures 3 and 4, respectively, for the three aforementioned SEI EoSs. The vertical shaded region in brown in the Figures correspond to the $K_0^{'}$ values predicted in Refs. [112,115]. The data for NS radii, obtained from various recent studies, are taken from Refs. [70,92,116–121] (for $R_{1.4}$) and [122,123] (for $R_{1.6}$) and displayed by different color areas in Figures 3 and 4, respectively. The LIGO-Virgo measurement leads to an upper limit of $R_{1.4}$ at 13.6 km [124], and that from the BNSM ascertained to be $R_{1.4} < 13.76$ km [125]. The minimum limit for an $R_{1.6}$ radius of non-rotating NS, constrained from GW170817 data by Bauswein et al., is $10.68^{+0.15}_{-0.04}$ km [126]. This value is shown as an orange band in Figure 4. $R_{1.4}$, $R_{1.6}$ and $K_0^{'}$ values predicted by 44-EoSs of Skyrme, RMF and microscopic interactions, which are taken from Table I of the Supplemental Material given in Alam et al., 2016 [112], are also shown in the two figures by green squares.

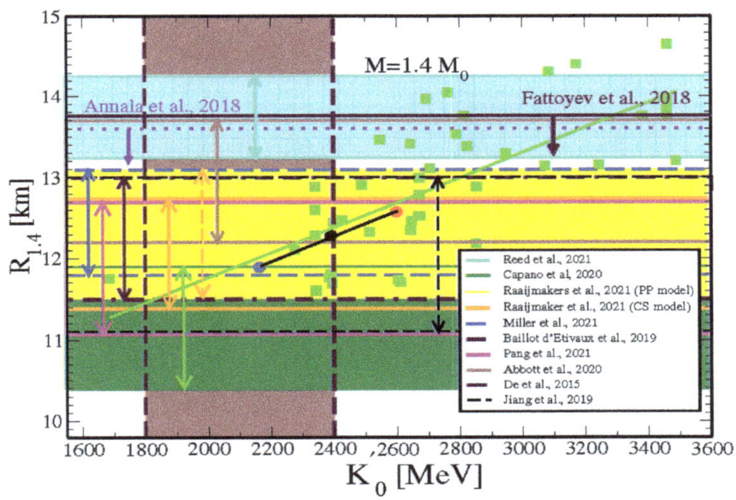

Figure 3. $R_{1.4}$ of 1.4 $M_\odot$ neutron stars versus the slope of the incompressibility obtained using different EoS of SEI-Y of γ = 1/3, 1/2, and 2/3. The green square are the results taken from supplementary material given in Alam et al., 2016 [112]. The horizontal shaded region data of $R_{1.4}$ are taken from: cyan [Reed et al., 2021 [70]], dark green [Capano et al., 2020 [116]], yellow and orange [Raaijmakers et al., 2021 (PP model and CS model), respectively [117]], blue [Miller et al., 2021 [118]], maroon [Baillot d' Etivaux et al., 2019 [92]], magenta [Pang et al., 2021 [119]], black dashed [Jiang et al., 2019 [120]], brown [Abbott et al., 2020 [121]], indigo dotted line [Annala et al., 2018 [124]] and Maroon line [Fattoyev et al., 2018 [125]]. The vertical shaded region in brown [De et al., 2015 [115]].

Using these 44-EoSs data, a moderate linear correlation between the NS radii and $K_0^{'}$ is obtained, as was also pointed out by Alam et al. in Ref. [112], whereas a rather strong correlation over the mass range 0.8 $M_\odot$–1.8 $M_\odot$ was obtained with a linear combination of $K_0^{'}$ and L_0, as we mentioned before. However, our three SEI-Y EoSs, with the γ parameter equal to 1/3, 1/2 and 2/3, show a strong correlation between the radii and the slope of the incompressibility alone for both NS masses, namely, 1.4 $M_\odot$ and 1.6 $M_\odot$. We have also verified that another strong linear correlation exists between $R_{1.4}$ and $R_{1.6}$ and the linear combination of K_0' and L_0, which is in agreement with the previous findings in [112].

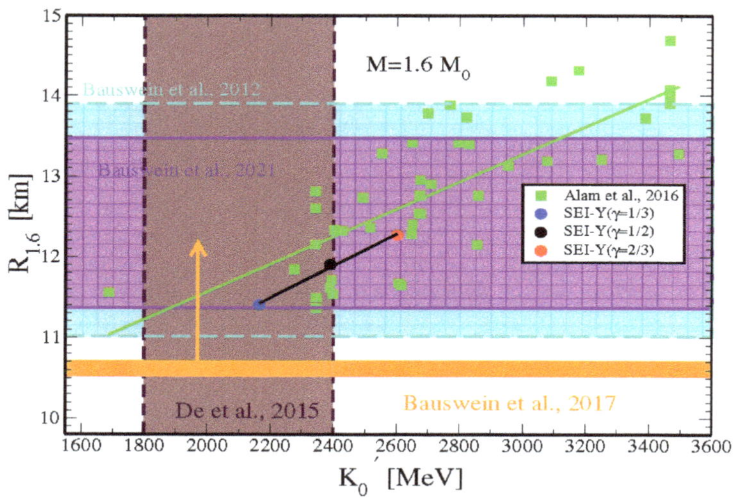

Figure 4. $R_{1.6}$ of 1.6 $M_\odot$ neutron stars versus the slope of the incompressibility obtained using different EoS of SEI-Y of γ = 1/3, 1/2, and 2/3. The green square are the results taken from supplementary material given in Alam et al., 2016 [112]. The violet and cyan horizontal shaded region data of $R_{1.6}$ are calculated from Table IX of [122] and Table I of [123]. The orange band refers to the minimum limit of $R_{1.6}$ in the work of Bauswein et al., 2017 [126]. The vertical shaded region in brown [De et al., 2015 [115]].

4.2. Neutron Star Merger and Incompressibility of Asymmetric Nuclear Matter

The incompressibility of ANM in Equation (9) depends on both density and isospin asymmetry, and it is found to have important implications in BNSM studies [127,128]. The threshold mass M_{th} for prompt collapse (PC) to form a black hole (BH) in BNSM is scaled in terms of maximum mass M_{max} of the non-rotating NS as $M_{th} = \kappa M_{max}$, where the scaling parameter κ is EoS dependent [129,130]. Bauswein et al. [127], from a simulation study of the BNSM for symmetric binary NS, found that using temperature-dependent nuclear EoSs, there was a strong correlation of κ with the compactness $C_{max} = \frac{GM_{max}}{c^2 R_{max}}$ of the TOV configuration (M_{max}, R_{max}) of the NS, where c and G are the speed of light and the gravitational constant, respectively. A universal ansatz proposed by Bauswein et al. (2013) is

$$\kappa = aC_{max} + b, \tag{23}$$

which is independent of the EoS. Such a linear ansatz represents a reasonable first approximation to the data, but it is not the most general one [131]. By using weighted averaged values of the linear-fit constants a and b of different works, given in Table II of Ref. [132], we computed the threshold mass M_{th} predicted by the three EoSs of SEI-Y (γ = 1/3, 1/2, and 2/3), which are reported in Table 8. In the case of delayed/no collapse, the estimated total binary mass of GW170817 provides a lower bound on the threshold mass for direct BH formation, $M_{th} > M_{tot}^{GW170817} = 2.74^{+0.04}_{-0.01}$ $M_\odot$ [126]. SEI-Y predictions of M_{th} for the three sets of values of a and b in Table 8 closely conform to this limiting value.

Table 8. Threshold mass M_{th} for the three EoSs of SEI-Y (γ = 1/3, 1/2, and 2/3) using the values of constants a and b from the literature, given in Table II of Ref. [132].

			SEI-Y (γ = 2/3)			
Ref	a	b	R_{max}^{TOV}	C_{max}^{TOV}	k	M_{th}
[127]	−3.342	2.42	10.523	0.275	1.499	2.937
[126]	−3.38	2.43	10.523	0.275	1.498	2.936
[132]	$-3.36^{+0.20}_{-0.20}$	$2.35^{+0.06}_{-0.06}$	10.523	0.275	$1.424^{+0.115}_{-0.115}$	$2.790^{+0.225}_{-0.225}$
			SEI-Y (γ = 1/2)			
Ref	a	b	R_{max}^{TOV}	C_{max}^{TOV}	k	M_{th}
[127]	−3.342	2.42	10.243	0.272	1.5095	2.846
[126]	−3.38	2.43	10.243	0.272	1.5091	2.845
[132]	$-3.36^{+0.20}_{-0.20}$	$2.35^{+0.06}_{-0.06}$	10.243	0.272	$1.434^{+0.114}_{-0.114}$	$2.705^{+0.215}_{-0.215}$
			SEI-Y (γ = 1/3)			
Ref	a	b	R_{max}^{TOV}	C_{max}^{TOV}	k	M_{th}
[127]	−3.342	2.42	9.943	0.267	1.5252	2.7437
[126]	−3.38	2.43	9.943	0.267	1.5250	2.7434
[132]	$-3.36^{+0.20}_{-0.20}$	$2.35^{+0.06}_{-0.06}$	9.943	0.267	$1.45^{+0.113}_{-0.113}$	$2.609^{+0.204}_{-0.204}$

In recent work, Perego et al. [128] performed a BNSM simulation study taking asymmetric masses in the NS binary. These authors have shown that the nuclear incompressibility at the central density ρ_c of M_{max}, $K_{max} = K(\rho_c^{max}, \delta)$, contains information on M_{th} for PC in the BNS merger. Consequently, if M_{th} is known, then K_{max} can potentially be predicted, which is not possible nowadays in any laboratory experiment. In Ref. [128], the authors examined the correlation between K_{max} and compactness C_{max} of maximum mass NS considering a large sample of EoSs comprising the nucleonic part, as well as containing hyperons and the transition to the quark phase.

A strong power law correlation has been obtained among these data. We have computed the density dependence of $K(\rho, \delta)$ for the three SEI-Y EoSs, where δ for each ρ is obtained by solving the charge neutral β-equilibrated NSM, and the results are shown as a function of the density in panel (a) of Figure 5. The values of the K_{max} computed at the central density of M_{max} predicted by the SEI-Y EoSs are shown as a function of the compactness C_{max} in panel (b) together with the data of the 66-EoSs taken from the supplementary material of Ref. [128]. The values of the C_{max}, shown in panel (b) of Figure 5 for different EoSs, lie below the empirical limit of compactness allowed by general relativity, C = 4/9 [133], and the Tolman VII analytical solution of the TOV equation, C = 0.3428. These limiting values are the universal upper bounds for compactness, as corroborated by the incorporation of realistic EoS [134,135]. The three SEI data points lie in the tighter threshold region of $K_{max} \approx 12$ GeV of Perego et al., 2022 (Figure 4 of [128]). They have also suggested that the information of M_{th} at different mass asymmetries q of the two NSs in the binary can provide constraints on the velocity of sound v_s close to the central density ρ_c of M_{max}.

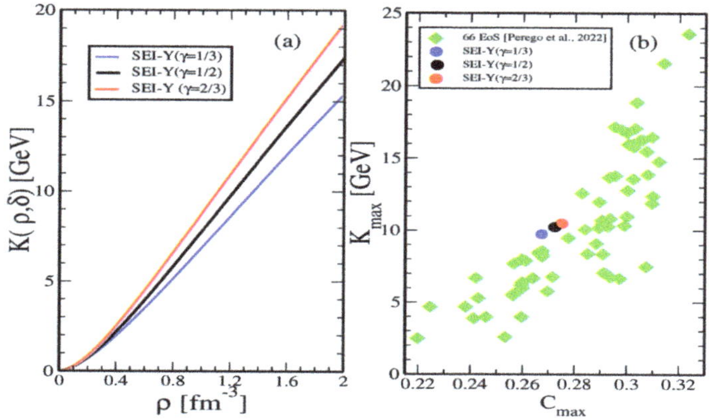

Figure 5. (**a**) K (ρ, δ) as a function of density in NSM for the SEI-Y ($\gamma = 1/3$), SEI-Y ($\gamma = 1/2$), and SEI-Y ($\gamma = 2/3$) EoS, (**b**) K_{max} as a function of the compactness of the heaviest NS for the three EoSs of SEI-Y. The green diamonds are the 66 EoS results taken from [128].

4.3. Sound Speed in Neutron Star Matter

The adiabatic speed of sound in ANM evaluated at constant entropy is given by [97,136,137].

$$\frac{v_s^2}{c^2} = \left(\frac{\partial P}{\partial H}\right)_s = \frac{K(\rho,\delta)}{9(mc^2 + e(\rho,\delta) + P(\rho,\delta)/\rho)} \tag{24}$$

where H is the energy density given by Equation (2), and m is the average nucleon mass.

The square of the speed of the sound v_s in NSM, where δ for each ρ in Equation (24) is the equilibrium value obtained from the β-stability condition, predicted by the three SEI-Y EoSs $\gamma = 1/3, 1/2,$ and $2/3$, is displayed as a function of density in panel (a) of Figure 6. From this figure, we can see that the square of the speed of sound increases with the density without exceeding the causality limit, and it also increases linearly with the incompressibility of nuclear matter, as predicted by Equation (24). The magenta line in panel (a) of Figure 6 represents the conformal limit ($\frac{v_s}{c} \leqslant \frac{1}{\sqrt{3}}$) [138]. More recently, by studying maximally rotating neutron stars, Margaritis et al., 2020, claimed that the sound speed likely exceeds the conformal limit [134,139].

In panel (b) of Figure 6, we display the speed of sound in NSM as a function of nucleonic pressure at density $1.85\rho_0$ computed with the SEI-Y ($\gamma = 1/2$ and $2/3$) EoSs with different slopes of the symmetry energy L in the range 60–110 MeV, together with the results of the EoSs from Bauswein et al., 2012 [123]. The SEI-Y predictions show a nice linear behavior between the speed of the sound and the pressure, with a correlation coefficient r = 0.998 for both EoSs. This linear behavior is, however, weaker with the set of EoSs selected by Bauswein et al. [123], probably due to the different origin and fitting protocols of these EoSs.

Panels (c), (d) and (e) of Figure 6 show, as a function of the NS radius, the square of the speed of sound, v_s^2, computed at the central densities, ρ_c, of the 1.8 $M_\odot$, 1.6 $M_\odot$ and 1.4 $M_\odot$ NSs obtained by solving the TOV equations using the SEI-Y ($\gamma = 1/2$ and $2/3$) EoSs with different values of the slope parameter L in the range 70–110 MeV. For the three considered masses, the square of the speed of the sound shows an inverse linear relationship with the radius of NS, with correlation coefficients close to unity, which show a moderate decreasing trend as the NS mass decreases. The inverse linear relationship in each given mass NS is due to the following. When the slope of the symmetry energy L increases in an EoS of given γ, the $\frac{M}{R}$ ratio decreases owing to the growth of the radius R, and therefore, the compactness also decreases, which implies a reduction in the incompressibility $K(\rho, \delta)$.

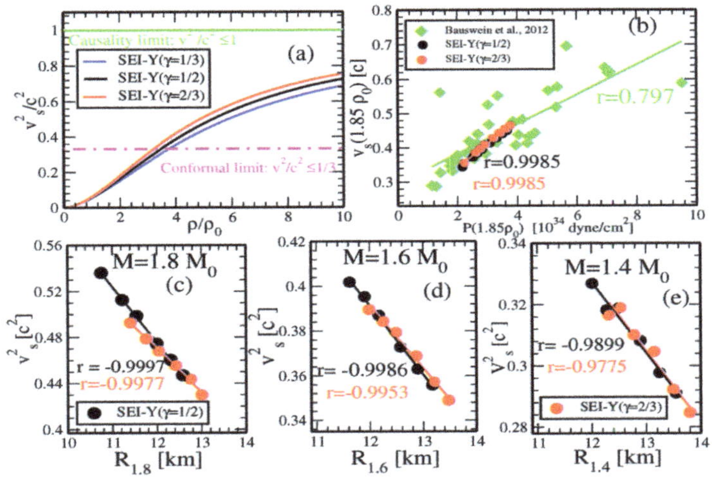

Figure 6. (**a**) Speed of sound in NSM as a function density, where δ for each ρ is the $\beta-$equilibrium value, obtained for the three EoS corresponding to $\gamma = 1/3, 1/2$, and $2/3$ of SEI-Y. The magenta and green lines are the conformal and casual limit, respectively. (**b**) Speed of sound as a function of pressure at density ($1.85\rho_0$) in NSM for SEI-Y (1/2 and 2/3) EoS compared with the results of Bauswein et al., 2012 [123]. The square speed of sound at the central densities of 1.8 $M_\odot$, 1.6 $M_\odot$ and 1.4 $M_\odot$ NSs as a function of radius $R_{1.8}$, $R_{1.6}$ and $R_{1.4}$ for the SEI (1/2 and 2/3) EoSs corresponding to different values of L in the range 70–110 MeV are shown in panels (**c**–**e**), respectively.

4.4. Gravitational Redshift

The gravitational redshift of a signal from the star's surface can be written as,

$$Z_{surf} = \left(1 - \frac{2GM}{c^2 R}\right)^{-1/2} - 1. \quad (25)$$

Measurements of the gravitational redshift of spectral lines can provide direct insights into the stellar compactness parameter and, as a result, can constrain the EoS for dense matter. The Z_{surf} calculated using the SEI-Y ($\gamma = 1/2$) and SEI-Y ($\gamma = 2/3$) EoSs as a function of gravitational mass is shown in Figure 7. From this figure, we can see that Z_{surf} increases as the mass of the NS rises for both considered EoSs. In the lower mass range, the Z_{surf} values for both EoS models are almost the same, but they diverge notably in the higher mass range. EoS having a higher value of incompressibility predicts a lower value of Z_{surf}.

The magenta horizontal line in Figure 7 corresponds to Z_{surf} = 0.35. This value was obtained by Cottam et al. from the X-ray burst source in the low-mass X-ray binary EXO 07482-676 [140]. The gravitational redshift of RBS 1223, RX J0720.4-3125, and RX J1856.5-3754, which are members of the so-called "The Magnificent Seven", are $0.16^{+0.03}_{-0.02}$ [green shaded region], $0.205^{+0.006}_{-0.003}$ [orange shaded region], and $0.22^{+0.06}_{-0.12}$ [maroon shaded region] [141,142] with masses $1.08^{+0.2}_{-0.11}$ $M_\odot$, $1.23^{+0.10}_{-0.05}$ $M_\odot$, and $1.24^{+0.29}_{-0.29}$ $M_\odot$, respectively, at 68% confidence level [143]. These observational data, which are also displayed in Figure 7, are well reproduced by our theoretical calculation using the SEI-Y EoSs, which pass well through the shaded areas representing the uncertainties in the respective observed data.

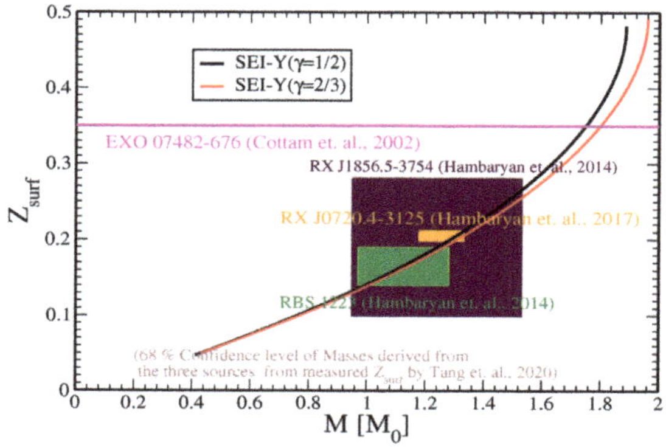

Figure 7. Gravitational redshift at the neutron star surface as a function of the stellar gravitational mass for the SEI-Y ($\gamma = 1/2$) and SEI-Y ($\gamma = 2/3$) EoSs. The magenta horizontal line corresponds to the results of Cottam et al., 2002 from the X-ray burst source in the low-mass X-ray binary EXO 07482-676 [140]. The extracted ranges for the three members of the so-called "The Magnificent Seven" NSs are RBS 1223 [Green shaded region], RX J0720.4-3125 [orange shaded region], and RX J1856.5-3754 [maroon shaded region] [141–143].

The gravitational redshift Z_{surf} in different masses, NSs, namely (a) 1.8 $M_\odot$, (b) 1.6 $M_\odot$, and (c) 1.4 $M_\odot$, are shown as a function of the auxiliary parameter $\eta = (K_0 L^2)^{1/3}$ in Figure 8 for the SEI-Y ($\gamma = 1/2$) and SEI-Y ($\gamma = 2/3$) EoSs. The auxiliary parameter η, which was proposed in Refs. [144,145], is a combination of incompressibility in SNM, K_0, and the slope of the symmetry energy, L. The values of L and K_0 extracted from different nuclear experiments and observations constrain the value of η in the range $60.8 \leq \eta \leq 174.5$ MeV from SπRIT, RCNP and PREX-II data, as can be seen in Figure 8. From this figure, we can also observe a strong anti-correlation between Z_{surf} and η predicted by the two considered SEI-Y EoSs for the three NS masses analyzed, namely 1.8 $M_\odot$, 1.6 $M_\odot$, and 1.4 $M_\odot$. The correlation coefficient results for the two EoSs are displayed in the respective panels, which predict a relatively strong correlation coefficient for the stiffer EoS. The Z_{surf} values as a function of η, predicted by the 44 EoSs of Skyrme, RMF and microscopic interactions used by Alam et al. [112], have been computed for the same masses and shown in the respective panels of Figure 8 as green diamonds. The inverse relation between Z_{surf} and η is also observed for these EoSs (although it is much weaker) can be assigned due to the different origins of the EoSs considered in [112].

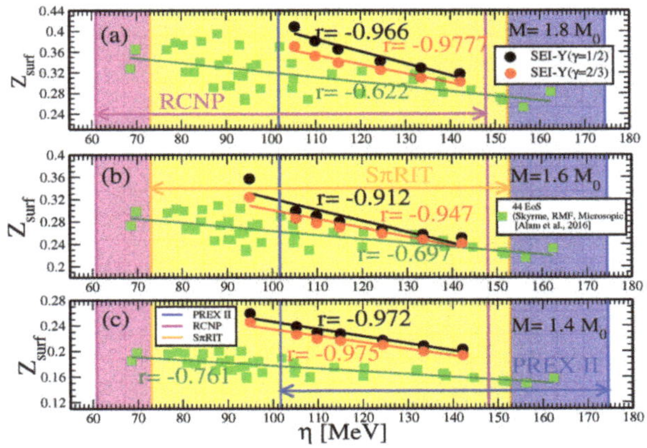

Figure 8. Z_{surf} as a function of η for (**a**) 1.8 M$_\odot$, (**b**) 1.6 M$_\odot$, and (**c**) 1.4 M$_\odot$ NS for SEI-Y ($\gamma = 1/2$) and SEI-Y ($\gamma = 2/3$) EoSs. The shaded region is the constrained value of η for PREX II [Blue], RCNP [magenta], and SπRIT [yellow] [145]. The green diamonds are the data for the 44-EoSs of Ref. [112].

4.5. Neutron Star Mass, Radius and Gravitational Redshift at Different Central Densities

The correlations between the mass M, radius R, and gravitational redshift Z_{surf} of NSs with central densities of ρ_0, $2\rho_0$, $3\rho_0$, $4\rho_0$, and $6\rho_0$ are studied as a function of the slope L in the range 60–110 MeV, using the SEI EoS (γ = 2/3). The corresponding results are shown in the three panels of Figure 9, which correspond to masses, radii and gravitational redshift from left to right, respectively.

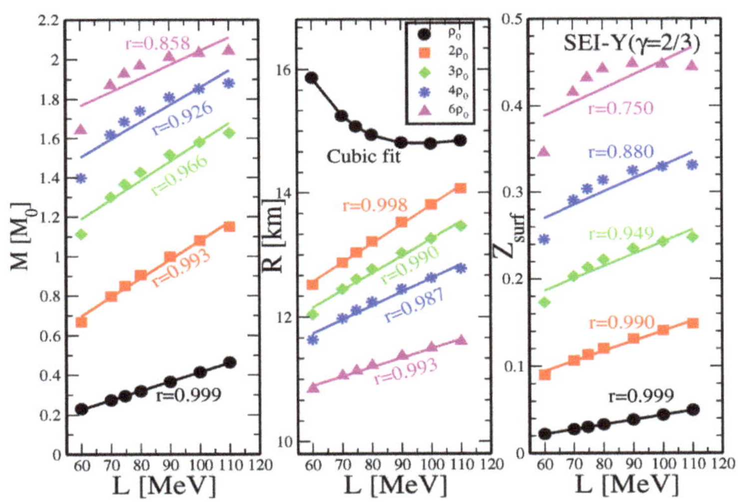

(**a**) (**b**) (**c**)

Figure 9. (**a**) Neutron star masses, (**b**) neutron star radius (**c**) Z_{surf} corresponding to central densities of ρ_0, $2\rho_0$, $3\rho_0$, $4\rho_0$, and $6\rho_0$ as a function of L.

A linear correlation has been observed between L and mass M for the NSs at all central densities ρ_0, $2\rho_0$, $3\rho_0$, $4\rho_0$, and $6\rho_0$. The correlation coefficients are r = 0.999, 0.993, 0.966, 0.926, and 0.858, respectively. The correlation becomes weaker as the density increases—a trend similar to the one found in the work of Ref. [146]—which reflects the role of other empirical parameters governing the density dependence of the EoS [146]. For NSs with central densities $2\rho_0$, $3\rho_0$, $4\rho_0$, and $6\rho_0$, a strong linear correlation between the NSs radii and L is also found. However, for a central density $\rho_c = \rho_0$, the radius decreases as the slope of the symmetry energy L increases. The reason for this behavior is given in Ref. [146]. A higher value of L implies a softer EoS for densities below ρ_0, which explains the anti-correlation observed in the central panel of Figure 9 for NSs having a central density ρ_0. In contrast, at higher densities, a larger value of the slope parameter L results in a stiffer EoS above ρ_0 and, consequently, the radius, which is correlated with L in this region, shows an increasing trend. In the right panel of Figure 9, we also found a strong linear correlation between the Z_{surf} and L for NSs with central densities ρ_0, $2\rho_0$, $3\rho_0$, $4\rho_0$, and $6\rho_0$, which gradually degrades as the central density increases and is a feature similar to the correlation between mass M and L displayed in the left panel of Figure 9.

5. Summary and Outlook

We have used the so-called finite-range simple effective interaction with a Yukawa form-factor to study some non-standard properties of symmetric and asymmetric nuclear matter, such as the Landau parameters associated with this interaction and the high-order derivatives of the energy per particle in symmetric matter and the symmetry energy at saturation density. In addition, we have explored the predictive power of SEI-Y in the high-density neutron-rich domain in describing recent neutron star phenomenology associated with a binary neutron star merger and gravitational redshift.

SEI-Y is a phenomenological effective interaction whose parameters, except the one fitted to finite nuclei data, are systematically fitted under very generic considerations to experimental or empirical data of symmetric nuclear matter and pure neutron matter, which provides a satisfactory account of the nuclear matter properties. An important characteristic of SEI-Y is the fact that the parameters that determine the momentum dependence of the mean-field are decoupled from the ones that fix their density dependence in such a way that each part in the isovector sector can be studied independently of the other without affecting the isoscalar predictions.

We have computed the Landau parameters for the SEI-Y EoSs that give an overall satisfactory account of the nuclear matter saturation properties as well as the sum rules. The nuclear matter properties predicted from the Fermi liquid formulation given in Table 5 are reproduced within a relative difference of approximately ~ 4 % by the corresponding values computed directly with the parameters of the SEI-Y interaction (see, in this respect, Table 2 for comparisons).

The properties of the higher-order derivatives of the energy per particle $e_0(\rho)$ and the symmetry energy $E_{sym}(\rho)$ in nuclear matter at saturation density, namely, Q_0, L, K_{sym} and Q_{sym}, have remained unconstrained, and their extraction from theoretical calculations, various terrestrial laboratory experiments and astrophysical observations predict values with largely differing uncertainties, as can be seen in Table 7. The results obtained with the different SEI-Y EoSs considered in this work are $-478 \lesssim Q_0 \lesssim -437$ MeV, $-103 \lesssim K_{sym} \lesssim -99$ MeV, $-418 \lesssim K_\tau \lesssim -388$ MeV and $234 \lesssim Q_{sym} \lesssim 273$ MeV, at it can be seen in Table 2. These values are found to lie within the range of values extracted from the different studies summarized in Table 7.

However, the main aim of this study is to discuss some recent phenomenology of neutron stars related to the binary neutron star merger and gravitational redshift. The compactness C_{max} of the maximum mass of a neutron star predicted by the three SEI-Y EoSs lies in the range of $0.267 \lesssim C_{max} \lesssim 0.275$, which approximately covers the range of incompressibilities between 220 and 260 MeV, predicting the threshold mass M_{th} for prompt collapse in the range between 2.61 and 2.94 $M_\odot$, which satisfies the minimum threshold mass constraint assessed from the binary masses in the GW170817 event. The K_{max} of these SEI-Y EoSs lies in the tighter threshold region of $K_{max} \approx 12$ GeV of Perego et al.'s, 2022, analysis using 34 EoSs of different types. The velocity of sound, v_s, computed with the SEI-Y model EoSs, is also found to remain causal in neutron star matter and does not exceed the velocity of light. Using SEI-Y EoSs with given symmetry stiffness (given γ) but different slope parameter L in a given neutron star, we find an antilinear relationship between its radius, R, and the square of its sound speed, ${v_s}^2$, computed at its central density. These results are shown in panels (c), (d) and (e) of Figure 6 for neutron stars with masses 1.8 $M_\odot$, 1.6 $M_\odot$ and 1.4 $M_\odot$. We have also used the SEI-Y EoSs to study the gravitational redshift at the surface of a neutron star, Z_{surf}, which is intrinsically connected to the compactness parameter, as a function of the mass of the neutron star. The SEI-Y predictions of Z_{surf} conform to the values constrained from the astrophysical observations on the three neutron stars, namely, RBS 1223, RX J0720.4-3125, and RX J1856.5-3754 in the X-ray binary shown in Figure 7 and whose observed redshifts are $0.16^{+0.03}_{-0.02}$, $0.205^{+0.006}_{-0.003}$, and $0.22^{+0.06}_{-0.12}$, respectively. In a neutron star of a given mass, the redshift parameter Z_{surf} shows a decreasing trend if the slope parameter, L, increases when the incompressibility modulus, K_0, is kept fixed. This implies that for a neutron star of a given mass, when the slope parameter L increases, the compactness, C, decreases. This behavior is also verified using the 44 EoSs of Skyrme, RMF and microscopic type given in the work of Alam et al. shown in Figure 8. On the other hand, Z_{surf} in a neutron star also grows with increasing central density, which in turn increases the compactness parameter that is a relevant parameter in order to understand the EoS of dense neutron-rich matter. Our immediate objective is to use SEI-Y to study the neutron star phenomenology at finite temperature and finite nuclei properties at non-zero temperature, as well as taking deformation into account.

Author Contributions: Conceptualization, X.V., P.B., Z.N. and T.R.R.; methodology, X.V., P.B., Z.N. and T.R.R.; formal analysis, X.V., P.B., Z.N. and T.R.R.; investigation, X.V., P.B., Z.N. and T.R.R.; data curation, X.V., P.B., Z.N. and T.R.R.; writing—original draft, T.R.R.; writing—review and editing, X.V., P.B., Z.N. and T.R.R.; supervision, X.V., Z.N. and T.R.R. All authors have read and agreed to the published version of the manuscript.

Funding: X.V. availed partial support from Grants No. PID2020-118758GB-I00 and No. CEX2019-000918-M (through the "Unit of Excellence María de Maeztu 2020–2023" award to ICCUB) from the Spanish MCIN/AEI/10.13039/501100011033.

Data Availability Statement: The original contributions presented in the study are included in the article, further inquiries can be directed to the corresponding author.

Acknowledgments: P.B. acknowledges support from the MANF Fellowship of UGC, India. Useful discussions with M. Centelles and J.N. De are also warmly acknowledged. T.R.R. offers sincere thanks to B. Behera for useful discussions.

Conflicts of Interest: The authors declare no conflicts of interest. The funders had no role in the design of the study; in the collection, analyses, or interpretation of data; in the writing of the manuscript; or in the decision to publish the results.

Appendix A. Relations between the Six Strength Parameters and the Interaction Parameters

The relationships between the six strength parameters in Equation (2) and the interaction parameters in Equation (1) are as follows:

$$\varepsilon_0^l = \rho_0\left[\frac{t_0}{2}(1-x_0) + \left(W + \frac{B}{2} - H - \frac{M}{2}\right)(4\pi\alpha^3)\right] \tag{A1}$$

$$\varepsilon_0^{ul} = \rho_0\left[\frac{t_0}{2}(2+x_0) + \left(W + \frac{B}{2}\right)(4\pi\alpha^3)\right] \tag{A2}$$

$$\varepsilon_\gamma^l = \frac{t_3}{12}\rho_0^{\gamma+1}(1-x_3) \tag{A3}$$

$$\varepsilon_\gamma^{ul} = \frac{t_3}{12}\rho_0^{\gamma+1}(2+x_3) \tag{A4}$$

$$\varepsilon_{ex}^{l} = \rho_0\left(M + \frac{H}{2} - B - \frac{W}{2}\right)(4\pi\alpha^3) \tag{A5}$$

$$\varepsilon_{ex}^{ul} = \rho_0\left(M + \frac{H}{2}\right)(4\pi\alpha^3) \tag{A6}$$

where the superscript indices l and ul denote the contributions resulting from the nucleon interactions between a like-pair and an unlike-pair. By inverting this set of equations and replacing the strength parameters with their values given in Table 1, one obtains the interaction parameters that are reported in Table A1.

Table A1. The twelve parameters for the SEI-Y EoSs corresponding to γ = 1/3, 1/2, and 2/3.

γ	**b** $[fm^3]$	α **[fm]**	x_3	t_3 **[MeVfm$^{3(\gamma+1)}$]**	**W [MeV]**
1/3	0.4161	0.4232	−0.0630	9536.129	−1380.539
1/2	0.5880	0.4242	−0.112	9277.281	−1321.847
2/3	0.7796	0.4250	−0.153	10228.257	−1214.475
B [MeV]	H [MeV]	M [MeV]	t_0 $[MeVfm^3]$	x_0	W_0 [MeV]
128.0918	−630.968	−808.871	333.5	1.151	119.3
100.950	−575.215	−832.339	566.7	0.664	118.4
49.094	−470.284	−881.144	647.4	0.520	118.2

Appendix B. Landau Parameters

The expression of Landau parameters for SEI-Y EoS are given as follows:

$$\begin{aligned} F_0 &= N_0\Bigg\{\frac{3}{4}t_0 + \frac{t_3}{16}\left(\frac{\rho}{1+b\rho}\right)^{\gamma}\left[(\gamma+1)(\gamma+2) - 2\gamma(\gamma+2)\left(\frac{b\rho}{1+b\rho}\right) + \gamma(\gamma+1)\left(\frac{b\rho}{1+b\rho}\right)^2\right] \\ &+ 4\pi\alpha^3\left(W + \frac{B}{2} - \frac{H}{2} - \frac{M}{4}\right) + \left(\frac{\pi\alpha}{k_F^2}\right)\left(M + \frac{H}{2} - \frac{B}{2} - \frac{W}{4}\right)\ln(1+4\alpha^2k_F^2)\Bigg\} \end{aligned} \tag{A7}$$

$$F_1 = N_0\left\{\left(\frac{3\pi\alpha}{k_F^2}\right)\left(M + \frac{H}{2} - \frac{B}{2} - \frac{W}{4}\right)\left[\left(1 + \frac{1}{2\alpha^2k_F^2}\right)\ln(1+4\alpha^2k_F^2) - 2\right]\right\} \tag{A8}$$

$$\begin{aligned} F_2 &= N_0\left(\frac{5\pi\alpha}{2k_F^2}\right)\left(M + \frac{H}{2} - \frac{B}{2} - \frac{W}{4}\right)\Bigg\{\left[3\left(1 + \frac{1}{2\alpha^2k_F^2}\right)^2 - 1\right]\ln(1+4\alpha^2k_F^2) \\ &- 6\left(1 + \frac{1}{2\alpha^2k_F^2}\right)\Bigg\} \end{aligned} \tag{A9}$$

$$\begin{aligned} F_0' &= N_0\Bigg\{-\frac{t_0}{4}(1+2x_0) - \frac{t_3}{24}(1+2x_3)\left(\frac{\rho}{1+b\rho}\right)^{\gamma} - 4\pi\alpha^3\left(\frac{H}{2} + \frac{M}{4}\right) \\ &- \left(\frac{\pi\alpha}{k_F^2}\right)\left(\frac{B}{2} + \frac{W}{4}\right)\ln(1+4\alpha^2k_F^2)\Bigg\} \end{aligned} \tag{A10}$$

$$F_1' = N_0\left\{-\left(\frac{3\pi\alpha}{k_F^2}\right)\left(\frac{B}{2} + \frac{W}{4}\right)\left[\left(1 + \frac{1}{2\alpha^2k_F^2}\right)\ln(1+4\alpha^2k_F^2) - 2\right]\right\} \tag{A11}$$

$$F_2' = N_0\left\{-\left(\frac{5\pi\alpha}{2k_F^2}\right)\left(\frac{B}{2}+\frac{W}{4}\right)\left[\left[3\left(1+\frac{1}{2\alpha^2k_F^2}\right)^2-1\right]\ln(1+4\alpha^2k_F^2)-6\left(1+\frac{1}{2\alpha^2k_F^2}\right)\right]\right\} \tag{A12}$$

$$\begin{aligned} G_0 &= N_0\left\{-\frac{t_0}{4}(1-2x_0)-\frac{t_3}{24}(1-2x_3)\left(\frac{\rho}{1+b\rho}\right)^\gamma+4\pi\alpha^3\left(\frac{B}{2}-\frac{M}{4}\right)\right. \\ &+ \left.\left(\frac{\pi\alpha}{k_F^2}\right)\left(\frac{H}{2}-\frac{W}{4}\right)\ln(1+4\alpha^2k_F^2)\right\} \end{aligned} \tag{A13}$$

$$G_1 = N_0\left\{\left(\frac{3\pi\alpha}{k_F^2}\right)\left(\frac{H}{2}-\frac{W}{4}\right)\left[\left(1+\frac{1}{2\alpha^2k_F^2}\right)\ln(1+4\alpha^2k_F^2)-2\right]\right\} \tag{A14}$$

$$G_2 = N_0\left\{\left(\frac{5\pi\alpha}{2k_F^2}\right)\left(\frac{H}{2}-\frac{W}{4}\right)\left[\left[3\left(1+\frac{1}{2\alpha^2k_F^2}\right)^2-1\right]\ln(1+4\alpha^2k_F^2)-6\left(1+\frac{1}{2\alpha^2k_F^2}\right)\right]\right\} \tag{A15}$$

$$G_0' = N_0\left\{-\frac{t_0}{4}-\frac{t_3}{24}\left(\frac{\rho}{1+b\rho}\right)^\gamma-4\pi\alpha^3\left(\frac{M}{4}\right)-\left(\frac{\pi\alpha}{k_F^2}\right)\left(\frac{W}{4}\right)\ln(1+4\alpha^2k_F^2)\right\} \tag{A16}$$

$$G_1' = N_0\left\{-\left(\frac{3\pi\alpha}{k_F^2}\right)\left(\frac{W}{4}\right)\left[\left(1+\frac{1}{2\alpha^2k_F^2}\right)\ln(1+4\alpha^2k_F^2)-2\right]\right\} \tag{A17}$$

$$G_2' = N_0\left\{-\left(\frac{5\pi\alpha}{2k_F^2}\right)\left(\frac{W}{4}\right)\left[\left[3\left(1+\frac{1}{2\alpha^2k_F^2}\right)^2-1\right]\ln(1+4\alpha^2k_F^2)-6\left(1+\frac{1}{2\alpha^2k_F^2}\right)\right]\right\}, \tag{A18}$$

where the normalization constant $N_0 = \frac{2k_F m_s^*}{\hbar^2\pi^2}$ is the level density at the Fermi surface. The numerical values of the Landau parameters reported in Table 4 are obtained with the help of the interaction parameters given in Table A1 and the effective mass and Fermi momentum at saturation taken from Table 2.

References

1. Haar, B.T.; Malfliet, R. Nucleons, mesons and deltas in nuclear matter a relativistic Dirac-Brueckner approach. *Phys. Rep.* **1987**, *149*, 207. [CrossRef]
2. Van Dalen, E.N.E.; Fuchs, C.; Faessler, A. The relativistic Dirac–Brueckner approach to asymmetric nuclear matter. *Nucl. Phys. A* **2004**, *744*, 227–248. [CrossRef]
3. de Jong, F.; Lenske, H. Asymmetric nuclear matter in the relativistic Brueckner-Hartree-Fock approach. *Phys. Rev. C* **1998**, *57*, 3099. [CrossRef]
4. Cugnon, J.; Deneye, P.; Lejeune, A.Z. Neutron matter properties in an extended Brueckner approach. *Zeitschrift für Physik A-Atomic Nuclei* **1987**, *328*, 409–415. [CrossRef]
5. Zuo, W.; Lejeune, A.; Lombardo, U.; Mathiot, J.F. Microscopic three-body force for asymmetric nuclear matter. *Eur. Phys. J. A* **2002**, *14*, 469–475. [CrossRef]
6. Bombaci, I.; Lombardo, U. Asymmetric nuclear matter equation of state. *Phys. Rev. C* **1991**, *44*, 1892. [CrossRef] [PubMed]
7. Day, B.D.; Wiringa, R.B. Brueckner-Bethe and variational calculations of nuclear matter. *Phys. Rev. C* **1985**, *32*, 1057. [CrossRef] [PubMed]
8. Baldo, M.; Burgio, G.F.; Schulze, H.-J. Hyperon stars in the Brueckner-Bethe-Goldstone theory. *Phys. Rev. C* **2000**, *61*, 055801 [CrossRef]

9. Vidaña, I.; Polls, A.; Ramos, A.; Engvik, L.; Hjorth-Jensen, M. Hyperon-hyperon interactions and properties of neutron star matter. *Phys. Rev. C* **2000**, *62*, 035801. [CrossRef]
10. Akmal, A.; Pandharipande, V.R.; Ravenhall, D.G. Equation of state of nucleon matter and neutron star structure. *Phys. Rev. C* **1998**, *58*, 1804. [CrossRef]
11. Mukherjee, A.; Pandharipande, V.R. Variational theory of hot nucleon matter. *Phys. Rev. C* **2007**, *75*, 035802. [CrossRef]
12. Hebeler, K.; Schwenk, A. Chiral three-nucleon forces and neutron matter. *Phys. Rev. C* **2010**, *82*, 014314. [CrossRef]
13. Hebeler, K.; Lattimer, J.M.; Pethick, C.J.; Schwenk, A. Constraints on Neutron Star Radii Based on Chiral Effective Field Theory Interactions. *Phys. Rev. Lett.* **2010**, *105*, 161102. [CrossRef] [PubMed]
14. Haensel, P.; Kutschera, M.; Prószyński, M. Uncertainty in the saturation density of nuclear matter and neutron star models. *Astron. Astrophys.* **1981**, *102*, 299–302. Available online: http://adsabs.harvard.edu/abs/1981A%26A...102..299H (accessed on 27 January 2024).
15. Skyrme, T.H.R. CVII. The Nuclear Surface. *Philos. Mag.* **1956**, *1*, 1043. [CrossRef]
16. Decharge, J.; Gogny, D. Hartree-Fock-Bogolyubov calculations with the D1 effective interaction on spherical nuclei. *Phys. Rev. C* **1980**, *21*, 1568. [CrossRef]
17. Bertsch, G.; Borysowicz, J.; McManus, H.; Love, W.G. Interactions for inelastic scattering derived from realistic potentials. *Nucl. Phys. A* **1977**, *284*, 399–419. [CrossRef]
18. Behera, B.; Routray, T.R.; Satpathy, R.K. Momentum and density dependence of the mean field in nuclear matter. *J. Phys. G Nucl. Part. Phys.* **1998**, *24*, 2073. [CrossRef]
19. Behera, B.; Routray, T.R.; Pradhan, A.; Patra, S.K.; Sahu, P.K. Momentum and density dependence of the isospin part of nuclear mean field and equation of state of asymmetric nuclear matter. *Nucl. Phys. A* **2005**, *753*, 367–386. [CrossRef]
20. Behera, B.; Viñas, X.; Routray, T.R.; Centelles, M. Study of spin polarized nuclear matter and finite nuclei with finite range simple effective interaction. *J. Phys. G Nucl. Part. Phys.* **2015**, *42*, 045103. Available online: http://iopscience.iop.org/0954-3899/42/4/045103 (accessed on 27 January 2024). [CrossRef]
21. Behera, B.; Viñas, X.; Bhuyan, M.; Routray, T.R.; Sharma, B.K.; Patra, S.K. Simple effective interaction: infinite nuclear matter and finite nuclei. *J. Phys. G Nucl. Part. Phys.* **2013**, *40*, 095105. Available online: http://iopscience.iop.org/0954-3899/40/9/095105 (accessed on 27 January 2024). [CrossRef]
22. Bertsch, G.F.; Das Gupta, S. A guide to microscopic models for intermediate energy heavy ion collisions. *Phys. Rep.* **1988**, *160*, 189. [CrossRef]
23. Gale, C.; Bertsch, G.F.; Das Gupta, S. Heavy-ion collision theory with momentum-dependent interactions. *Phys. Rev. C* **1987**, *35*, 1666. [CrossRef] [PubMed]
24. Gale, C.; Welke, G.M.; Prakash, M.; Lee, S.J.; Das Gupta, S. Transverse momenta, nuclear equation of state, and momentum-dependent interactions in heavy-ion collisions. *Phys. Rev. C* **1990**, *41*, 1545. [CrossRef]
25. Csernai, L.P.; Fai, G.; Gale, C.; Osnes, E. Nuclear equation of state with momentum-dependent interactions. *Phys. Rev. C* **1992**, *46*, 736. [CrossRef]
26. Behera, B.; Routray, T.R.; Satpathy, R.K. Causal violation of the speed of sound and the equation of state of nuclear matter. *J. Phys. G* **1997**, *23*, 445. Available online: http://iopscience.iop.org/0954-3899/23/4/005 (accessed on 27 January 2024). [CrossRef]
27. Danielewicz, P.; Lacey, R.; Lynch, W.G. Determination of the Equation of State of Dense Matter. *Science* **2002**, *298*, 1592. [CrossRef]
28. Behera, B.; Routray, T.R.; Tripathy, S.K. Neutron–proton effective mass splitting and thermal evolution in neutron-rich matter. *J. Phys. G Nucl. Part. Phys.* **2011**, *38*, 115104. Available online: http://iopscience.iop.org/0954-3899/38/11/115104 (accessed on 27 January 2024). [CrossRef]
29. Dutra, M.; Lourenço, O.; Martins, J.S.S.; Delfino, A.; Stone, J.R.; Stevenson, P.D. Skyrme interaction and nuclear matter constraints. *Phys. Rev. C* **2012**, *85*, 035201. [CrossRef]
30. Behera, B.; Routray, T.R.; Pradhan, A.; Patra, S.K.; Sahu, P.K. Nuclear mean field and equation of state of asymmetric nuclear matter. *Nucl. Phys. A* **2007**, *794*, 132–148. [CrossRef]
31. Wiringa, R.B. Single-particle potential in dense nuclear matter. *Phys. Rev. C* **1988**, *38*, 2967. [CrossRef]
32. Sammarruca, F. The Microscopic Approach to Nuclear Matter and Neutron Star Matter. *Int. J. Mod. Phys. E* **2010**, *19*, 1259–1313. [CrossRef]
33. Behera, B.; Routray, T.R.; Tripathy, S.K. Temperature dependence of the nuclear symmetry energy and equation of state of charge neutral $n + p + e + \mu$ matter in beta equilibrium. *J. Phys. G Nucl. Part. Phys.* **2009**, *36*, 125105. [CrossRef]
34. Sammarruca, F.; Krastev, P.G. Spin polarized neutron matter within the Dirac-Brueckner-Hartree-Fock approach. *Phys. Rev. C* **2007**, *75*, 034315. [CrossRef]
35. Behera, B.; Viñas, X.; Routray, T.R.; Robledo, L.M.; Centelles, M.; Pattnaik, S.P. Deformation properties with a finite-range simple effective interaction. *J. Phys. G Nucl. Part. Phys.* **2016**, *43*, 045115. Available online: http://iopscience.iop.org/0954-3899/43/4/045115 (accessed on 27 January 2024). [CrossRef]
36. Li, B.A.; Ramos, À.; Verde, G.; Vidaña, I. Topical issue on nuclear symmetry energy. *Eur. Phys. J. A* **2014**, *50*, 9. [CrossRef]
37. Chen, L.W.; Cai, B.J.; Ko, C.M.; Li, B.A.; Shen, C.; Xu, J. Higher-order effects on the incompressibility of isospin asymmetric nuclear matter. *Phys. Rev. C* **2009**, *80*, 014322. [CrossRef]
38. Prakash, M.; Bedell, K.S. Incompressibility of neutron-rich nuclear matter. *Phys. Rev. C* **1985**, *32*, 1118(R). [CrossRef]

39. Shlomo, S.; Kolomietz, V.M.; Colò, G. Deducing the nuclear-matter incompressibility coefficient from data on isoscalar compression modes. *Eur. Phys. J. A* **2006**, *30*, 23–30. [CrossRef]
40. Garg, U.; Colò, G. The compression-mode giant resonances and nuclear incompressibility. *Prog. Part. Nucl. Phys.* **2018**, *101*, 55. [CrossRef]
41. Tagami, S.; Wakasa, T.; Yahiro, M. Slope parameters determined from CREX and PREX2. *Res. Phys.* **2022**, *43*, 106037. [CrossRef]
42. Dutra, M.; Lourenço, O.; Avancini, S.S.; Carlson, B.V.; Delfino, A.; Menezes, D.P.; Providência, C.; Typel, S.; Stone, J.R. Relativistic mean-field hadronic models under nuclear matter constraints. *Phys. Rev. C* **2014**, *90*, 055203. [CrossRef]
43. Stone, J.R.; Stone, N.J.; Moszkowski, S.A. Incompressibility in finite nuclei and nuclear matter. *Phys. Rev. C* **2014**, *89*, 044316. [CrossRef]
44. Avogadro, P.; Bertulani, C.A. Role of pairing in the description of giant monopole resonances. *Phys. Rev. C* **2013**, *88*, 044319. [CrossRef]
45. Khan, E.; Margueron, J.; Vidaña, I. Constraining the Nuclear Equation of State at Subsaturation Densities. *Phys. Rev. Lett.* **2012**, *109*, 092501. [CrossRef]
46. Vretenar, D.; Nikšić, T.; Ring, P. A microscopic estimate of the nuclear matter compressibility and symmetry energy in relativistic mean-field models. *Phys. Rev. C* **2003**, *68*, 024310. [CrossRef]
47. Landau, L.D. The Theory of a Fermi Liquid. *J. Exp. Theor. Phys.* **1957**, *3*, 920. Available online: http://jetp.ras.ru/cgi-bin/e/index/e/3/6/p920?a=list (accessed on 27 January 2024).
48. Landau, L.D. Oscillations in a Fermi Liquid. *J. Exp. Theor. Phys.* **1957**, *5*, 101. Available online: http://jetp.ras.ru/cgi-bin/e/index/e/5/1/p101?a=list (accessed on 27 January 2024).
49. Landau, L.D. On the theory of the Fermi liquid. *J. Exp. Theor. Phys.* **1959**, *8*, 70. Available online: http://jetp.ras.ru/cgi-bin/e/index/e/8/1/p70?a=list (accessed on 27 January 2024).
50. Vautherin, D.; Brink, D.M. Hartree-Fock calculations with Skyrme's interaction. I. Spherical nuclei. *Phys. Rev. C* **1972**, *5*, 626. [CrossRef]
51. Bäckman, S.O.; Källman, C.G.; Sjöberg, O. Calculation of Landau's fermi-liquid parameters in pure neutron matter. *Phys. Lett. B* **1973**, *43*, 263–266. [CrossRef]
52. Bäckman, S.O.; Jackson, A.D.; Speth, J. Landau parameters for nuclear matter using the Skyrme interaction. *Phys. Lett. B* **1975**, *56*, 209–211. [CrossRef]
53. Cao, L.-G.; Colò, G.; Sagawa, H. Spin and spin-isospin instabilities and Landau parameters of Skyrme interactions with tensor correlations. *Phys. Rev. C* **2010**, *81*, 044302. [CrossRef]
54. Chabanat, E.; Bonche, P.; Haensel, P.; Meyer, J.; Schaeffer, R. A Skyrme parametrization from subnuclear to neutron star densities Part II. Nuclei far from stabilities. *Nucl. Phys. A* **1998**, *635*, 231–256. [CrossRef]
55. Chabanat, E.; Bonche, P.; Haensel, P.; Meyer, J.; Schaeffer, R. A Skyrme parametrization from subnuclear to neutron star densities. *Nucl. Phys. A* **1997**, *627*, 710–746. [CrossRef]
56. Taqi, A.H.; Khidher, E.G. Nuclear multipole excitations in the framework of self-consistent Hartree–Fock random phase approximation for Skyrme forces. *Pramana J. Phys.* **2019**, *93*, 60. [CrossRef]
57. Goriely, S.; Tondeur, F.; Pearson, J.M. A Hartree-Fock nuclear mass table. *At. Data Nucl. Data Tables* **2001**, *77*, 311–381. [CrossRef]
58. Tselyaev, V.; Lyutorovich, N.; Speth, J.; Reinhard, P.-G.; Smirnov, D. Low-energy M1 excitations in ^{208}Pb and the spin channel of the Skyrme energy-density functional. *Phys. Rev. C* **2019**, *99*, 064329. [CrossRef]
59. Agrawal, B.K.; Shlomo, S.; Kim Au, V. Determination of the parameters of a Skyrme type effective interaction using the simulated annealing approach. *Phys. Rev. C* **2005**, *72*, 014310. [CrossRef]
60. Lahiri, J.; Atta, D.; Basu, D.N. Properties of glitching pulsars in the Skyrme-Hartree-Fock framework. *arXiv* **2022**, arXiv:2207.13384v3. [CrossRef]
61. Keh-Fei, L.; Hongde, L.; Ma, Z.; Shen, Q.; Moszkowski, S.A. Skyrme-Landau parameterization of effective interactions (I). Hartree-Fock ground states. *Nucl. Phys. A* **1991**, *534*, 1–24. [CrossRef]
62. Vuong, A.K. New Skyrme Nucleon-Nucleon Interaction for the Mean-Field Approximation. Doctoral Dissertation, Texas A & M University, College Station, TX, USA, 2003. Available online: http://hdl.handle.net/1969.1/5996 (accessed on 27 January 2024).
63. Dutra, M.; Lourenço, O.; Delfino, A. Consistent Skyrme parametrization and its critical parameter values. *J. Phys. Conf. Ser.* **2019**, *1291*, 012040. Available online: https://iopscience.iop.org/article/10.1088/1742-6596/1291/1/012040/meta (accessed on 27 January 2024). [CrossRef]
64. Nakada, H. Hartree-Fock approach to nuclear matter and finite nuclei with M3Y-type nucleon-nucleon interactions. *Phys. Rev. C* **2003**, *68*, 014316. [CrossRef]
65. Idini, A.; Bennaceur, K.; Dobaczewski, J. Landau parameters for energy density functionals generated by local finite-range pseudopotentials. *J. Phys. G Nucl. Part. Phys.* **2017**, *44*, 064004. [CrossRef]
66. Margueron, J.; Navarro, J.; Van Giai, N. Instabilities of infinite matter with effective Skyrme-type interactions. *Phys. Rev. C* **2002**, *66*, 014303. [CrossRef]
67. Goriely, S.; Chamel, N.; Pearson, J.M. Further explorations of Skyrme-Hartree-Fock-Bogoliubov mass formulas. XII. Stiffness and stability of neutron-star matter. *Phys. Rev. C* **2010**, *82*, 035804. [CrossRef]
68. Li, B.A.; Han, X. Constraining the neutron–proton effective mass splitting using empirical constraints on the density dependence of nuclear symmetry energy around normal density. *Phys. Lett. B* **2013**, *727*, 276–281. [CrossRef]

69. Oertel, M.; Hempel, M.; Klähn, T.; Typel, S. Equations of state for supernovae and compact stars. *Rev. Mod. Phys.* **2017**, *89*, 015007. [CrossRef]
70. Reed, B.T.; Fattoyev, F.J.; Horowitz, C.J.; Piekarewicz, J. Implications of PREX-2 on the Equation of State of Neutron-Rich Matter. *Phys. Rev. Lett.* **2021**, *126*, 172503. [CrossRef]
71. Estee, J.; Lynch, W.G.; Tsang, C.Y.; Barney, J.; Jhang, G.; Tsang, M.B.; Wang, R.; Kaneko, M.; Lee, J.W.; Isobe, T.; et al. Probing the Symmetry Energy with the Spectral Pion Ratio. *Phys. Rev. Lett.* **2021**, *126*, 162701. [CrossRef]
72. Danielewicz, P.; Singh, P.; Lee, J. Symmetry energy III: Isovector skins. *Nucl. Phys. A* **2017**, *958*, 147. [CrossRef]
73. Xie, W.-J.; Li, B.-A. Bayesian Inference of High-density Nuclear Symmetry Energy from Radii of Canonical Neutron Stars. *Astrophys. J.* **2019**, *883*, 174. [CrossRef]
74. Russotto, P.; Gannon, S.; Kupny, S.; Lasko, P.; Acosta, L.; Adamczyk, M.; Al-Ajlan, A.; Al-Garawi, M.; Al-Homaidhi, S.; Amorini, F.; et al. Results of the ASY-EOS experiment at GSI: The symmetry energy at suprasaturation density. *Phys. Rev. C* **2016**, *94*, 034608. [CrossRef]
75. Senger, P. Probing Dense Nuclear Matter in the Laboratory: Experiments at FAIR and NICA. *Universe* **2021**, *7*, 171. [CrossRef]
76. Zhang, Y.; Liu, M.; Xia, C.J.; Li, Z.; Biswal, S.K. Constraints on the symmetry energy and its associated parameters from nuclei to neutron stars. *Phys. Rev. C* **2020**, *101*, 034303. [CrossRef]
77. Zhang, N.B.; Li, B.A. Extracting nuclear symmetry energies at high densities from observations of neutron stars and gravitational waves. *Eur. Phys. J. A* **2019**, *55*, 39. [CrossRef]
78. Tong, H.; Zhao, P.; Meng, J. Symmetry energy at supra-saturation densities via the gravitational waves from GW170817. *Phys. Rev. C* **2020**, *101*, 035802. [CrossRef]
79. Li, B.-A.; Cai, B.-J.; Xie, W.-J.; Zhang, N.-B. Progress in Constraining Nuclear Symmetry Energy Using Neutron Star Observables Since GW170817. *Universe* **2021**, *7*, 182. [CrossRef]
80. Nakazato, K.; Suzuki, H. Cooling Timescale for Protoneutron Stars and Properties of Nuclear Matter: Effective Mass and Symmetry Energy at High Densities. *Astrophys. J.* **2019**, *878*, 25. [CrossRef]
81. Yue, T.-G.; Chen, L.-W.; Zhang, Z.; Zhou, Y. Constraints on the symmetry energy from PREX-II in the multimessenger era. *Phys. Rev. Res.* **2022**, *4*, L022054. [CrossRef]
82. Xie, W.J.; Li, B.A. Bayesian Inference of the Symmetry Energy of Superdense Neutron-rich Matter from Future Radius Measurements of Massive Neutron Stars. *Astrophys. J.* **2020**, *899*, 4. [CrossRef]
83. Zhou, Y.; Chen, L.-W.; Zhang, Z. Equation of state of dense matter in the multimessenger era. *Phys. Rev. D* **2019**, *99*, 121301(R). [CrossRef]
84. Drischler, C.; Furnstahl, R.J.; Melendez, J.A.; Phillips, D.R. How Well Do We Know the Neutron-Matter Equation of State at the Densities Inside Neutron Stars? A Bayesian Approach with Correlated Uncertainties. *Phys. Rev. Lett.* **2020**, *125*, 202702. [CrossRef]
85. Lonardoni, D.; Tews, I.; Gandolfi, S.; Carlson, J. Nuclear and neutron-star matter from local chiral interactions. *Phys. Rev. Res.* **2020**, *2*, 022033(R). [CrossRef]
86. Chen, L.-W. Symmetry energy systematics and its high density behavior. *EPJ Web Conf.* **2015**, *88*, 00017. [CrossRef]
87. Lynch, W.G.; Tsang, M.B. Decoding the density dependence of the nuclear symmetry energy. *Phys. Lett. B* **2022**, *830*, 137098. [CrossRef]
88. Lattimer, J.M.; Lim, Y. Constraining the symmetry parameters of the nuclear interaction. *Astrophys. J.* **2013**, *771*, 51. [CrossRef]
89. Chen, L.W.; Ko, C.M.; Li, B.A. Nuclear matter symmetry energy and the neutron skin thickness of heavy nuclei. *Phys. Rev. C* **2005**, *72*, 064309. [CrossRef]
90. Bano, P.; Pattnaik, S.P.; Centelles, M.; Viñas X.; Routray, T.R. Correlations between charge radii differences of mirror nuclei and stellar observables. *Phys. Rev. C* **2023**, *108*, 015802. [CrossRef]
91. Li, B.-A.; Magno, M. Curvature-slope correlation of nuclear symmetry energy and its imprints on the crust-core transition, radius, and tidal deformability of canonical neutron stars. *Phy. Rev. C* **2020**, *102*, 045807. [CrossRef]
92. d'Etivaux, N.B.; Guillot, S.; Margueron, J.; Webb, N.; Catelan, M.; Reisenegger, A. New Constraints on the Nuclear Equation of State from the Thermal Emission of Neutron Stars in Quiescent Low-mass X-ray Binaries. *Astrophys. J.* **2019**, *887*, 48. [CrossRef]
93. Carson, Z.; Steiner, A.W.; Yagi, K. Constraining nuclear matter parameters with GW170817. *Phys. Rev. D* **2019**, *99*, 043010. [CrossRef]
94. Choi, S.; Miyatsu, T.; Cheoun, M.; Saito, K. Constraints on Nuclear Saturation Properties from Terrestrial Experiments and Astrophysical Observations of Neutron Stars. *Astrophys. J.* **2021**, *909*, 156. [CrossRef]
95. Drischler, C.; Hebeler, K.; Schwenk, A. Asymmetric nuclear matter based on chiral two- and three-nucleon interactions. *Phys. Rev. C* **2016**, *93*, 054314. [CrossRef]
96. Newton, W.G.; Crocombe, G. Nuclear symmetry energy from neutron skins and pure neutron matter in a Bayesian framework. *Phy. Rev. C* **2021**, *103*, 064323. [CrossRef]
97. Grams, G.; Somasundaram, R.; Margueron, J.; Khan, E. Nuclear incompressibility and speed of sound in uniform matter and finite nuclei. *Phys. Rev. C* **2022**, *106*, 044305. [CrossRef]
98. Sagawa, H.; Yoshida, S.; Cao, L.G. EoS from terrestrial experiments: Static and dynamic polarizations of nuclear density. *AIP Conf. Proc.* **2019**, *2127*, 020002. [CrossRef]
99. Tews, I.; Lattimer, J.M.; Ohnishi, A.; Kolomeitsev, E.E. Symmetry Parameter Constraints from a Lower Bound on Neutron-matter Energy. *Astrophys. J.* **2017**, *848*, 105. [CrossRef]

100. Zhang, N.B.; Cai, B.J.; Li, B.A.; Newton, W.G.; Xu, J. How tightly is the nuclear symmetry energy constrained by a unitary Fermi gas? *Nucl. Sci. Tech.* **2017**, *28*, 181. [CrossRef]
101. Mondal, C.; Agrawal, B.K.; De, J.N.; Samaddar, S.K.; Centelles, M.; Viñas, X. Interdependence of different symmetry energy elements. *Phy. Rev C* **2017**, *96*, 021302(R). [CrossRef]
102. Sagawa, H.; Yoshida, S.; Zeng, G.M.; Gu, J.Z.; Zhang, X.Z. Isospin dependence of incompressibility in relativistic and nonrelativistic mean field calculations. *Phys. Rev. C* **2007**, *76*, 034327. [CrossRef]
103. Li, T.; Garg, U.; Liu, Y.; Marks, R.; Nayak, B.K.; Rao, P.V.M.; Fujiwara, M.; Hashimoto, H.; Kawase, K.; Nakanishi, K.; et al. Isotopic Dependence of the Giant Monopole Resonance in the Even-A $^{112-124}$Sn Isotopes and the Asymmetry Term in Nuclear Incompressibility. *Phys. Rev. Lett.* **2007**, *99*, 162503. [CrossRef]
104. Li, T.; Garg, U.; Liu, Y.; Marks, R.; Nayak, B.K.; Rao, P.V.M.; Fujiwara, M.; Hashimoto, H.; Kawase, K.; Nakanishi, K.; et al. Isoscalar giant resonances in the Sn nuclei and implications for the asymmetry term in the nuclear-matter incompressibility. *Phys. Rev. C* **2010**, *81*, 034309. [CrossRef]
105. Cozma, M.D. Feasibility of constraining the curvature parameter of the symmetry energy using elliptic flow data. *Eur. Phys. J. A* **2018**, *54*, 40. [CrossRef]
106. Centelles, M.; Roca-Maza, X.; Viñas, X.; Warda, M. Nuclear Symmetry Energy Probed by Neutron Skin Thickness of Nuclei. *Phys. Rev. Lett.* **2009**, *102*, 122502. [CrossRef]
107. Cai, B.J.; Chen, L.W. Constraints on the skewness coefficient of symmetric nuclear matter within the nonlinear relativistic mean field model. *Nucl. Sci. Tech.* **2017**, *28*, 185. [CrossRef]
108. Farine, M.; Pearson, J.M.; Tondeur, F. Nuclear-matter incompressibility from fits of generalized Skyrme force to breathing-mode energies. *Nucl. Phy. A* **1997**, *615*, 135. [CrossRef]
109. Xie, W.-J.; Li, B.-A. Bayesian inference of the incompressibility, skewness and kurtosis of nuclear matter from empirical pressures in relativistic heavy-ion collisions. *J. Phys. G Nucl. Part. Phys.* **2021**, *48*, 025110. [CrossRef]
110. Steiner, A.W.; Lattimer, J.M.; Brown, E.F. The Equation of State from Observed Masses and Radii of Neutron Stars. *Astrophys. J.* **2010**, *722*, 33–54. [CrossRef]
111. Alam, N.; Agrawal, B.K.; De, J.N.; Samaddar, S.K.; Colò, G. Equation of state of nuclear matter from empirical constraints. *Phys. Rev. C* **2014**, *90*, 054317. [CrossRef]
112. Alam, N.; Agrawal, B.K.; Fortin, M.; Pais, H.; Providência, C.; Raduta, A.R.; Sulaksono, A. Strong correlations of neutron star radii with the slopes of nuclear matter incompressibility and symmetry energy at saturation. *Phys. Rev. C* **2016**, *94*, 052801(R). [CrossRef]
113. Baym. G.; Bethe, H.A.; Pethick, C.J. Neutron star matter. *Nucl. Phys. A* **1971**, *175*, 225–271. [CrossRef]
114. Baym, G.; Pethick, C.J.; Sutherland, P. The Ground State of Matter at High Densities: Equation of State and Stellar Models. *Astrophys. J* **1971**, *170*, 299–317. [CrossRef]
115. De, J.N.; Samaddar, S.K.; Agrawal, B.K. Reassessing nuclear matter incompressibility and its density dependence. *Phys. Rev. C* **2015**, *92*, 014304. [CrossRef]
116. Capano, C.D.; Tews, I.; Brown, S.M.; Margalit, B.; De, S.; Kumar, S.; Brown, D.A.; Krishnan, B.; Reddy, S. Stringent constraints on neutron-star radii from multimessenger observations and nuclear theory. *Nat. Astron.* **2020**, *4*, 625–632. [CrossRef]
117. Raaijmakers, G.; Greif, S.K.; Hebeler, K.; Hinderer, T.; Nissanke, S.; Schwenk, A.; Riley, T.E.; Watts, A.L.; Lattimer, J.M.; Ho, W.C.J. Constraints on the Dense Matter Equation of State and Neutron Star Properties from NICER's Mass–Radius Estimate of PSR J0740+6620 and Multimessenger Observations. *Astrophys. J. Lett.* **2021**, *918*, L29. [CrossRef]
118. Miller, M.C.; Lamb, F.K.; Dittmann, A.J.; Bogdanov, S.; Arzoumanian, Z.; Gendreau, K.C.; Guillot, S.; Ho, W.C.G.; Lattimer, J.M.; Loewenstein, M.; et al. The Radius of PSR J0740 + 6620 from NICER and XMM-Newton Data. *Astrophys. J. Lett.* **2021**, *918*, L28. [CrossRef]
119. Pang, P.T.H.; Tews, I.; Coughlin, M.W.; Bulla, M.; Broeck, C.V.D.; Dietrich, T. Nuclear Physics Multimessenger Astrophysics Constraints on the Neutron Star Equation of State: Adding NICER's PSR J0740 + 6620 Measurement. *Astrophys. J.* **2021**, *922*, 14. [CrossRef]
120. Jiang, J.-L.; Tang, S.-P.; Shao, D.-S.; Han, M.-Z.; Li, Y.-J.; Wang, Y.-Z.; Jin, Z.-P.; Fan, Y.-Z.; Wei, D.-M. The Equation of State and Some Key Parameters of Neutron Stars: Constraints from GW170817, the Nuclear Data, and the Low-mass X-Ray Binary Data. *Astrophys. J.* **2019**, *885*, 39. [CrossRef]
121. Abbott, R.; Abbott, T.D.; Abraham, S.; Acernese, F.; Ackley, K.; Adams, C.; Adhikari, R.X.; Adya, V.B.; Affeldt, C.; Agathos, M.; et al. GW190814: Gravitational Waves from the Coalescence of a 23 Solar Mass Black Hole with a 2.6 Solar Mass Compact Object. (LIGO Scientific Collaboration and Virgo Collaboration). *Astrophys. J.* **2020**, *896*, L44. [CrossRef]
122. Bauswein, A.; Blacker, S.; Lioutas, G.; Soultanis, T.; Vijayan, V.; Stergioulas, N. Systematics of prompt black-hole formation in neutron star mergers. *Phys. Rev. D* **2021**, *103*, 123004. [CrossRef]
123. Bauswein, A.; Janka, H.-T. Hebeler, K.; Schwenk, A. Equation-of-state dependence of the gravitational-wave signal from the ring-down phase of neutron-star mergers. *Phys. Rev. D* **2012**, *86*, 063001. [CrossRef]
124. Annala, E.; Gorda, T.; Kurkela, A.; Vuorinen, A. Gravitational-Wave Constraints on the Neutron-Star-Matter Equation of State. *Phys. Rev. Lett.* **2018**, *120*, 172703. [CrossRef] [PubMed]
125. Fattoyev, F.J.; Piekarewicz, J.; Horowitz, C.J. Neutron Skins and Neutron Stars in the Multimessenger Era. *Phys. Rev. Lett.* **2018**, *120*, 172702. [CrossRef]

126. Bauswein, A.; Just, O.; Janka, H.-T.; Stergioulas, N. Neutron-star Radius Constraints from GW170817 and Future Detections. *Astrophys. J. Lett.* **2017**, *850*, L34. [CrossRef]
127. Bauswein, A.; Baumgarte, T.W.; Janka, H.T. Prompt Merger Collapse and the Maximum Mass of Neutron Stars. *Phys. Rev. Lett.* **2013**, *111*, 131101. [CrossRef]
128. Perego, A.; Logoteta, D.; Radice, D.; Bernuzzi, S.; Kashyap, R.; Das, A.; Padamata, S.; Prakash, A. Probing the Incompressibility of Nuclear Matter at Ultrahigh Density through the Prompt Collapse of Asymmetric Neutron Star Binaries. *Phys. Rev. Lett.* **2022**, *129*, 032701. [CrossRef] [PubMed]
129. Shibata, M.; Taniguchi, K.; Uryū, K. Merger of binary neutron stars with realistic equations of state in full general relativity. *Phys. Rev. D* **2005**, *71*, 084021. [CrossRef]
130. Hotokezaka, K.; Kyutoku, K.; Okawa, H.; Shibata, M.; Kiuchi, K. Binary neutron star mergers: Dependence on the nuclear equation of state. *Phys. Rev. D* **2011**, *83*, 124008. [CrossRef]
131. Köppel, S.; Bovard, L.; Rezzolla, L. A General-relativistic Determination of the Threshold Mass to Prompt Collapse in Binary Neutron Star Mergers. *Astrophys. J. Lett.* **2019**, *872*, L16. [CrossRef]
132. Kashyap, R.; Das, A.; Radice, D.; Padamata, S.; Prakash, A.; Logoteta, D.; Perego, A.; Godzieba, D.A.; Bernuzzi, S.; Bombaci, I.; et al. Numerical relativity simulations of prompt collapse mergers: Threshold mass and phenomenological constraints on neutron star properties after GW170817. *Phys. Rev. D* **2022**, *105*, 103022. [CrossRef]
133. Glendenning, N.K. *Compact Stars: Nuclear Physics, Particle Physics, and General Relativity*; Springer: Berlin/Heidelberg, Germany, 2000. [CrossRef]
134. Margaritis, C.; Koliogiannis, P.S.; Moustakidis, C.C. Speed of sound constraints on maximally rotating neutron stars. *Phys. Rev. D* **2020**, *101*, 043023. [CrossRef]
135. Koliogiannis, P.S.; Moustakidis, C.C. Constraints on the equation of state from the stability condition of neutron stars. *Astrophys. Space Sci.* **2019**, *364*, 52. [CrossRef]
136. Blaizot, J.P. Nuclear compressibilities. *Phys. Rep.* **1980**, *64*, 171–248. [CrossRef]
137. Zhang, N.B.; Li, B.A. Impact of symmetry energy on sound speed and spinodal decomposition in dense neutron-rich matter. *Eur. Phys. J. A* **2023**, *59*, 86. [CrossRef]
138. Hartle, J.B. Bounds on the mass and moment of inertia of non-rotating neutron stars. *Phys. Rep.* **1978**, *46*, 201–247. [CrossRef]
139. Reed, B.; Horowitz, C.J. Large sound speed in dense matter and the deformability of neutron stars. *Phys. Rev. C* **2020**, *101*, 045803. [CrossRef]
140. Cottam, J.; Paerels, F.; Mendez, M. Gravitationally redshifted absorption lines in the X-ray burst spectra of a neutron star. *Nature* **2002**, *420*, 51–54. [CrossRef]
141. Hambaryan, V.; Neuhäuser, R.; Suleimanov, V.; Werner, K. Observational constraints of the compactness of isolated neutron stars. *J. Phys. Conf. Ser.* **2014**, *496*, 012015. [CrossRef]
142. Hambaryan, V.; Suleimanov, V.; Haberl, F.; Schwope, A.D.; Neuhäuser, R.; Hohle, M.; Werner, K. The compactness of the isolated neutron star RXJ0720.4-3125. *Astron. Astrophys.* **2017**, *601*, A108. [CrossRef]
143. Tang, S.P.; Jiang, J.L.; Gao, W.-H.; Fan, Y.-Z.; Wei, D.-M. The Masses of Isolated Neutron Stars Inferred from the Gravitational Redshift Measurements. *Astrophys. J.* **2020**, *888*, 45. [CrossRef]
144. Sotani, H.; Iida, K.; Oyamatsu, K.; Ohnishi, A. Mass and radius formulas for low-mass neutron stars. *Prog. Theor. Exp. Phys.* **2014**, *2014*, 051E01. [CrossRef]
145. Sotani, H.; Nishimura, N.; Naito, T. New constraints on the neutron-star mass and radius relation from terrestrial nuclear experiments. *Prog. Theor. Exp. Phys.* **2022**, *2022*, 041D01. [CrossRef]
146. Carlson, B.V.; Dutra, M.; Lourenço, O.; Margueron, J. Low-energy nuclear physics and global neutron star properties. *Phys. Rev. C* **2023**, *107*, 035805. [CrossRef]

symmetry

MDPI

Article

Ghost Stars in General Relativity

Luis Herrera [1,*,†], Alicia Di Prisco [2,†] and Justo Ospino [3,†]

1 Instituto Universitario de Física Fundamental y Matemáticas, Universidad de Salamanca, 37007 Salamanca, Spain
2 Escuela de Física, Facultad de Ciencias, Universidad Central de Venezuela, Caracas 1050, Venezuela; alicia.diprisco@ucv.ve
3 Departamento de Matemática Aplicada and Instituto Universitario de Física Fundamental y Matemáticas, Universidad de Salamanca, 37007 Salamanca, Spain; j.ospino@usal.es
* Correspondence: lherrera@usal.es
† These authors contributed equally to this work.

Abstract: We explore an idea put forward many years ago by Zeldovich and Novikov concerning the existence of compact objects endowed with arbitrarily small mass. The energy density of such objects, which we call "ghost stars", is negative in some regions of the fluid distribution, producing a vanishing total mass. Thus, the interior is matched on the boundary surface to Minkowski space–time. Some exact analytical solutions are exhibited and their properties are analyzed. Observational data that could confirm or dismiss the existence of this kind of stellar object are discussed.

Keywords: relativistic fluids; interior solutions; spherically symmetric sources

PACS: 04.40.-b; 04.40.Nr; 04.40.Dg

1. Introduction

In their book on relativistic astrophysics, Zeldovich and Novikov (ZN) [1] (see also [2]), raise the question about the possibility of packaging the constituents of a self–gravitating fluid distribution in such a way that the total mass of the resulting compact object is arbitrarily small.

Specifically, they consider static spherically symmetric fluid distributions, for which the line element may be written as

$$ds^2 = e^{\nu}c^2dt^2 - e^{\lambda}dr^2 - r^2\left(d\theta^2 + \sin^2\theta d\phi^2\right), \tag{1}$$

where $\nu(r)$ and $\lambda(r)$ are functions of r, and c is the light velocity. In this section, we shall follow the notation of [1]; however, in the rest of the manuscript we shall use relativistic units, in which case we put $c = G = 1$.

The fluid distribution is bounded from the exterior by a surface, Σ, whose equation is $r = r_\Sigma =$ constant.

From (1) and the Einstein equations we may write

$$e^{-\lambda} = 1 - \frac{8\pi G}{rc^2}\int_0^r \mu r^2 dr, \tag{2}$$

and for the three-dimensional volume element we have

$$dV = 4\pi e^{\lambda/2} r^2 dr, \tag{3}$$

where μ denotes the energy density of the fluid.

Then, we have for the total mass (energy) the well-known expression

Citation: Herrera, L.; Di Prisco, A.; Ospino, J. Ghost Stars in General Relativity. *Symmetry* **2024**, *16*, 562. https://doi.org/10.3390/sym16050562

Academic Editor: Charalampos Moustakidis

Received: 12 April 2024
Revised: 29 April 2024
Accepted: 30 April 2024
Published: 5 May 2024

$$E = Mc^2 = 4\pi c^2 \int_0^{r_\Sigma} \mu r^2 dr. \tag{4}$$

ZN also introduce the rest energy of the constituent particles, E_0, given by

$$E_0 = M_0 c^2 = N m_0 c^2, \tag{5}$$

where m_0 is the particle mass and N denotes the total number of particles that may be expressed through the particle density, n, as

$$N = \int_V n dV. \tag{6}$$

Also, denoting by E_1 the rest energy, E_0, plus the kinetic energy and the interaction energy of the constituents (excluding the gravitational interaction) we may write

$$E_1 = M_1 c^2 = c^2 \int_V \mu dV = 4\pi c^2 \int_0^{r_\Sigma} e^{\lambda/2} \mu r^2 dr. \tag{7}$$

Since $e^{\lambda/2} \geq 1$, then the mass defect $\Delta M = M_1 - M$ should be positive.

Thus, the original question posed by ZN may be rephrased as: can the constituents of a star be packaged in such a way that the mass defect equals M_1?

They answer affirmatively to the above question, and illustrate their point by analyzing the case of an ideal Fermi gas. Although their analysis is flawed, as we shall see below, the case for the existence of stars with arbitrarily small total mass should not be dismissed.

Let us first reproduce the analysis of ZN, following strictly their line of arguments (with only slight changes in notation).

Thus, let us consider an ultra-relativistic Fermi gas, characterized by an equation of state given by

$$\mu = \beta n^{4/3}, \qquad \beta \equiv \frac{3}{8}\hbar(3\pi^2)^{1/3}, \tag{8}$$

where $\hbar$ is the Planck constant over 2π.

Next, ZN assume for the distribution of energy density the form

$$\mu = \frac{a}{r^2}, \quad a = constant. \tag{9}$$

It is worth emphasizing that the above choice is justified by the fact that it coincides with the well-known Tolman VI solution [3], whose equation of state for large values of μ approaches that for a highly compressed Fermi gas.

Then, using (9) in (4), it follows at once

$$M = 4\pi a r_\Sigma. \tag{10}$$

On the other hand, using (2), (3), (6), (8) and (9), we obtain for N

$$N = \frac{\alpha r_\Sigma^{3/2}}{\sqrt{1 - \frac{8\pi G a}{c^2}}}, \quad \alpha \equiv \frac{8\pi}{3}\left(\frac{a}{\beta}\right)^{3/4}, \tag{11}$$

implying

$$r_\Sigma = \alpha^{-2/3} N^{2/3} \left(1 - \frac{8\pi G a}{c^2}\right)^{1/3}. \tag{12}$$

Feeding back (12) into (10) produces

$$M \sim N^{2/3} \left(1 - \frac{8\pi G a}{c^2}\right)^{1/3}. \tag{13}$$

From (13), ZN conclude that, in the limit when $a \to \frac{c^2}{8\pi G}$, the total mass, M, tends to zero.

Such a conclusion is incorrect, as (11) and (12) imply that, in the limit $a \to \frac{c^2}{8\pi G}$, N diverges as $N^{2/3} \sim \frac{1}{(1-\frac{8\pi Ga}{c^2})^{1/3}}$, thereby canceling the term $(1 - \frac{8\pi Ga}{c^2})^{1/3}$ in (13). This is also evident from (10), which shows that M does not tend to zero for any value of a (different from zero).

In general, it should be clear from its very definition (4), that M cannot be zero for any positive defined energy-density function, μ. Thus, vanishing total mass is only possible if we accept the existence of fluid distributions allowing negative energy density, or in the trivial case $\mu = 0$.

The appearance of negative energy density (mass) in general relativity has been considered in the past by several researchers, starting with a paper by Bondi [4]. This issue also appears in relation to the Reissner–Nordstrom solution and classical electron models (see [5–8] and references therein). More recently, negative masses have been invoked in the construction of some cosmological models (see [9,10] and references therein). Also, it is worth mentioning that negative energy density appears in hyperbolically symmetric fluids (see [11,12] and references therein). In all the cases above, quantum effects were not taken into account. However, in spite of these examples, we believe that it is fair to say that the assumption of positive energy density is well justified, at the classic level, for any realistic fluid.

Notwithstanding, the situation is quite different in the quantum regime. Indeed, as it has been argued in the recent past (see [13–17] and references therein), the appearance of negative energy density is possible, whenever quantum effects are expected to be relevant.

Thus, the idea of compact objects with arbitrarily small total mass is still feasible, if we accept the possibility of negative energy density. We call such objects "ghost stars", in analogy with a somehow similar situation observed in some Einstein–Dirac neutrinos (named ghost neutrinos), which do not produce a gravitational field but still are characterized by non-vanishing current density [18–20].

In this work, we shall explore such a possibility by presenting explicit analytical models of ghost stars.

2. The Einstein Equations for Static Locally Anisotropic Fluids

In what follows, we shall briefly summarize the definitions and main equations required for describing spherically symmetric static anisotropic fluids. We shall heavily rely on [21], and therefore we shall omit many steps in the calculations, details of which the reader may find in that reference.

We consider a spherically symmetric distribution of static fluid, bounded by a spherical surface, Σ. The fluid is assumed to be locally anisotropic (principal stresses unequal).

The justification to consider anisotropic fluids, instead of isotropic ones, is provided by the fact that pressure anisotropy is produced by many different physical phenomena of the kind expected in a gravitational collapse scenario (see [22] and references therein). In particular, we expect that the final stages of stellar evolution should be accompanied by intense dissipative processes, which, as shown in [23], should produce pressure anisotropy.

In curvature coordinates (using relativistic units), the line element reads (please notice that we are using signature -2, instead $+2$, as in [21])

$$ds^2 = e^{\nu(r)}dt^2 - e^{\lambda(r)}dr^2 - r^2(d\theta^2 + r^2 sin^2\theta d\phi^2), \tag{14}$$

which has to satisfy the Einstein equations. For a locally anisotropic fluid they are

$$8\pi\mu = \frac{1}{r^2} - e^{-\lambda}\left(\frac{1}{r^2} - \frac{\lambda'}{r}\right), \tag{15}$$

$$8\pi P_r = -\frac{1}{r^2} + e^{-\lambda}\left(\frac{1}{r^2} + \frac{\nu'}{r}\right), \tag{16}$$

$$8\pi P_\perp = \frac{e^{-\lambda}}{4}\left(2\nu'' + \nu'^2 - \lambda'\nu' + 2\frac{\nu' - \lambda'}{r}\right), \tag{17}$$

where primes denote derivatives with respect to r, and μ, P_r and $P_\perp$ are proper energy density, radial pressure and tangential pressure, respectively.

The above is a system of three ordinary differential equations for the five unknown functions ν, λ, μ, P_r and $P_\perp$, and accordingly their solutions would depend on two arbitrary functions.

From the above field equations, the Tolman–Oppenheimer–Volkof equation follows

$$P_r' = -\frac{(m + 4\pi P_r r^3)}{r(r - 2m)}(\mu + P_r) + \frac{2(P_\perp - P_r)}{r}, \tag{18}$$

where we have introduced the mass function, m [24], defined by

$$e^{-\lambda} = 1 - \frac{2m(r)}{r}. \tag{19}$$

In [21], a general algorithm to express any solution for anisotropic fluids in terms of two generating functions was proposed (see also [25]). It generalizes a previous work by Lake for isotropic fluids [26].

Specifically, it was shown that the general line element corresponding to any solution to the system (15)–(17) may be written as

$$\begin{aligned} ds^2 &= e^{\int(2z(r)-2/r)dr}dt^2 - \frac{z^2(r)e^{\int(\frac{4}{r^2 z(r)}+2z(r))dr}}{r^6\left(-2\int \frac{z(r)(1+\Pi(r)r^2)e^{\int(\frac{4}{r^2 z(r)}+2z(r))dr}}{r^8}dr + C\right)}dr^2 \\ &- r^2 d\theta^2 - r^2 sin^2\theta d\phi^2. \end{aligned} \tag{20}$$

with $\Pi(r) = 8\pi(P_r - P_\perp)$ and

$$e^{\nu(r)} = e^{\int(2z(r)-2/r)dr} \tag{21}$$

where C is a constant of integration.

The expression (20) follows from (21) and the formal integration of $\Pi(r) = 8\pi(P_r - P_\perp)$, after replacing P_r and $P_\perp$ by their expressions in (16) and (17) (see [21] for details).

In the next sections, z will be obtained from specific restrictions on the fluid distribution (e.g., conformal flatness, vanishing complexity factor).

The physical variables may be written as

$$4\pi P_r = \frac{z(r - 2m) + m/r - 1}{r^2}, \tag{22}$$

$$4\pi\mu = \frac{m'}{r^2}, \tag{23}$$

and

$$4\pi P_\perp = \left(1 - \frac{2m}{r}\right)\left(z' + z^2 - \frac{z}{r} + \frac{1}{r^2}\right) + z\left(\frac{m}{r^2} - \frac{m'}{r}\right). \tag{24}$$

In order to match smoothly the metric (14) with the Schwarzschild metric on the boundary surface $r = r_\Sigma = constant$, we require the continuity of the first and the second fundamental forms across that surface, producing

$$e^{\nu_\Sigma} = 1 - \frac{2M}{r_\Sigma}, \tag{25}$$

$$e^{-\lambda_\Sigma} = 1 - \frac{2M}{r_\Sigma}, \tag{26}$$

$$[P_r]_\Sigma = 0, \tag{27}$$

where subscript Σ indicates that the quantity is evaluated on the boundary surface, Σ.

The above conditions hold for any value of M, including $M = 0$.

For configurations with $M = 0$, we obtain from (22) and (27)

$$z_\Sigma = \frac{1}{r_\Sigma}. \tag{28}$$

We shall next present solutions describing fluid spheres with vanishing total mass. To do that, we shall resort to a variety of assumptions, some of which are usually invoked in the modeling of relativistic stars.

3. Conformally Flat Ghost Stars

The Weyl tensor is known to play a very important role in the structure and evolution of compact objects (see [27] and references therein), which explains why the vanishing Weyl tensor condition (conformal flatness) has been used so frequently in the study of self-gravitating objects.

If we assume the space–time within the fluid distribution to be conformally flat, then the two generating functions read

$$z = \frac{2}{r} \pm \frac{e^{\frac{\lambda}{2}}}{r} tanh\left(\int \frac{e^{\frac{\lambda}{2}}}{r} dr\right). \tag{29}$$

and

$$\Pi = r\left(\frac{1 - e^{-\lambda}}{r^2}\right)'. \tag{30}$$

In (29), we shall choose the minus sign, since the plus sign leads (in this case) to a model not satisfying the boundary condition (28).

We shall present two conformally flat models of ghost star. For that purpose, we shall complement the conformal flatness condition with some additional restrictions.

3.1. Ghost Star with a Given Density Profile

Let us assume a density profile of the form

$$4\pi\mu = \sum_{i=0}^{n} a_i r^{i-2}, \tag{31}$$

which using (23) produces

$$m = \sum_{i=0}^{n} \frac{a_i}{i+1} r^{i+1}. \tag{32}$$

Since the total mass is assumed to vanish, then the following condition has to be satisfied

$$\sum_{i=0}^{n} \frac{\bar{a}_i}{i+1} = 0, \tag{33}$$

with $\bar{a}_i = a_i r^i_\Sigma$.

In order to describe a specific model, let us restrict the expression (31) to $n = 2$.

Thus, we obtain for the energy density and the mass function

$$4\pi\mu = -\frac{3}{2r^2} + \frac{a_1}{r} + a_2, \tag{34}$$

and

$$m = -\frac{3}{2}r + \frac{a_1}{2}r^2 + \frac{a_2}{3}r^3, \quad (35)$$

where we have chosen $a_0 = -\frac{3}{2}$ to simplify the calculation of the second term on the right of (29).

Then the condition (33) reads

$$a_2 = \frac{9}{2r_\Sigma^2} - \frac{3a_1}{2r_\Sigma}, \quad (36)$$

and using (35) and (36) in (19) we obtain

$$e^{-\lambda} = 4 - \frac{3r^2}{r_\Sigma^2} - a_1 r\left(1 - \frac{r}{r_\Sigma}\right). \quad (37)$$

With the expression for λ given by (37), the two generating functions for this case become

$$z = \frac{5}{2r} - \sqrt{\frac{a_2}{-24 + 6a_1 r + 4a_2 r^2}}, \quad (38)$$

and

$$\Pi = \frac{6}{r^2} - \frac{a_1}{r}. \quad (39)$$

The constant a_1 may be easily obtained from (38), and, using condition (28), it reads

$$a_1 = \frac{12}{r_\Sigma}, \quad (40)$$

which combined with (36) produces

$$a_2 = -\frac{27}{2r_\Sigma^2}. \quad (41)$$

With the two expressions above, we finally obtain for z and m

$$z = \frac{6r - 5r_\Sigma}{r(3r - 2r_\Sigma)}, \quad (42)$$

$$m = -\frac{3}{2}r + \frac{6r^2}{r_\Sigma} - \frac{9r^3}{2r_\Sigma^2}, \quad (43)$$

and using using (22), (34) and (39) we obtain for the energy density, the radial pressure and Π

$$4\pi\mu = -\frac{3}{2r^2} + \frac{12}{r_\Sigma r} - \frac{27}{2r_\Sigma^2}, \quad (44)$$

$$4\pi P_r = \frac{27}{2r_\Sigma^2} - \frac{21}{r r_\Sigma} + \frac{15}{2r^2}, \quad (45)$$

$$\Pi = \frac{6}{r^2} - \frac{12}{r r_\Sigma}. \quad (46)$$

Using (44) and (46), the reader can easily check that the condition of conformal flatness (see Equation (29) in [28])

$$P_r - P_\perp = \frac{1}{r^3}\int_0^r r^3 \mu' dr, \quad (47)$$

is satisfied.

From (44), we see that μ is negative in the intervals $0 < r \lessapprox 0.15 r_\Sigma$ and $r \gtrapprox 0.73 r_\Sigma$. As it is apparent from the expressions of the physical variables, the fluid distribution has a singularity at the origin ($r = 0$), and therefore the center should be excluded from the discussion. The best way to handle this drawback consists in assuming that a vacuum cavity surrounds the center. Denoting the equation of the boundary of the cavity by $r = r_i = constant$, we obtain from (43) $r_i = \frac{r_\Sigma}{3}$, which ensures the continuity of the mass function on that surface. However, the radial pressure is discontinuous on that surface, and therefore it is a thin shell, endowed with a singular matter distribution satisfying the Israel conditions [29].

3.2. Ghost Star with the Gokhroo and Mehra Ansatz

We shall now complement the conformal flatness condition with an ansatz proposed by Gokhroo and Mehra [30]. Its virtue consists in providing physically satisfactory models for compact objects.

Thus, we shall assume for λ the condition

$$e^{-\lambda} = 1 - \alpha r^2 + \frac{3K\alpha r^4}{5r_\Sigma^2}, \tag{48}$$

producing, because of (15) and (19),

$$\mu = \mu_0\left(1 - \frac{Kr^2}{r_\Sigma^2}\right), \tag{49}$$

and

$$m(r) = \frac{4\pi\mu_0 r^3}{3}\left(1 - \frac{3Kr^2}{5r_\Sigma^2}\right), \tag{50}$$

where K is a constant, μ_0 is the central density and

$$\alpha \equiv \frac{8\pi\mu_0}{3}. \tag{51}$$

Since we must impose $m(r_\Sigma) = 0$, then $K = \frac{5}{3}$.

Feeding back this value of K into (48)–(50), we obtain

$$4\pi\mu = \frac{6}{r_\Sigma^2}\left(1 - \frac{5r^2}{3r_\Sigma^2}\right), \tag{52}$$

$$m = \frac{2r^3}{r_\Sigma^2}\left(1 - \frac{r^2}{r_\Sigma^2}\right), \tag{53}$$

and

$$e^{-\lambda} = 1 - \frac{4r^2}{r_\Sigma^2} + \frac{4r^4}{r_\Sigma^4}, \tag{54}$$

where we have chosen $\alpha = \frac{4}{r_\Sigma^2}$, in order to facilitate the calculation of the second term on the right of (29). Thus, we obtain for z

$$z = \frac{3}{r} - \frac{2r}{2r^2 - r_\Sigma^2}, \tag{55}$$

whereas for Π we obtain from (30)

$$\Pi = -\frac{8r^2}{r_\Sigma^4}, \tag{56}$$

and from (22) we obtain the expression for P_r

$$8\pi P_r = \frac{4}{r^2}\left(1 - \frac{4r^2}{r_\Sigma^2} + \frac{3r^4}{r_\Sigma^4}\right). \tag{57}$$

As it follows from (52), the energy density becomes negative for $r \gtrapprox 0.77 r_\Sigma$.

As in the precedent model, there appears a singularity at the center, which could be embedded in a vacuum cavity bounded by a thin shell.

4. Ghost Stars with Vanishing Complexity Factor

The complexity factor, usually denoted by Y_{TF}, is a scalar function intended to measure the degree of complexity of a given fluid distribution, and was introduced in [28] for static spherically symmetric configurations. A rigorous definition of complexity has been the goal of many scientists in different branches of sciences, with such interest being motivated by the intuitive idea that complexity should, somehow, measure a basic property describing the structures existing within a system.

Mathematically, the complexity factor describes the trace-free part of the electric Riemann tensor and may be written as (see [28] for details)

$$Y_{TF} = \Pi - \frac{4\pi}{r^3}\int_0^r \tilde{r}^3 \mu' d\tilde{r}, \tag{58}$$

and, accordingly, the vanishing complexity factor condition reads

$$\Pi = \frac{4\pi}{r^3}\int_0^r \tilde{r}^3 \mu' d\tilde{r}, \tag{59}$$

and please notice that the symbol Π here differs from the one in [28] by a factor 8π.

Using (22)–(24) and (59), we are led to the following differential equation for z

$$2\left(1-\frac{2m}{r}\right)\left(z'+z^2\right) - \left(\frac{2}{r}-\frac{5m}{r^2}+\frac{m'}{r}\right)\left(2z-\frac{1}{r}\right) + \frac{2}{r}-\frac{4m}{r^3} = 0. \tag{60}$$

The first integral of the above equation (for m) reads

$$1-\frac{2m}{r} = e^{\int_r^{r_\Sigma} \frac{4(r^2 z' + r^2 z^2 - 2rz + 2)}{2r^2 z - r} dr}, \tag{61}$$

from which we see that, for any z satisfying (28), we have a model with a vanishing complexity factor. However, we shall follow here a different strategy, and we shall present two models of ghost stars satisfying the vanishing complexity factor condition, by imposing two different additional restrictions.

4.1. A Model with a Given Energy-Density Profile

In order to specify this model, we shall propose the following energy-density profile,

$$8\pi\mu = \frac{1-9(\frac{r}{r_\Sigma})^8}{r^2}, \tag{62}$$

producing for m

$$m = \frac{r}{2}\left[1-\left(\frac{r}{r_\Sigma}\right)^8\right], \tag{63}$$

the reason behind this choice being simply that it allows the integration of (60).

Indeed, feeding back (63) into (60), we may easily integrate this equation for z, obtaining

$$z = \frac{1}{c_1 r^2 - r}, \tag{64}$$

where c_1 is a constant of integration, which according to (28) reads

$$c_1 = \frac{2}{r_\Sigma}. \tag{65}$$

Having obtained the two generators of the solution, we may write for P_r and $P_\perp$

$$8\pi P_r = -\frac{1}{r^2} + \frac{r^6}{r_\Sigma^8}\left(\frac{3 - \frac{2r}{r_\Sigma}}{\frac{2r}{r_\Sigma} - 1}\right), \tag{66}$$

$$8\pi P_\perp = \frac{4r^7}{r_\Sigma^9\left(\frac{2r}{r_\Sigma} - 1\right)}. \tag{67}$$

In this model, the energy density becomes negative for values of r in the interval $[0.76 r_\Sigma < r, r = r_\Sigma]$. As in the previous model, the fluid presents a singularity at the origin, which could be surrounded by a cavity bounded by a thin shell.

4.2. Ghost Star with Vanishing Active Gravitational Mass

For this model, we shall additionally assume that the active gravitational (Tolman) mass [31] vanishes.

This last condition implies (see Equation (7.30) in [22])

$$m + 4\pi P_r r^3 = 0. \tag{68}$$

Feeding the above condition into (18) and using (59), we obtain

$$P_r' + \frac{\Pi}{4\pi r} = 0, \tag{69}$$

which can be easily transformed into

$$P_r'' + \frac{4P_r'}{r} + \frac{\mu'}{r} = 0. \tag{70}$$

In order to find a solution to the above equation, we shall split it in two equations, as

$$P_r'' + \frac{3P_r'}{r} = 0, \tag{71}$$

and

$$\frac{P_r'}{r} + \frac{\mu'}{r} = 0, \tag{72}$$

whose solutions reads

$$P_r = b\left(\frac{1}{r^2} - \frac{1}{r_\Sigma^2}\right), \tag{73}$$

and

$$\mu = b\left(\frac{3}{r_\Sigma^2} - \frac{1}{r^2}\right), \tag{74}$$

where boundary condition (27) has been used and b is a constant of integration.

Using (73) in (68), we obtain for the mass

$$m = 4\pi r^3 b\left(\frac{1}{r_\Sigma^2} - \frac{1}{r^2}\right), \tag{75}$$

while using (74) in (59), we obtain for Π

$$\Pi = \frac{8\pi b}{r^2}. \tag{76}$$

In this model, the energy density becomes negative in the interval $0 < r \lessapprox 0.58 r_\Sigma$ (if we assume $b > 0$). As in the previous models, the physical variables exhibit a singular behavior at the center, and any surface delimiting a vacuum cavity surrounding the center would be a thin shell.

5. Discussion

Exploring the possibility of the existence of compact objects endowed with vanishing total mass (energy), we have presented four exact solutions to Einstein equations for static spherical distribution of anisotropic fluids, sharing this property. Such solutions must, within some regions of the distribution, be endowed with negative energy density. Negative energy-density values appear indistinctly in outer or in inner regions, depending on the model, not a universal pattern of distribution having been detected.

Although some of the assumptions adopted to obtain the presented solutions (e.g., the vanishing complexity factor or the conformal flatness) are physically meaningful, the obtained solutions are intended only to illustrate the above-mentioned possibility but not to describe any specific astrophysical scenario. A pending problem regarding this issue consists in finding exact solutions for ghost stars, directly related to relevant astrophysical data.

In the same order of ideas, an important open question concerning ghost stars is related to possible astrophysical observations that could confirm (or dismiss) the existence of this kind of object. We have in mind, for example, a new trend of investigations based on the recent observations of shadow images of the gravitationally collapsed objects at the center of the elliptical galaxy *M87* and at the center of the Milky Way galaxy by the Event Horizon Telescope (EHT) Collaboration (see [32–35] and references therein). More specifically, we wonder if it could be possible to establish the existence of a ghost star by its shadow.

The solutions we have presented should be considered as the final state of collapsing stars, where quantum effects become relevant during the evolution process. Accordingly, it is of utmost interest to describe the process leading to the final stage with vanishing total mass. To do that, we should find non-static exact solutions describing such a process. Additionally, a detailed description of the mechanism by means of which quantum effects allow negative energy density should be provided. These two problems are out of the scope of this manuscript, but remain among the most relevant questions to solve concerning the physical viability of ghost stars.

Regarding the formation of a ghost star, it should be clear from elementary physical considerations that, as a final product of gravitational collapse, the formation of such configurations must be preceded by an intense radiative process. The problem regarding the efficiency of energy release in gravitational collapse has been discussed by several authors (see [36–38] and references therein). Some of these authors conclude that a 100% efficiency (all the mass is radiated away) is possible under rather mild restrictions [36,38], while others [37] claim that 100% efficiency is forbidden under physically meaningful conditions, among which positive energy density plays a relevant role. Thus, the violation of such a condition, as it happens in our models, is a strong argument to believe that 100% efficiency could be a likely possibility. In such a case, the detection of a strong emission of radiation might indicate the location of a ghost star.

We would like to conclude with five remarks oriented to encourage future research on this issue

- We have explored the possibility of ghost stars within the context of general relativity. It would be interesting to explore such a possibility under some of the extended theories of gravity [39].
- For reasons exposed before, we have considered anisotropic fluids. However, it seems clear that ghost star models described by isotropic fluids should also exist. It could be interesting to find some models of this kind.

- All the models presented here exhibit a singularity at the origin. In order to exclude such a region, we have proposed to surround the center with a vacuum cavity. However, in all examples analyzed, the boundary surface of such a cavity appears to be a thin shell. It would be interesting to find singularity-free solutions and/or singular solutions whose center could be embedded in a vacuum cavity delimited by a regular boundary.
- We would like to insist on the importance of finding exact (analytical or numerical) solutions describing the evolution leading to a ghost star.
- Alternatively, it could be also of interest to find solutions describing the evolution of an initial ghost star leading to a $M > 0$ object, by absorbing radiation. As strange as this scenario might look like (compact object absorbing radiation), it is worth noticing that it has been invoked in the past to explain the origin of gas in quasars [40]. A semi-numerical example for such a model is described in [41].

Author Contributions: All authors contributed equally to this work. Conceptualization, L.H.; methodology, L.H., A.D.P. and J.O.; software, J.O.; formal analysis, L.H., A.D.P. and J.O.; writing—original draft preparation, L.H. writing—review and editing, L.H., A.D.P. and J.O.; funding acquisition, L.H. and J.O. All authors have read and agreed to the published version of the manuscript.

Funding: This work was partially supported by the Spanish Ministerio de Ciencia, Innovación, under Research Project No. PID2021-122938NB-I00.

Data Availability Statement: Data are contained within the article.

Conflicts of Interest: The authors declare no conflicts of interest. The funders had no role in the design of the study; in the collection, analyses, or interpretation of data; in the writing of the manuscript; or in the decision to publish the results.

References

1. Zeldovich, Y.B.; Novikov, I.D. *Relativistic Astrophysics. Vol. I. Stars and Relativity*; University of Chicago Press: Chicago, IL, USA, 1971.
2. Zeldovich, Y.B. The Collapse of a Small Mass in the General Theory of Relativity. *Sov. Phys. JETP* **1962**, *15*, 446.
3. Tolman, R. Static Solutions of Einstein's Field Equations for Spheres of Fluid. *Phys. Rev.* **1939**, *55*, 364. [CrossRef]
4. Bondi, H. Negative Mass in General Relativity. *Rev. Mod. Phys.* **1957**, *29*, 423–428. [CrossRef]
5. Cooperstock, F.I.; Rosen, N. A nonlinear gauge-invariant field theory of leptons. *Int. J. Theor. Phys.* **1989**, *28*, 423–440. [CrossRef]
6. Bonnor, W.B.; Cooperstock, F.I. Does the electron contain negative mass? *Phys. Lett. A.* **1989**, *139*, 442–444. [CrossRef]
7. Papapetrou, A. *Lectures on General Relativity*; D. Reidel: Dordrecht, The Netherlands; Boston, MA, USA, 1974.
8. Herrera, L.; Varela, V. Negative energy density and classical electron models. *Phys. Lett. A* **1994**, *189*, 11–14. [CrossRef]
9. Najera, S.; Gamboa, A.; Aguilar-Nieto, A.; Escamilla-Rivera, C. On Negative Mass Cosmology in General Relativity. *arXiv* **2021**, arXiv:2105.110041v1.
10. Farnes, J.S. A unifying theory of dark energy and dark matter: Negative masses and matter creation within a modified λ cdm framework. *Astron. Astrophys.* **2018**, *620*, A92. [CrossRef]
11. Maciel, A.; Delliou, M.L.; Mimoso, J.P. New perspectives on the TOV equilibrium from a dual null approach. *Class. Quantum Gravity* **2020**, *37*, 125005. [CrossRef]
12. Herrera, L. Nonstatic hyperbolically symmetric fluids. *Int. J. Mod. Phys. D* **2022**, *31*, 2240001. [CrossRef]
13. Capozziello, S.; Lobo, F.S.N.; Mimoso, J.P. Energy conditions in modified gravity. *Phys. Lett. B* **2014**, *730*, 280–283. [CrossRef]
14. Capozziello, S.; Lobo, F.S.N.; Mimoso, J.P. Generalized energy conditions in extended theories of gravity. *Phys. Rev. D* **2015**, *91*, 124019. [CrossRef]
15. Barcelo, C.; Visser, M. Twilight for the Energy Conditions? *Int. J. Mod. Phys. D* **2002**, *11*, 1553–1560. [CrossRef]
16. Kontou, E.A.; Sanders, K. Energy conditions in general relativity and quantum field theory. *Class. Quantum Gravity* **2020**, *37*, 193001. [CrossRef]
17. Pavsic, M. On negative energies, strings, branes, and braneworlds: A review of novel approaches. *Int. J. Mod. Phys. A* **2020**, *35*, 2030020. [CrossRef]
18. Davis, T.; Ray, J. Ghost neutrinos in general relativity. *Phys. Rev. D* **1974**, *9*, 331. [CrossRef]
19. Collinson, C.D.; Morris, P.B. Space-times admitting non-unique neutrino fields. *Nuovo Cim. B* **1973**, *16*, 273–284. [CrossRef]
20. Collinson, C.D.; Morris, P.B. Space-times admitting neutrino fields with zero energy and momentum. *J. Phys. A Math. Gen.* **1973**, *6*, 915–916. [CrossRef]
21. Herrera, L.; Ospino, J.; Prisco, A.D. All static spherically symmetric anisotropic solutions of Einstein's equations. *Phys. Rev. D* **2008**, *77*, 027502 . [CrossRef]

22. Herrera, L.; Santos, N.O. Local anisotropy in self–gravitating systems. *Phys. Rep.* **1997**, *286*, 53–130. [CrossRef]
23. Herrera, L. Stabilty of the isotropic pressure condition. *Phys. Rev. D* **2020**, *101*, 104024. [CrossRef]
24. Misner, C.; Sharp, D. Relativistic Equations for Adiabatic, Spherically Symmetric Gravitational Collapse. *Phys. Rev.* **1964**, *136*, B571. [CrossRef]
25. Lake, K. Galactic Potentials. *Phys. Rev. Lett.* **2004**, *92*, 051101. [CrossRef]
26. Lake, K. All static spherically symmetric perfect-fluid solutions of Einstein's equations. *Phys. Rev. D* **2003**, *67*, 104015. [CrossRef]
27. Herrera, L.; Prisco, A.D.; Ospino, J.; Fuenmayor, E. Conformally flat anisotropic spheres in general relativity. *J. Math Phys.* **2001**, *42*, 2129–2143. [CrossRef]
28. Herrera, L. New definition of complexity for self–gravitating fluid distributions: The spherically symmetric static case. *Phys. Rev. D* **2018**, *97*, 044010. [CrossRef]
29. Israel, W. Singular hypersurfaces and thin shells in general relativity. *Nuovo Cim. B* **1966**, *44*, 1–14. [CrossRef]
30. Gokrhoo, M.K.; Mehra, A.L. Anisotropic spheres with variable energy density in general relativity. *Gen. Relativ. Gravit.* **1994**, *26*, 75. [CrossRef]
31. Tolman, R. On the use of the energy-momentum principle in general relativity. *Phys. Rev* **1930**, *35*, 875. [CrossRef]
32. Akiyama, K.; Alberdi, A.; Alef, W.; Asada, K.; Azulay, R.; Baczko, A.; Ball, D.; Baloković, M.; Barret, J.; Bintley, D.; et al. First M87 Event Horizon Telescope Results. I. The Shadow of the Supermassive Black Hole. *Astrophys. J. Lett.* **2019**, *875*, L1.
33. Akiyama, K.; Alberdi, A.; Alef, W.; Algaba, J.C.; Anantua, R.; Asada, K.; Azulay, R.; Bach, U.; Baczko, A.; Ball, D.; et al. First Sagittarius A Event Horizon Telescope Results. I. The Shadow of the Supermassive Black Hole in the Center of the Milky Way. *Astrophys. J. Lett.* **2022**, *930*, L12.
34. Psaltis, D. Testing general relativity with the Event Horizon Telescope. *Gen. Relativ. Gravit.* **2019**, *51*, 137. [CrossRef]
35. Gralla, S.E. Can the EHT M87 results be used to test general relativity? *Phys. Rev. D* **2021**, *103*, 024023. [CrossRef]
36. Dyson, F.J. The Efficiency of Energy Release in Gravitational Collapse. *Comments Astrophys. Space Phys.* **1969**, **1**, 75.
37. Leibovitz, C.; Israel, W. Maximum Efficiency of Energy Release in Spherical Collapse. *Phys. Rev. D* **1970**, *1*, 3226. [CrossRef]
38. Mitra, A.; Singh, K. On the Maximum Energy Release from Formation of Static Compact Objects. *Galaxies* **2023**, *11*, 116. [CrossRef]
39. Capozziello, S.; Laurentis, M.D. Extended theories of gravity. *Phys. Rep.* **2011**, *509*, 167. [CrossRef]
40. Mathews, W.G. Reverse stellar evolution, stellar ablation, and the origin of gas in quasars. *Astrophys. J.* **1983**, *272*, 390–399. [CrossRef]
41. Esculpi, M.; Herrera, L. Conformally symmetric radiating spheres in general relativity. *J. Math. Phys.* **1987**, *27*, 2087–2096. [CrossRef]

MDPI AG
Grosspeteranlage 5
4052 Basel
Switzerland
Tel.: +41 61 683 77 34

Symmetry Editorial Office
E-mail: symmetry@mdpi.com
www.mdpi.com/journal/symmetry

www.ingramcontent.com/pod-product-compliance
Lightning Source LLC
Chambersburg PA
CBHW061552210926
51850CB00009B/2085
9783725815999